Finite Element Computations in Mechanics with R

Finite Element Computations in Mechanics with R

A Problem-Centered Programming Approach

Khameel Bayo Mustapha

CRC Press
Taylor & Francis Group
Boca Raton London New York

CRC Press is an imprint of the
Taylor & Francis Group, an **informa** business

CRC Press
Taylor & Francis Group
6000 Broken Sound Parkway NW, Suite 300
Boca Raton, FL 33487-2742

Printed on acid-free paper

International Standard Book Number-13: 978-1-1385-0162-1 (Hardback)

Library of Congress Cataloging-in-Publication Data

Names: Mustapha, Khameel Bayo, author.
Title: Finite element computations in mechanics with R : a problem-centered programming approach / Khameel Bayo Mustapha.
Description: First edition. | Boca Raton, FL : CRC Press/Taylor & Francis Group, 2018. | "A CRC title, part of the Taylor & Francis imprint, a member of the Taylor & Francis Group, the academic division of T&F Informa plc." | Includes bibliographical references and index.
Identifiers: LCCN 2017056045 | ISBN 9781138501621 (hardback : acid-free paper) | ISBN 9781315144474 (ebook)
Subjects: LCSH: Mechanics--Data processing. | Engineering mathematics--Data processing. | R (Computer program language)
Classification: LCC QA808 .M87 2018 | DDC 620.100285/5133--dc23
LC record available at https://lccn.loc.gov/2017056045

Visit the Taylor & Francis Web site at
http://www.taylorandfrancis.com

and the CRC Press Web site at
http://www.crcpress.com

To my parent, for their love and sacrifice.

Contents

Preface

The finite element method (FEM) occupies a prominent position in the simulations of many intractable multiphysics problems. Across a wide range of engineering fields (mechanical, civil, biomedical, electromagnetic, etc.), the method remains a technical tool of choice in the workflow of simulation-driven development and design of a variety of complex systems. The essentiality of the knowledge of FEM is recognized in engineering education curriculum, and it undergirds the proliferation of commercial finite element analysis software in academia and industries. Yet the teaching of the FEM is not without its challenges. Finding it difficult to strike a balance between theory and hands-on practice, courses in the teaching of the FEM are either too deep in the mathematical framework of the method or biased toward a heavy reliant on commercial FEM software. The former approach presents challenges for students with shallow mathematics knowledge, whereas the latter tends to create expert software users with hazy understanding of the rudimentary activities required for conducting reliable finite element analyses.

Although not discouraging the use of commercial FEM software, a modest aim of this book is to contribute to the attempts, instigated by respected figures in this field, to address this issue by encouraging the use of programming languages to enhance the understanding of the procedural steps of FEM. Indeed, in the past decades, programming approaches to the FEM have been presented by several authors using FORTRAN, JAVA, MATLAB, MATHEMATICA, and Python. Nonetheless, it is not the author's motivation to write a book just for the sole purpose of using R for the FEM. The desire arises out of a deep belief in the suitability of the language along with its growing ecosystem as another versatile computational tool in the arsenal of engineers. In this spirit, this book provides an introductory coverage of the FEM with the R programming language in a manner that emphasizes link between theory and the procedural steps of implementing finite element solutions for problems in engineering mechanics. A multiplatform open-source language, the R programming language is matured with thousands of library and functions developed by experienced developers. It is also attractive as students and practitioners can download and install the software free of charge.

The primary target of this book is readers who are studying structural and solid mechanics with interests in an interactive exploration of the FEM. Therefore, it is assumed that the reader is familiar with the fundamentals of basic engineering mechanics and the theory of elasticity. Besides, although knowledge of a computer language is helpful, it is not essential. The book adopts a programming-by-example philosophy and provides automated R functions that will allow readers to build the relevant skill set of understanding the essential conceptual basis of finite element computations, without being burdened by the tedious longhand calculations associated with the method. It also should appeal to readers who may already be familiar with R and wish to get started on the fundamentals of FEM. Throughout the chapters of the book, the convenience of the language for implementing vectorized finite element routines for structural, thermal, and dynamic analyses of elementary structural components is illustrated. The use of the implemented functions is also demonstrated with numerous example problems, most of which are derived from authoritative sources. In all, this book contains 12 chapters. Chapter 1 presents a brief introduction to the R programming language and the RStudio (a user-friendly GUI for writing and editing of R scripts). In Chapter 2, we explored the functions available for the manipulations of vectors and matrices in R. The remaining chapters cover the finite element models and implemented functions for a variety of problems with practical implications in components design. Specifically, the other chapters contain problems that relate to

- Discrete systems (springs and assemblage of springs).
- One-dimensional structural elements (bars and beams).
- Two-dimensional truss and space truss.
- Two-dimensional frame and space frame.
- Plane stress and axisymmetric problems.
- Thermal stress problems.
- Free vibration of bars, beams, and truss.

Within each chapter (starting from Chapter 3), we start with a cursory account of the finite element model and then present the relevant code snippet in the form of functions that implement specific aspects of the FEM procedure. Demonstrations of the functions are then presented with specific examples. The implemented functions are all tested and validated with exercise problems from authoritative sources. The codes for all the chapters can be downloaded from the book's website, can be used freely, and can also easily be extended by readers.

As a final point, although attempts have been made to check for errors, the author welcomes any suggestions or corrections by email KhameelB.Mustapha@nottingham.edu.my. I leave the readers with the following quote:

> If you really wish to learn you must mount a machine and become acquainted with its tricks by actual trial.—Wilbur Wright

Khameel Bayo Mustapha

Acknowledgments

I profoundly appreciate all my teachers for their mentorship. Indeed, I have gained over many years from the explanations and writings of many brilliant minds in the field of finite element method. I especially remain indebted to Prof. A.F.M. Arif of the King Fahd University of Petroleum and Minerals, Dhahran, Saudi Arabia, whose excellent lectures on *Finite Element Analysis in Mechanical Design* whetted my appetites and prodded me to dig deeper into the field. I appreciate the support of my friends as well as the many students (in Swinburne University of Technology Malaysia Campus and University of Nottingham Malaysia Campus) whose questions have helped to improve parts of this book. The institutional support of the University of Nottingham Malaysia Campus is crucial for the completion of this book.

Special thanks to the editorial and production staff of CRC Press. I specifically express gratitude to Jonathan Plant for his trust and commitments in bringing this project to fruition. I also acknowledge the constructive suggestions obtained from the reviewers of the manuscript.

I cannot thank my wife, Aminah Oziohu, enough for her unremitting encouragement and for being a beacon of love. Her engineering background came handy in the preparation of this manuscript. To my children Hibatullah Adenike, AbdulAlim Adewale, and Hameedah Aderonke, I am thankful for your encouragement and love in the face of the demanding process of writing a book. Your regular probing of "What chapter of the book are you writing now?" was a motivating force throughout the writing process.

Finally, all praise is due to Allah, the Most Exalted. He lavishes His blessings on us, apparent and non-apparent.

Author

Dr. Khameel Bayo Mustapha is an assistant professor in the department of mechanical, materials, and manufacturing engineering at the University of Nottingham, (Malaysia Campus). He teaches mechanics of solids, advanced solid mechanics (stress analysis techniques), computer modeling techniques, and additive manufacturing and 3D printing. He was previously a lecturer at Swinburne University of Technology, (Malaysia Campus), where he convened and taught the following related courses: solid mechanics, structural mechanics, computer-aided engineering, and computer modeling, analysis and visualization. He holds a PhD degree in mechanical engineering from Nanyang Technological University (NTU, Singapore), a BEng (mechanical engineering) from University of Ilorin (Ilorin, Nigeria), and a Graduate Certificate in Learning and Teaching from Swinburne University of Technology (Melbourne, Australia). His other previous roles include the post of a research scholar with NTU, research assistant at the King Fahd University of Petroleum and Minerals (Saudi Arabia), and a very brief stint as an in-house R&D engineer with a manufacturing firm in Singapore. His research interest involves mechanics of advanced materials, vibration analysis of systems, and computational modeling and simulation of mechanical systems across different length scales and complexities.

1

Overview of the R Programming Environment, Installations, and Basic Syntax

1.1 Introduction

The choice of R for the implementation of the finite element method (FEM) will appear eccentric to those familiar with the language as a byword for statistical computing. Remarkably, within the wider engineering community, the language has not received the attention it deserves for computational analyses. But this is expected to change as more and more practitioners get familiar with the status of R as a general-purpose programming language with top-notch supports for a wide range of computational tasks.

Originally designed for statistical computations and visualization, R is an open-source software environment that has gained reputation for intensive data analyses. Yet, the language offers an enabling platform for computational tasks in mechanics, particularly for the coding of the FEM. For one, it is free (no cost for download, installation, and usage). Further, the language is designed with a rigorous support for matrix algebra (the backbone of FEM), and it is endowed with comprehensive ecosystems for tackling of essential modern computational tasks (such as parallel computations, big data analytics, etc.). As an added bonus, a swelling rank of expert users, most of whom are eminent academics and researchers, keeps pushing the boundary of its application into domains beyond the historical conception.

As part of our journey to deploy the R programming language for computations related to the FEM, we will need two tools: the R installer and the RStudio®. Sections 1.2 and 1.3 explain how to download and install these two tools.

1.2 Downloading and Installing R

The R software is made available under the terms of the Free Software Foundation's GNU. The following steps explain how to download and install the precompiled binary distributions of the installer for your operating system:

- Visit the R project website (https://www.r-project.org/) and follow the link to download R.
- While at the CRAN web page, select a mirror location closest to you (https://cran.r-project.org/mirrors.html).
- Download the appropriate R installer for your operating system.
- You can either run the installer on the fly or save it to a directory from which to install later.

The R project website (https://www.r-project.org/) provides more information about the history of R, including an extensive documentation about the language. Technically, the R programming language has its own user interface. However, this interface is somewhat inadequate for large programming projects. Thus, this text adopts a user-friendlier R editor called RStudio. Be aware that RStudio does not come with a preinstalled R. Therefore, the R installer must be downloaded and installed separately.

1.3 Downloading and Installing RStudio

As a matured integrated development environment (IDE), RStudio facilitates an effective venturing into the world of R programming. It provides a console for the writing and editing of commands, a well-integrated interface for many behind-the-hood tasks and elegant project management features.

RStudio is a part of the affiliated projects of the Foundation for Open Access Statistics, and it can be downloaded for free from the RStudio website (https://www.rstudio.com/). Once installed, clicking the RStudio icon either from the Window's Start menu or desktop shortcut opens the interface as shown in Figure 1.1.

An in-depth introduction to many features of RStudio can be found in Reference 1. However, such a comprehensive tomb may not be necessary for users not interested in software engineering. We therefore present a short guide to RStudio in Section 1.4. For clarity, in the rest of this book, programming commands are typeset in a distinctive font `like this`. This is to distinguish between the programming syntax and the regular English words within a chunk of texts.

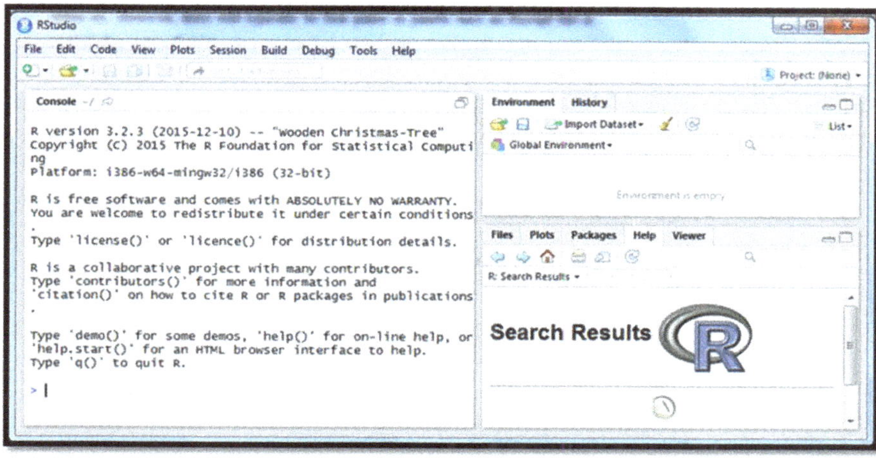

FIGURE 1.1
The RStudio interface.

1.4 Getting Started with R and RStudio

Most of the in-built R functions that are of relevance to our journey are covered in Chapter 2. A comprehensive guide to the language is also provided in the online manual (*An Introduction to R*, found at https://cran.r-project.org/index.html). One can also learn the language from a host of other books. I have personally benefitted from References 2–4.

In this book, all interactions with the R language are channeled through the interface provided by RStudio. With the RStudio installed, invoking it is as simple as double-clicking the RStudio icon via the desktop, or clicking its icon from the Start menu. As shown in Figure 1.1, once invoked, the RStudio's interface comes up with three panes: (a) *console*, for interactive execution of R commands; (b) *workspace browser*, to show a record of past variables; and (c) *History interface*, for displaying accessed files, plots, used packages, help menu, and so on.

1.4.1 Basic Interaction with RStudio

Once RStudio is started, you should see the console pane displaying:

```
>
```

where the symbol > is a prompt. At the prompt, one can enter any R command or expression, hit the *Enter* key, and the R interpreter will discern the command and spits out the result accordingly in the **`Console`**. A series of simple calculations are shown in the next block of code to demonstrate this.

```
> 2*5
[1] 10

> sqrt(16)
[1] 4

> 2^4
[1] 16

> log10(1000)
[1] 3

> log(1000)
[1] 6.907755

> exp(1)
[1] 2.718282
```

The first expression in the preceding code snippet shows the multiplication of 2 and 5, and below the expression **`> 2*5`** is the result. The rest of the expressions show the use of the square root command (**`sqrt()`**), base 10 logarithm (**`log10()`**), natural logarithm (**`log()`**), and exponential (**`exp()`**).

Like in most programming languages, the results of an execution in an R session can be stored using a ***variable***. It is easy to create a variable by using an assignment statement. Variables in R can be stored by using one of these assignment operators:

- `=`
- `<-`
- `->`

as illustrated in the ***Console pane*** of Figure 1.2 (the window on the left-hand side). Typically, the workspace browser shows the name and value(s) of variables that have been created as shown in the ***Workspace*** pane of Figure 1.2.

As revealed in Figure 1.2, R allows a variable name to be on the left of the assignment operators, for = and <- (which is the conventional way). Remarkably, the language also supports the case of placing the variable name on the right if the right-pointing operator (i.e., ->) is used (as demonstrated in the third variable definition shown in Figure 1.2). This latter approach is not common with other programming languages such as MATLAB®, for instance.

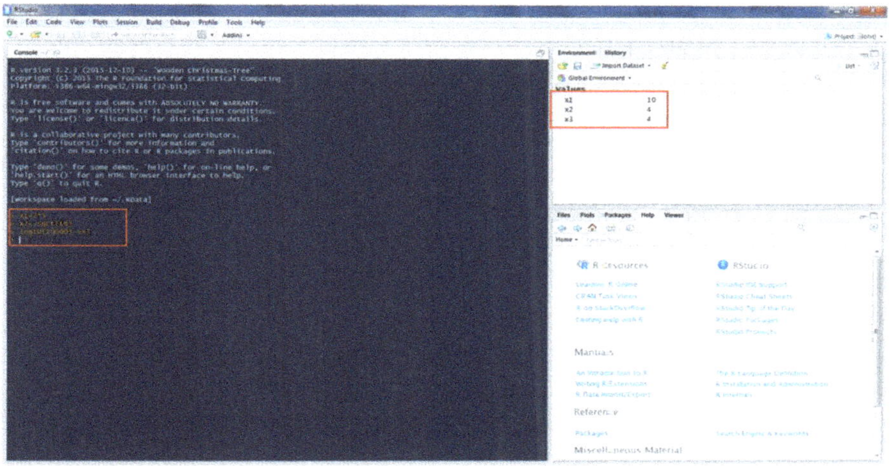

FIGURE 1.2
A demonstration of the use of assignment operators.

1.4.2 Getting Help in R

R offers a comprehensive built-in documentation that can be queried for help, and knowing how to use this documentation is fundamental to one's versatility with the powerful built-in functions offered by the language.

Table 1.1 demonstrates the various ways to search for help within the programming environment. In principle, you can always search the R

TABLE 1.1
Approaches for Searching the R Documentation

Search Format	Output
`help('sum')`	Search result for the function **sum** within the R documentation.
`help('%%')`	Search result for the operator **%%** within the R documentation.
`help("break")`	Search result for control flow **break** within the R documentation.
`help.search("decomposition")`	Search result, typically hyperlinked pages of topics containing the word "**decomposition**".
`help.search("random number")`	Search result, typically hyperlinked pages of topic containing the phrase "**random number**".
`?sum`	Search result for the function **sum** within the R documentation.
`??"decomposition"`	Search for hyperlinked pages of topics containing the word "**decomposition**".

documentation by using either the functions **help()** or **help.search()**. As illustrated in the table, these two functions require the term you are searching to be enclosed within a quotation mark. Alternatively, you can type **?** (i.e., a question mark), followed by the name of the function with no quotation mark or (**??**) followed by a keyword related to the search term (see rows 6 and 7). However, note that if a search term is a member of the special characters, reserved words, or compound words, it must be put in quotes.

1.4.3 In-Built Examples and Demos

The R environment provides several examples for many in-built functions that one can run to get a hang of how the functions work. Some of these examples can be accessed by using the **example()** function. For instance, if you are interested in how the **matrix()** function works, you can use:

```
example(matrix)
```

A list of demonstrations is also available, and one can explore these demonstrations by simply typing and running:

- **demo()**
- **demo(graphics)**

Note that **demo()** will list all available demonstrations, whereas **demo(graphics)** will list demonstrations related to the graphics capability of R.

1.4.4 Packages and Modularity in R

R is a modular programming platform. This modularity is influenced by its open license structure that allows developers—many of whom are professors, professional programmers, and data hobbyists—to design a modular collection of tools that ease the process of data, numerical, and graphical analyses. These tools, typically a collection of well thought-out functions, and objects, are called **packages** in R. They are made freely available for anyone to download. Most of the R packages are hosted at http://cran.r-project.org. A specific package can be installed, provided a system is connected to the Internet. For instance, a popular package for numerical analysis is the **Matrix** package. This or any other package can be installed as follows:

- Open RStudio.
- Run **install.packages("package_name")** at the command line.

To download any package, just replace **package_name** with the name of the package of your interest (e.g., the **Matrix** package) and R will connect your computer to the CRAN website to get the **Matrix** package, and have it installed in the directory that hosts your R interpreter. As at the time of writing this book, the CRAN website holds over 10000 packages for different tasks.

References

1. J. Verzani, *Getting Started with RStudio*, O'Reilly Media: Sebastopol, CA, 2011.
2. V. A. Bloomfield, *Using R for Numerical Analysis in Science and Engineering*, Taylor & Francis Group: Boca Raton, FL, 2014.
3. G. Grolemund, *Hands-On Programming with R: Write Your Own Functions and Simulations*, O'Reilly Media: Sebastopol, CA, 2014.
4. O. Jones, R. Maillardet, A. Robinson, *Introduction to Scientific Programming and Simulation Using R, Second Edition*, Taylor & Francis Group: Boca Raton, FL, 2014.

2

Vectors and Matrices

2.1 Introduction

The finite element method (FEM) is a mathematical procedure that relies heavily on matrix algebra. This chapter therefore introduces some basic ideas related to matrix algebra necessary to make sense of the chapters that follow. Among other things, we will look at data types such as vectors and matrices (the hallmark of matrix algebra and the FEM) along with the R built-in functions available to perform operations on them.

2.2 Creating Vectors

We start with the treatment of vectors here because R has the distinctive feature of having the vector data type as the fundamental data type. This means that anything we evaluate in R is treated as a vector (i.e., an ordered set of n elements).

As stated in Chapter 1, when you enter an expression into the R console and press the Enter key, the expression will be evaluated and the results (if there are any) will be displayed. Bearing this in mind, we list a few ways to create a vector in R.

2.2.1 The Sequence Operator

seq() function

This function creates an ordered sequence of numbers, with an added advantage that the interval of the sequence can be controlled. Few examples, evaluated within the R console, are used to demonstrate the usage of this function.

```
#A sequence with the default interval of 1
> seq(1,10)
 [1]  1  2  3  4  5  6  7  8  9 10
#A sequence with a specfied interval of 2
> seq(1,10,by=2)
[1] 1 3 5 7 9
#A sequence of n equally spaced numbers between 2 numbers:
> seq(1,2,length.out = 5)
[1] 1.00 1.25 1.50 1.75 2.00
#A sequence with a specfied interval of -1:
> seq(10,1,by=-1)
 [1] 10  9  8  7  6  5  4  3  2  1
```

Couple of comments is necessary about the preceding block of code. First, R regards any line headed by the symbol "#" as a comment, hence the first line of the above-mentioned block of code will not be evaluated. Second, each line that starts with a prompt symbol (>) is the expression we wish to evaluate. When the expression is evaluated, after pressing *Enter* on the keyboard, R returns a value. The returned value is preceded by "[1]", which refers to the index of the first item displayed in the row.

2.2.2 The Colon Operator

```
:
```

The colon operator is a binary operator that creates an ordered vector with a unit increment between the number on its left and right sides.

```
#Integer numbers
> 5:10
[1]  5  6  7  8  9 10
#Non-integer numbers
> pi:10
[1] 3.142 4.142 5.142 6.142 7.142 8.142 9.142
```

2.2.3 The Combine Function

```
c()
```

This function concatenates its argument into a vector. It can also coerce different data types into a vector.

```
#Arguments of the same data type:
> c(4,8,9,12,11,27)
[1]  4  8  9 12 11 27
#Arguments of different data types:
> c(12,13,18,"A","D","F")
[1] "12" "13" "18" "A"  "D"  "F"
```

2.2.4 Other Sequence Functions

There are other generalized functions for creating sequences.

```
seq_along()
seq_len()
```

The first function creates a sequence from 1 up to the length of its input, whereas the second creates a sequence from 1 up to its input. Some examples of the usage of this are given as follows:

```
#Using the seq_along() function
> studentnames=c("Michael","Bayo","John","Hibatullah",
"AbdulAlim")
> for(i in seq_along(studentnames)) print(studentnames[i])
[1] "Michael"
[1] "Bayo"
[1] "John"
[1] "Hibatullah"
[1] "AbdulAlim"

#Using the seq_len() function
> seq_len(12)
 [1]  1  2  3  4  5  6  7  8  9 10 11 12
```

From this, you will notice that **seq_along()** requires a vector pointer as an input.

2.2.5 The Repeat Function

```
rep()
```

This function replicates its argument by an amount that must be specified. It works on various data types.

```
> #rep(0,N) to produce a vector containing zeros, but of
length N
> rep(0,10)
 [1] 0 0 0 0 0 0 0 0 0 0
> #rep(j,N) to produce a vector containing a number, but
of length N
> rep(pi,10)
 [1] 3.142 3.142 3.142 3.142 3.142 3.142 3.142 3.142
3.142 3.142
> #rep() can also replicate a vector
> v=c(10,20,30)
> rep(v,3)
[1] 10 20 30 10 20 30 10 20 30
> #We can also repeat each element of a vector v a number
of times
> rep(v,each=3)
[1] 10 10 10 20 20 20 30 30 30
```

2.2.6 The Vector Function

```
vector()
```

This function is used to create a vector of certain length and mode, and it is particularly useful in creating a placeholder for a vector whose content or elements are not known ahead of time.

```
> #Creating a numeric vector of zero length
> v=vector(mode="numeric",length=0);v
numeric(0)

> #Creating a numeric vector of zeros of length five
> v=vector(mode="numeric",length=5);v
[1] 0 0 0 0 0

> #Creating a logical vector of specified length
> vlogical=vector(mode="logical",length=7);vlogical
[1] FALSE FALSE FALSE FALSE FALSE FALSE FALSE
```

2.3 Operations on Vectors

2.3.1 Assessing the Element of a Vector

R provides many ways to pick out or refer to a member (or members) of a vector. One can refer to an element (or elements) by location in a vector as shown in the following:

```
#Consider a vector named vectora=c(1,2,3,4,5,5,7,8)
> vector_a=c(1,2,3,4,5,6,7,8)
#We can extract the nth item (e.g. 3rd item) in the
vector_a as
> vector_a[3]
[1] 3
#Extract items 2 to 6 as
> vector_a[2:6]
[1] 2 3 4 5 6
#Extract elements of vector_a that are divisible by e.g. 2
> vector_a[vector_a%%2==0]
[1] 2 4 6 8
#Extract elements 1,3,7
> vector_a[c(1,3,7)]
[1] 1 3 7

> #Extract all but the 1st and 8th element
> vector_a=c(1,2,3,4,5,6,7,8)
> vector_a[-c(1,8)]
[1] 2 3 4 5 6 7
```

The examples in the foregoing blocks of code must have revealed that R starts counting from 1. Hence, the first element of a vector stored in a vector variable **x** will be **x[1]**. This is different from the Python programming language, for instance, in which the first element of such a vector is extracted by **x[C]**.

2.3.2 Length of a Vector

One can find the length of a vector by using the **length()** function.

```
> vector_b=c(12,15,18,21,24,27,30,33,36,39)
> length(vector_b)
[1] 10
```

2.3.3 Creating Row and Column Vectors

```
cbind()
```

This function is used to construct column matrix from a row vector (it basically transforms the vector into a matrix). In addition, **cbind()** can be used to construct a row matrix if an unconcatenated array of numbers is passed to it directly. See the following examples:

```
#turns vector c(12,14,16,18,20) into a column matrix
> cbind(c(12,14,16,18,20))
     [,1]
[1,]   12
[2,]   14
[3,]   16
[4,]   18
[5,]   20

> cbind(12,14,16,18,20)
     [,1] [,2] [,3] [,4] [,5]
[1,]   12   14   16   18   20
```

A related function to **cbind()** is **rbind()**, which is explained in the following:

```
rbind()
```

This function is used to explicitly construct a row matrix from a vector. However, this function will also construct a column matrix if an unconcatenated array of numbers is passed to it directly. See the following examples:

```
#turns vector c(12,14,16) into a row matrix
> rbind(c(12,14,16))
     [,1] [,2] [,3]
[1,]   12   14   16

> rbind(12,14,16)
     [,1]
[1,]   12
[2,]   14
[3,]   16
```

2.3.4 Transpose of Vectors

We can find the transpose of a vector by using the **t()** function.

```
#Consider a column matrix z
> z=cbind(c(10,20,30,40,50,60,70,80));z
     [,1]
[1,]   10
[2,]   20
[3,]   30
[4,]   40
[5,]   50
[6,]   60
[7,]   70
[8,]   80
> #Find the transpose as
> t(z)
     [,1] [,2] [,3] [,4] [,5] [,6] [,7] [,8]
[1,]   10   20   30   40   50   60   70   80
```

2.3.5 Assigning and Modifying the Elements of Vectors

Assuming we have declared a vector of zeros with length 4, we can assign values via the assignment operator as shown in the following:

```
#Assigning values to a zero vector of length 4
> y=vector(mode="numeric",length=4);y
[1] 0 0 0 0

#Assign a value to the first element
> y[1]=6;y
[1] 6 0 0 0

#Assign values to elements 2 to 4 of vector y
> y[2:4]=c(12,14,16);y
[1]  6 12 14 16

#Modify the third item of vector y
> y[3]=90;y
[1]  6 12 90 16
```

2.3.6 Adding of Vectors

Vectors can be added by using the plus operator +.

```
#Adding vectors together
> x=c(2,5,7,9)
> y=c(3,6,8,10)
> x+y
[1]  5 11 15 19
```

2.3.7 Multiplication of Vectors

Vectors can be multiplied by using the operator *, which returns an element-wise multiplication.

```
#Multiply vector x and y
> x=c(2,5,7,9)
> y=c(3,6,8,10)
> x*y
[1]  6 30 56 90
```

2.3.8 Reversing and Ordering a Vector

The order of elements within a vector can be reversed and sorted using the **rev()** and **sort()** functions, respectively.

```
#Consider a vector y=c(10,20,30,40,50,60,70,80
> y=c(10,20,30,40,50,60,70,80)
#We can reverse it order by
> rev(y)
[1] 80 70 60 50 40 30 20 10
#Sorting
> y=c(12,3,4,90,10,2,19)
> sort(y)
[1]  2  3  4 10 12 19 90
```

2.3.9 Recycling

When applying an operation on two vectors that are required to be of equal length, but in which one of them is shorter, R will recycle (i.e., repeat) the shorter vector until its length matches the longer vector. For instance, if we attempt to add two vectors x (with a length of 4) and y (with a length of 8):

$$x = [12,13,14,15] \tag{2.1}$$

$$y = [10,20,30,40,50,60,70,80] \tag{2.2}$$

As illustrated next, *x* will be repeated to match the length of *y*.

```
#Adding two vectors of unequal lengths
> x=c(12,13,14,15)
> y=c(10,20,30,40,50,60,70,80)
> x+y
[1] 22 33 44 55 62 73 84 95
```

2.4 Vectorized Functions of R

A good number of arithmetic operators and functions in R are vectorized, which makes it easy to work on arrays, such as vectors and matrices, without the use of loops. Some selected, but not in-depth examples, are presented in the following:

```
#Square root of a vector
> y=c(10,20,30,40,50,60,70,80)
> sqrt(y)
[1] 3.162 4.472 5.477 6.325 7.071 7.746 8.367 8.944
#Base 10 logarithm of a vector
> log10(y)
[1] 1.000 1.301 1.477 1.602 1.699 1.778 1.845 1.903
#Natural logarithm of a vector
> log(y)
[1] 2.303 2.996 3.401 3.689 3.912 4.094 4.248 4.382

#Trigonometric functions
> y=c(10,20,30,40,50,60,70,80)
> cos(y)
[1] -0.8391  0.4081  0.1543  -0.6669  0.9650  -0.9524
0.6333  -0.1104

#Element-wise comparison of x and y
> y=c(10,20,30,40,50,60,70,80)
> x=c(1,2,3,45,80,90,50,21)
> x>y
[1] FALSE FALSE FALSE  TRUE  TRUE  TRUE FALSE FALSE

#Rounding of
> y=pi:10;y
[1] 3.142 4.142 5.142 6.142 7.142 8.142 9.142
> round(y,2)
[1] 3.14 4.14 5.14 6.14 7.14 8.14 9.14
```

There are hundreds of other functions that can be seen in the light of the preceding examples. Interested readers can pick up one of the many references for a deep dive into the exploration of some of these other functions.

2.5 Creating Matrices

In the preceding sections, we have looked at vector variables, which can be described as one-dimensional objects. Vectors are only characterized by their length. On the other hand, a matrix is a two-dimensional array with two attributes: the number of rows and columns. R provides very good support for dealing with matrices.

2.5.1 The Matrix Function

The most common function to create an $m \times n$ matrix in R is the **matrix()** function. You control the shape of the matrix by specifying two parameters: the number of columns (**ncol**) or the number of rows (**nrow**). If neither **ncol** nor **nrow** is specified, the vector passed to the matrix function is turned into a column matrix.

```
#Consider the vector y below which we wish to turn into
a 3 by 2 matrix
> y=c(4,8,15,18,26,34)
#Turn y into a matrix of 2 columns
> matrix(y,ncol=2)
     [,1] [,2]
[1,]    4   18
[2,]    8   26
[3,]   15   34
#Turn y into a matrix of 2 rows
> matrix(y,nrow=2)
     [,1] [,2] [,3]
[1,]    4   15   26
[2,]    8   18   34

#Not specifying ncol or nrow turns the vector
into a column matrix
> matrix(y)
     [,1]
[1,]    4
[2,]    8
[3,]   15
[4,]   18
[5,]   26
[6,]   34
```

Essentially, in forming a matrix, R typically arranges the numbers columnwise. However, this default behavior can be overwritten as highlighted in the following example:

```
#byrow=TRUE
> y=c(4,8,15,18,26,34)
> matrix(y,nrow=2, byrow = TRUE)
     [,1] [,2] [,3]
[1,]    4    8   15
[2,]   18   26   34
#byrow=FALSE
> matrix(y,nrow=2, byrow = FALSE)
     [,1] [,2] [,3]
[1,]    4   15   26
[2,]    8   18   34

#Both nrow and ncol can be supplied at the same time tc
form
> y=seq(12)
> matrix(y,ncol=3,nrow=4)
     [,1] [,2] [,3]
[1,]    1    5    9
[2,]    2    6   10
[3,]    3    7   11
[4,]    4    8   12
> matrix(y,ncol=4,nrow=3)
     [,1] [,2] [,3] [,4]
[1,]    1    4    7   10
[2,]    2    5    8   11
[3,]    3    6    9   12
```

Further, an *m* by *n* matrix of zeros can be created as done in the next block of code.

```
#Creating an m by n matrix with zeros
> matrix(0,4,4)
     [,1] [,2] [,3] [,4]
[1,]    0    0    0    0
[2,]    0    0    0    0
[3,]    0    0    0    0
[4,]    0    0    0    0
```

2.5.2 The Diagonal Function

You can use the **diag()** function to create (i) an identity matrix of any size and (ii) a diagonal matrix from a vector.

```
#Build an n×n identity matrix
> diag(6)
     [,1] [,2] [,3] [,4] [,5] [,6]
[1,]    1    0    0    0    0    0
[2,]    0    1    0    0    0    0
[3,]    0    0    1    0    0    0
[4,]    0    0    0    1    0    0
[5,]    0    0    0    0    1    0
[6,]    0    0    0    0    0    1
#Build a diagonal matrix from a vector
> y=c(10,20,30,40,50,60)
> diag(y)
     [,1] [,2] [,3] [,4] [,5] [,6]
[1,]   10    0    0    0    0    0
[2,]    0   20    0    0    0    0
[3,]    0    0   30    0    0    0
[4,]    0    0    0   40    0    0
[5,]    0    0    0    0   50    0
[6,]    0    0    0    0    0   60
```

2.5.3 The Vector-Binding Functions

One may form a matrix by using the **rbind()** or **cbind()** functions as shown in the next code snippet:

```
> x = 1:5; y = 1:5
#Using rbind
> rbind(x,y)
  [,1] [,2] [,3] [,4] [,5]
x    1    2    3    4    5
y    1    2    3    4    5
#Using cbind
> cbind(x,y)
     x y
[1,] 1 1
[2,] 2 2
[3,] 3 3
[4,] 4 4
[5,] 5 5
```

2.5.4 Building a Matrix by Repeating a Vector

It is possible to combine the **rep()** function explored in Section 2.2.5 with the **matrix()** function to build a matrix as demonstrated in the following:

```
> x = 1:5
> matrix(rep(x,4),ncol=4)
     [,1] [,2] [,3] [,4]
[1,]    1    1    1    1
[2,]    2    2    2    2
[3,]    3    3    3    3
[4,]    4    4    4    4
[5,]    5    5    5    5
#A 5 by 5 matrix  of zeros
> matrix(rep(0,10),ncol=5,nrow=5)
     [,1] [,2] [,3] [,4] [,5]
[1,]    0    0    0    0    0
[2,]    0    0    0    0    0
[3,]    0    0    0    0    0
[4,]    0    0    0    0    0
[5,]    0    0    0    0    0
```

2.6 Operations on Matrices

2.6.1 Indexing

We can access a specific element of a matrix (say B) by specifying its row and column indices in square brackets (e.g., $B\left[i,j\right]$). Further, a particular row i in matrix B is accessed by using $B\left[i,\right]$, and a specific column j can be accessed by using $B\left[,j\right]$. See the following examples:

```
#Consider the matrix below
> a_matrix=matrix(c(12,13,14,10,20,30,15,25,35),nrow=3,ncol=3)
> a_matrix
     [,1] [,2] [,3]
[1,]   12   10   15
[2,]   13   20   25
[3,]   14   30   35

#Access row 1
> a_matrix[1,]
[1] 12 10 15
#Access column 1
> a_matrix[,1]
[1] 12 13 14

#Obtain the element in row 2, column 3
> a_matrix[2,3]
[1] 25
```

One can also assign a value to, or access, a single/multiple element of a matrix with the right index.

```
#Using a single index
> a_matrix[7]
[1] 15
#Let us change the above 7th element of the matrix
> a_matrix[7]=90
> a_matrix
     [,1] [,2] [,3]
[1,]   12   10   90
[2,]   13   20   25
[3,]   14   30   35
#Change the first row of the matrix
> a_matrix[1,]=c(100,150,250)
> a_matrix
     [,1] [,2] [,3]
[1,]  100  150  250
[2,]   13   20   25
[3,]   14   30   35
```

Notice that when a single number is used within the square bracket without a comma, only a single element of the matrix is being referenced. However, a comma before (after) the indexing number indicates a reference to a row (a column). Further, we may access a submatrix from a bigger matrix by specifying the desired row and column indices within the square brackets as revealed in the following lines of code:

```
> a_matrix=matrix(c(12,13,14,10,20,30,15,25,35),nrow=3,ncol=3)
> a_matrix
     [,1] [,2] [,3]
[1,]   12   10   15
[2,]   13   20   25
[3,]   14   30   35
#A submatrix formed by row 1 and 2 as well as column 1 and 2
> b_matrix=a_matrix[c(1,2),c(1,2)]
> b_matrix
     [,1] [,2]
[1,]   12   10
[2,]   13   20
```

2.6.2 Diagonal, Trace, and Dimension

One can obtain the diagonal of a matrix using the **diag()** function as shown next (different from the way it was used in Section 2.5.2). Besides, the **diag()**

function can also be used along with the **sum()** function to find the trace of a matrix. Use the **dim()** function to obtain the dimension of the matrix (i.e., its number of row and column).

```
> a_matrix=matrix(c(12,13,14,10,20,30,15,25,35),nrow=3,ncol=3)
> a_matrix
     [,1] [,2] [,3]
[1,]   12   10   15
[2,]   13   20   25
[3,]   14   30   35
> diag(a_matrix)
[1] 12 20 35
#Find the trace
> sum(diag(a_matrix))
[1] 67
#The dimension
> dim(a_matrix)
[1] 3 3
```

2.6.3 Determinant, Inverse, and Transpose

The three functions available in R to obtain the determinant, inverse, and transpose of a matrix are **det()**, **solve()**, and **t()**, respectively.

```
> matb=matrix(c(2,3,5,7,11,13,17,19,23), ncol=3,nrow=3,
byrow = T)
> matb
     [,1] [,2] [,3]
[1,]    2    3    5
[2,]    7   11   13
[3,]   17   19   23
#Determinant of matb
> det(matb)
[1] -78
#Inverse of matb
> solve(matb)
            [,1]       [,2]        [,3]
[1,] -0.07692308 -0.3333333  0.20512821
[2,] -0.76923077  0.5000000 -0.11538462
[3,]  0.69230769 -0.1666667 -0.01282051
#Transpose of matb
> t(matb)
     [,1] [,2] [,3]
[1,]    2    7   17
[2,]    3   11   19
[3,]    5   13   23
```

2.6.4 Eigenvalues and Eigenvectors

Eigenvalues appear frequently in the analyses of dynamic problems (vibration analyses, for instance) and stability problems (such as buckling and post-buckling analyses). Indeed, we will deal with eigenvalue problems in Chapter 12.

R provides the **eigen()** function to find the eigenvalues of a matrix.

```
> matb=matrix(c(2,-1,-1,2),byrow=T,nrow=2)
> matb
     [,1] [,2]
[1,]    2   -1
[2,]   -1    2
> eigen(matb)
$values
[1] 3 1

$vectors
           [,1]       [,2]
[1,] -0.7071068 -0.7071068
[2,]  0.7071068 -0.7071068
```

2.6.5 Arithmetic Operations on Matrices

Common operations such as addition, subtraction, division, and multiplication can be performed between matrix variables and a scalar.

```
> matrix_c=matrix(1:8,ncol=4,nrow=4)
> matrix_c
     [,1] [,2] [,3] [,4]
[1,]    1    5    1    5
[2,]    2    6    2    6
[3,]    3    7    3    7
[4,]    4    8    4    8
#Multiply by a scalar
> matrix_c*2
     [,1] [,2] [,3] [,4]
[1,]    2   10    2   10
[2,]    4   12    4   12
[3,]    6   14    6   14
[4,]    8   16    8   16
#Subtracting a scalar from the matrix
> matrix_c- 3
     [,1] [,2] [,3] [,4]
[1,]   -2    2   -2    2
[2,]   -1    3   -1    3
[3,]    0    4    0    4
[4,]    1    5    1    5
```

Further, given two or more matrices of the same order, we can perform element-by-element arithmetic operations involving addition, subtraction, division, and multiplication as shown in the following:

```
> matd=matrix(c(2,3,5,7,11,13,17,19,23),nrow=3,byrow=T)
> matf=matrix(c(0.2,0.3,0.5,0.7,1.1,1.3,1.7,1.9,2.3),nrow=3,
byrow=T)
> matd
     [,1] [,2] [,3]
[1,]    2    3    5
[2,]    7   11   13
[3,]   17   19   23
> matf
     [,1] [,2] [,3]
[1,]  0.2  0.3  0.5
[2,]  0.7  1.1  1.3
[3,]  1.7  1.9  2.3
#Adding together
> matd+matf
     [,1] [,2] [,3]
[1,]  2.2  3.3  5.5
[2,]  7.7 12.1 14.3
[3,] 18.7 20.9 25.3
#Subtracting
> matd-matf
     [,1] [,2] [,3]
[1,]  1.8  2.7  4.5
[2,]  6.3  9.9 11.7
[3,] 15.3 17.1 20.7
#Dividing
> matd/matf
     [,1] [,2] [,3]
[1,]   10   10   10
[2,]   10   10   10
[3,]   10   10   10
#Multiplication
> matd*matf
     [,1] [,2] [,3]
[1,]  0.4  0.9  2.5
[2,]  4.9 12.1 16.9
[3,] 28.9 36.1 52.9

#Exponentiating
> matd^matf
           [,1]       [,2]        [,3]
[1,]   1.148698   1.390389    2.236068
[2,]   3.904529  13.980798   28.062500
[3,] 123.527362 268.925411 1355.108355
```

2.6.6 Matrix Multiplication

When multiplying two matrices, say A and B, the multiplication obeys the rule that the column of B be equal to the row of A. The operator for matrix multiplication in R is `%*%`.

In the next code snippet, we will use the **`runif()`** function to generate the elements of two random matrices. Their matrix multiplication will then be carried out.

```
> set.seed(344)
> matrix(runif(9),3,3)
          [,1]      [,2]      [,3]
[1,] 0.8584339 0.2030884 0.8821016
[2,] 0.9462583 0.2324289 0.5857386
[3,] 0.9995209 0.5836422 0.6193479
> matrix(runif(9),3,3)
          [,1]       [,2]      [,3]
[1,] 0.3624611 0.15553808 0.7646178
[2,] 0.6464682 0.73466221 0.2300535
[3,] 0.9526003 0.05195411 0.9894010
> set.seed(344)
> matrix(runif(9),3,3)
          [,1]      [,2]      [,3]
[1,] 0.8584339 0.2030884 0.8821016
[2,] 0.9462583 0.2324289 0.5857386
[3,] 0.9995209 0.5836422 0.6193479
> set.seed(344)
> matA=matrix(runif(9),3,3)
> matA
          [,1]      [,2]      [,3]
[1,] 0.8584339 0.2030884 0.8821016
[2,] 0.9462583 0.2324289 0.5857386
[3,] 0.9995209 0.5836422 0.6193479
> set.seed(345)
> matB=matrix(runif(9),3,3)
> matB
          [,1]      [,2]      [,3]
[1,] 0.2162537 0.6557397 0.3856799
[2,] 0.2747640 0.4358664 0.8333017
[3,] 0.3899251 0.8034841 0.4730793
```

The product of the two matrices (**`matA`** and **`matB`**) and the cross products are then found using `%*%` and **`crossprod()`**, respectively.

```
> matA%*%matB
          [,1]     [,2]      [,3]
[1,] 0.5853944 1.360183 0.9176186
[2,] 0.4968891 1.192439 0.8357370
[3,] 0.6180132 1.407452 1.1648458
#Cross product
> crossprod(matA,matB)
          [,1]      [,2]      [,3]
[1,] 0.8353755 1.7784505 1.5924519
[2,] 0.3353584 0.7034283 0.5481195
[3,] 0.5931969 1.3313690 1.1213065
```

Many other functions related to vectors and matrices, not covered here, can be found in other references [1–3]. However, the functions explored in this chapter are sufficient for our journey in the coding of the FEM with R.

References

1. N. Matloff, *The Art of R Programming: A Tour of Statistical Software Design*, No Starch Press, San Francisco, CA, 2011.
2. V. A. Bloomfield, *Using R for Numerical Analysis in Science and Engineering*, Taylor & Francis Group, Boca Raton, FL, 2014.
3. N. Fieller, *Basics of Matrix Algebra for Statistics with R*, CRC Press, Boca Raton, FL, 2015.

3

Linear Spring Elements

3.1 Introduction

The philosophy of practical finite element computations rests on the matrix solution approach to structural analyses. A major focus of structural analyses is to find the relationships between loads and the corresponding deformations developed within properly constrained load-carrying members. In this book, we will focus mostly on the structural analyses of members under linear elastic deformations. Indeed, for the structures we will study, the load-deformation relationships may be obtained or available in analytical forms. However, a horde of practical cases exists where the impracticality of obtaining analytical solutions favors the use of numerical methods such as the finite element methods (FEM).

We will begin, in this chapter, our exploration of the concepts of FEM through the formulation of the model and computer implementation of the spring element, which is often used to idealize the mechanical behavior of linear elastic springs, by which we mean springs for which the direct proportionality of the force and deformation holds. The study of the analysis of linear springs provides a simple entrance into the theory of FEM for several reasons. Patently, springs provide simple means of storing potential energy in several mechanical systems, a usefulness that enables their suitability for use as mechanical links [1]. Besides, linear springs are at the heart of discrete modeling of mechanical systems.

The mathematical model of linear springs, which dates back to Robert Hooke (1635–1703), provides an intuitive relation that can be mapped, after careful mathematical manipulations, to other complex systems [2]. A common form of Hooke's law as it relates to linear elastic springs takes the form:

$$F = k\delta \tag{3.1}$$

where, according to Figure 3.1, F is the applied force that generates an extension δ, within a spring with stiffness k. From static analyses perspective, Figure 3.1b represents the free-body diagram of the extended spring with the reaction force F_R.

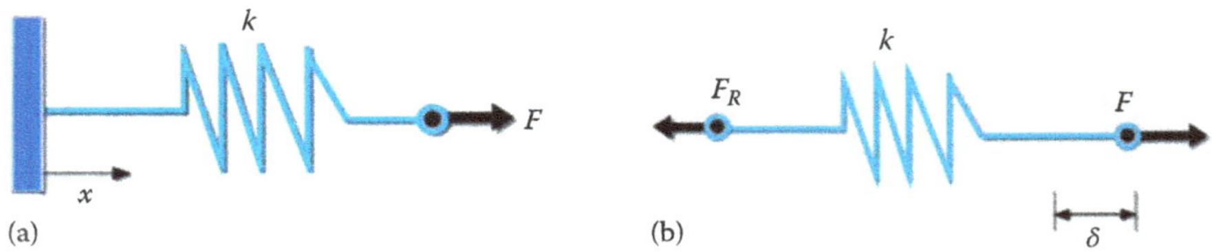

FIGURE 3.1
(a) A linear spring loaded with force F at the free end and (b) the free-body diagram of the spring.

3.2 The Spring Element

3.2.1 Element Matrix Equation Using the Direct Method of Formulation

For a single spring, Equation 3.1 is enough to estimate the displacement of the free end. However, in the case of a system of interconnected springs, obtaining the distribution of displacements and the reaction forces developed within the system in a closed-form version is not so straightforward. In this case, numerical solutions based on the FEM can be used for this purpose. The development of the spring element requires the following derivations in such situations.

The derivation of the mathematical expressions for the spring element for use in the finite element analysis of interconnected springs is detailed in a number of studies [3]. Procedurally, one can formulate the expressions for the spring element by relying on Figure 3.2, which shows an isolated linear spring (which may be initially part of a system of springs) with nodes i and j.

If the nodal displacements (u_1 and u_2) and nodal forces (f_1 and f_2) of the isolated linear spring element are as symbolized in Figure 3.2, then, consonant with [3], the equilibrium equations for the nodal forces can be written as:

$$f_1 = -f_2 = k(u_1 - u_2) \tag{3.2}$$

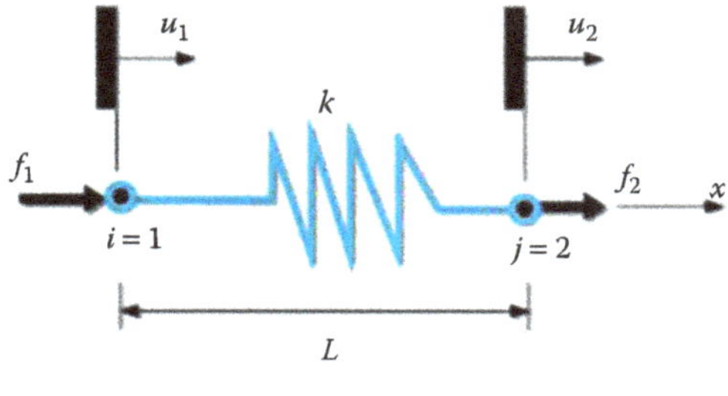

FIGURE 3.2
An isolated linear spring element with nodal force and nodal displacements.

From Equation 3.2, the finite element matrix equation for the spring element with a spring rate k can be derived in the form:

$$\begin{Bmatrix} f_1 \\ f_2 \end{Bmatrix}^e = \boldsymbol{f}^e = \begin{bmatrix} k & -k \\ -k & k \end{bmatrix}^e \begin{Bmatrix} u_1 \\ u_2 \end{Bmatrix}^e = \boldsymbol{k}^e \boldsymbol{u}^e \tag{3.3}$$

where the left-hand side of the matrix equation represents a vector of nodal forces, whereas the right-hand side is a product of the local stiffness matrix $\left(\boldsymbol{k}^e\right)$ and the vector of nodal displacements $\left(\boldsymbol{u}^e\right)$.

3.2.2 Element Matrix from the Principle of Minimum Potential Energy

For more advanced elements, we will be deriving their matrix equations via the powerful method that revolves around the *principle of minimum potential energy*. Using this principle, we can arrive at Equation 3.3 very easily. The *principle of minimum potential energy* states that [4]:

> For conservative systems, of all the kinematically admissible displacement fields, those corresponding to equilibrium extremize the total potential energy.

There are two main ingredients for the use of this principle. The first is identifying the total potential energy Π_{sp}, which mathematically is given by:

$$\Pi_{sp} = U - W \tag{3.4}$$

where U is the strain energy that develops from the deformation of the system and W is the work done by external forces (W). The second is minimizing the expression of the total potential energy with respect to the deformation measures. For the spring in Figure 3.2, we have:

$$U = \frac{1}{2}k\left(u_2 - u_1\right)^2 \tag{3.5}$$

$$W = f_1 u_1 + f_2 u_2 \tag{3.6}$$

The total potential energy for the linear spring (Π_{sp}) is therefore given by:

$$\Pi_{sp} = \frac{1}{2}k\left(u_2 - u_1\right)^2 - f_1 u_1 - f_2 u_2 \tag{3.7}$$

Minimizing Π_{sp} with respect to $u_k\,(k=1,2)$ gives:

$$\frac{\partial \Pi_{sp}}{\partial u_1} = k(u_1 - u_2) - f_1 \tag{3.8}$$

$$\frac{\partial \Pi_{sp}}{\partial u_2} = k(u_2 - u_1) - f_2 \tag{3.9}$$

Rearranging and combining Equations 3.8 and 3.9 lead to the same expression stated in Equation 3.3.

That is:

$$\begin{Bmatrix} f_1 \\ f_2 \end{Bmatrix}^e = \begin{bmatrix} k & -k \\ -k & k \end{bmatrix}^e \begin{Bmatrix} u_1 \\ u_2 \end{Bmatrix}^e \tag{3.10}$$

In principle, the size of the vector of nodal displacements (which in this case is 2) determines the total degree of freedom for the element. Further, half of the size of the vector of nodal displacements (which in this case is 1) determines the degree of freedom per node of the element. Moreover, we noticed that $[k]$ is a symmetric square matrix of dimension 2×2. This symmetricity carries over to the analyses of a combined system of connected springs.

For connected linear springs, one must form the global matrix equation from the elements of the connected system. Written in the global reference frame, the global matrix equation for a system of connected springs relates the vector of global nodal forces $\{\boldsymbol{F}\}$, the global stiffness matrix $[\boldsymbol{K}]$, and the vector of global nodal displacements $\{\boldsymbol{U}\}$ as:

$$\{\boldsymbol{F}\} = [\boldsymbol{K}]\{\boldsymbol{U}\} \tag{3.11}$$

We have highlighted three key properties of $\boldsymbol{K}$ in Equation 3.11 in the following:

- In terms of size, the global stiffness matrix for the system of connected springs with N nodes is $N \times N$.
- The matrix $\boldsymbol{K}$ is symmetric, which means it satisfies the conditions $\boldsymbol{K}^T = \boldsymbol{K}$.
- The matrix $\boldsymbol{K}$ is singular, which means we need to apply boundary conditions to the system's matrix equation to obtain the relevant nodal variables (forces and displacements).

3.3 Computer Implementation

3.3.1 Systematic Procedure for Solving Static Problems of Connected Linear Springs

For problems dealing with connected springs, the following steps are recommended.

3.3.1.1 Precomputation Phase

- Discretization of the system into individual springs
- Labeling of the nodes and elements, with no two elements having the same set of nodal labels

3.3.1.2 Computation Phase

1. Identify the elements' properties and write the finite element matrix equation for each using Equation 3.10.
2. Determine the global degree of freedom of the spring system and expand the matrix equation for each element (from step 1) to conform to the size of the global degree of freedom.
3. Establish the global matrix equation from the assemblage of the expanded matrices of the elements.
4. Identify the essential boundary condition(s) and apply the boundary condition to the global matrix equation to obtain a reduced matrix equation. Programmatically, this step involves two actions:
   a. Obtain the reduced force vector.
   b. Use the identified boundary condition to obtain the reduced stiffness matrix.
5. Determine the unknown nodal displacements by using the reduced matrix equation.
6. Obtain the nodal element forces using the nodal displacement.
7. Obtain the unknown global forces (which will include the reaction forces).

3.3.2 Implemented R Subroutines for Linear Spring Elements

Automated functions for dealing with the above-listed steps are presented here. In principle, programmed functions in the R programming language are equivalent to subroutines in some other languages such as FORTRAN.

```
#*****************************************#
# R functions for linear spring elements  #
# KB Mustapha                             #
#*****************************************#

LinearSpring_Element_Matrix = function(DOF=2,stiffk)
{
  ematrix=stiffk*matrix(c(1,-1,-1,1),nrow = DOF,byrow = T)
  return(ematrix)
}

LinearSpring_ExpandedElement_Matrix
= function(TDOF,eMatrix,i,j)
{
  bigMatrix = matrix(vector(l = TDOF * TDOF),nrow
  = TDOF,byrow = T);
  bigMatrix[c(i,j),c(i,j)] = eMatrix; return (bigMatrix)
}

LinearSpring_Element_ReducedStiffnessMatrix
= function(globalstiffnessmatrix,knownforcenodes)
{
  reducedstiffnessmatrix = globalstiffnessmatrix[c(knownf
  orcenodes),(knownforcenodes)]
  return(reducedstiffnessmatrix)
}

LinearSpring_Element_ReducedForceVector
= function(forcevector)
{
  reducedforcevector = matrix(forcevector,ncol = 1)
  return(reducedforcevector)
}

LinearSpring_Element_NodalDisplacement = function(reduced
matrix,reducedforcevec)
{
  return(solve(reducedmatrix,reducedforcevec))
}

LinearSpring_Element_Forces = function(DOF=2,stiffk,total
nodaldisp,i,j)
{
  nodaldisplacment=totalnodaldisp[c(i,j)]
  stiffnessmatrix = stiffk * matrix(c(1,-1,-1,1),nrow
= DOF,byrow = T)
  nodal_forces = stiffnessmatrix %*% nodaldisplacment
  return(nodal_forces)
}
```

```
LinearSpring_Global_Forces = function(globalstiffness,tot
alnodaldisp)
{
  global_forces = globalstiffness %*% totalnodaldisp
  return(round(global_forces))
}
```

We have highlighted the usage of each of the listed functions along with the parameters they accept in Tables 3.1 and 3.2.

TABLE 3.1

Functions for the Element's Matrices and Vectors

Description of Functions
LinearSpring_Element_Matrix(DOF=2,stiffk) – This function generates the stiffness matrix for a horizontally positioned linear spring element. It accepts two parameters. The first being **DOF**, which stands for degree of freedom of the spring (predefined as 2), and the second is **stiffk**, which represents the stiffness of a spring element.
LinearSpring_ExpandedElement_Matrix(TDOF,eMatrix,i,j) – This function forms the expanded stiffness matrix for a spring element. It accepts four parameters. The first is **TDOF**, which stands for total degree of freedom of an assemblage. The second is **eMatrix**, which represents the unexpanded stiffness matrix of a spring element. The third and fourth parameters relate to the nodal labels of a spring element.
LinearSpring_Element_ReducedForceVector(forcevector) – This function generates a reduced force vector needed to find the unknown nodal displacements. It accepts the parameter **forcevector**.
LinearSpring_Element_ReducedStiffnessMatrix(globalstiffnessmatrix, knownforcenodes) – This function generates a reduced stiffness matrix for a spring assemblage. Its parameters are the global stiffness matrix and a vector specifying the rows of known force terms.

TABLE 3.2

Functions for the FEA Solution Phase

Description of Functions
LinearSpring_Element_NodalDisplacement(reducedmatrix, reducedforcevec) – This function solves the unknown nodal displacements. It uses the parameters **reducedmatrix** (reduced stiffness matrix) and **reducedforcevec** (reduced force vector).
LinearSpring_Element_Forces(DOF=2,stiffk,totalnodaldisp,i,j) – This function determines the axial forces in a spring element. It accepts the default argument of 2 for the **DOF** parameter, **stiffk** (spring stiffness), **totalnodaldisp** (a vector of global nodal displacements), and **i,j** (nodal labels of the element).
LinearSpring_Global_Forces(globalstiffness,totalnodaldisp) – This function calculates the global forces (global reaction forces) in a spring assemblage. Its parameters are **globalstiffness** (the global stiffness matrix) and **totalnodaldisp** (a vector of the global nodal displacement).

We shall demonstrate the applications of the functions for the spring element in Examples 3.1 and 3.2.

3.4 Examples

3.4.1 Example 3.1

For the spring assemblage shown in Figure 3.3, (a) formulate the global stiffness matrix; (b) apply the requisite boundary conditions to obtain the reduced stiffness matrix; (c) determine the nodal element forces in each spring element; and (d) determine the reaction forces.

3.4.1.1 Solution

It is important to note that to use the codes provided in Section 3.3.2 for any problem, the functions must first be compiled to make them available to the computers' memory.

Step 1: Write the finite element matrix equation for each element.

The stiffness matrix is an important aspect of the finite element matrix equation. Therefore for each element, we obtain the stiffness matrix using the function **LinearSpring_Element_Matrix()** in the following manner:

```
springstiffness=20000;
k1=LinearSpring_Element_Matrix(2,springstiffness);k1
k2=LinearSpring_Element_Matrix(2,springstiffness);k2
k3=LinearSpring_Element_Matrix(2,springstiffness);k3
k4=LinearSpring_Element_Matrix(2,springstiffness);k4
```

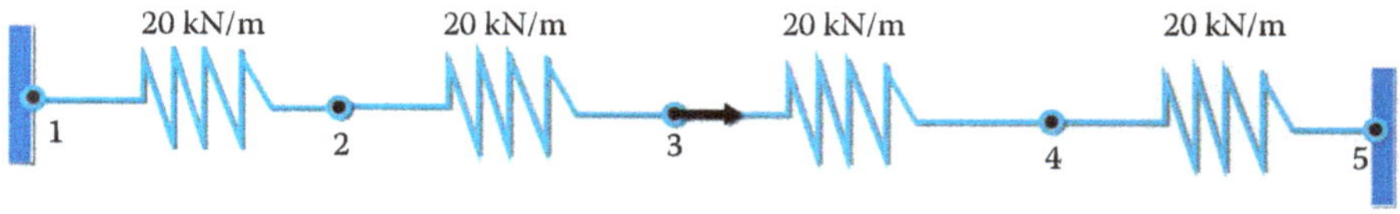

FIGURE 3.3
Interconnected serial springs.

The output from running the above reads as:

```
> springstiffness=20000;
> k1=LinearSpring_Element_Matrix(2,springstiffness);k1
       [,1]   [,2]
[1,]  20000 -20000
[2,] -20000  20000
> k2=LinearSpring_Element_Matrix(2,springstiffness);k2
       [,1]   [,2]
[1,]  20000 -20000
[2,] -20000  20000
> k3=LinearSpring_Element_Matrix(2,springstiffness);k3
       [,1]   [,2]
[1,]  20000 -20000
[2,] -20000  20000
> k4=LinearSpring_Element_Matrix(2,springstiffness);k4
       [,1]   [,2]
[1,]  20000 -20000
[2,] -20000  20000
```

Step 2: Determine the global degree of freedom of the spring assemblage and expand the matrix equation for each element (from step 1) to conform to the size of the global degree of freedom.

From Figure 3.3, it is noticed that there are five nodes in the assemblage. For the spring element, each node of this is endowed with just one degree of freedom. Thus, a five-noded system will have a total of five degrees of freedom. We use the function **LinearSpring_ExpandedElement_Matrix()** from Section 3.3.2 for the expansion of the element matrix.

```
n_nodes=5;Total_dof=n_nodes*1;
K1 = LinearSpring_ExpandedElement_Matrix(Total_
dof,k1,1,2);K1
K2 = LinearSpring_ExpandedElement_Matrix(Total_
dof,k2,2,3);K2
K3 = LinearSpring_ExpandedElement_Matrix(Total_
dof,k3,3,4);K3
K4 = LinearSpring_ExpandedElement_Matrix(Total_
dof,k4,4,5);K4
```

The output of this is obtained as follows:

```
> n_nodes=5;Total_dof=n_nodes*1;
> K1 = LinearSpring_ExpandedElement_Matrix(Total_
dof,k1,1,2);K1
       [,1]   [,2] [,3] [,4] [,5]
[1,]  20000 -20000    0    0    0
[2,] -20000  20000    0    0    0
[3,]      0      0    0    0    0
[4,]      0      0    0    0    0
[5,]      0      0    0    0    0
> K2 = LinearSpring_ExpandedElement_Matrix(Total_
dof,k2,2,3);K2
     [,1]   [,2]   [,3] [,4] [,5]
[1,]    0      0      0    0    0
[2,]    0  20000 -20000    0    0
[3,]    0 -20000  20000    0    0
[4,]    0      0      0    0    0
[5,]    0      0      0    0    0
> K3 = LinearSpring_ExpandedElement_Matrix(Total_
dof,k3,3,4);K3
     [,1] [,2]   [,3]   [,4] [,5]
[1,]    0    0      0      0    0
[2,]    0    0      0      0    0
[3,]    0    0  20000 -20000    0
[4,]    0    0 -20000  20000    0
[5,]    0    0      0      0    0
> K4 = LinearSpring_ExpandedElement_Matrix(Total_
dof,k4,4,5);K4
     [,1] [,2] [,3]   [,4]   [,5]
[1,]    0    0    0      0      0
[2,]    0    0    0      0      0
[3,]    0    0    0      0      0
[4,]    0    0    0  20000 -20000
[5,]    0    0    0 -20000  20000
```

Step 3: Establish the global stiffness matrix by adding the expanded matrices of the elements as shown in the following:

```
GlobalK = K1+K2+K3+K4;
GlobalK
```

The output of this gives the global stiffness matrix:

```
> GlobalK = K1+K2+K3+K4;
> GlobalK
       [,1]   [,2]   [,3]   [,4]   [,5]
[1,]  20000 -20000      0      0      0
[2,] -20000  40000 -20000      0      0
[3,]      0 -20000  40000 -20000      0
[4,]      0      0 -20000  40000 -20000
[5,]      0      0      0 -20000  20000
```

Step 4: We identify the essential boundary condition(s) and apply the boundary condition(s) to the global matrix equation to obtain a reduced stiffness matrix and reduced force vector. Given the global stiffness matrix obtained in Step 3, we write the global matrix equation as:

$$\begin{Bmatrix} F_1 \\ F_2 \\ F_3 \\ F_4 \\ F_5 \end{Bmatrix} = \begin{bmatrix} 20000 & -20000 & 0 & 0 & 0 \\ -20000 & 40000 & -20000 & 0 & 0 \\ 0 & -20000 & 40000 & -20000 & 0 \\ 0 & 0 & -20000 & 40000 & -20000 \\ 0 & 0 & 0 & -20000 & 20000 \end{bmatrix} \begin{Bmatrix} U_1 \\ U_2 \\ U_3 \\ U_4 \\ U_5 \end{Bmatrix} \tag{3.12}$$

As per the boundary condition, the fixing of nodes 1 and 5 (as seen from Figure 3.3) translates to zero displacements at these nodes. As a result, the boundary conditions will be:

$$U_1 = U_5 = 0 \tag{3.13}$$

Further, the vector of known nodal forces (in *Newton*) can be identified as:

$$\begin{Bmatrix} F_2 \\ F_3 \\ F_4 \end{Bmatrix} = \begin{Bmatrix} 0 \\ 5000 \\ 0 \end{Bmatrix} \tag{3.14}$$

Notice that the nodes at which the force values are known (Equation 3.14) are different from those at which the displacement values are known (Equation 3.13). This is a fundamental principle in the application of the FEM. Moving on, the reduced force vector in Equation 3.14 is used as an argument to the function **LinearSpring_Element_ReducedForceVector()**:

```
vec_values_of_knownloads=c(0,5000,0)
ReducedForceVector = LinearSpring_Element_
ReducedForceVector(vec_values_of_knownloads);
```

Next, from Equation 3.13, it is observed that the known nodal forces belong to rows 2, 3, and 4. This information is useful in finding the reduced stiffness matrix using the function **`LinearSpring_Element_ReducedStiffnessMatrix()`** as shown in the following:

```
vec_rows_of_knownloads=c(2,3,4)
ReducedK = LinearSpring_Element_ReducedStiffnessMatrix
(GlobalK,vec_rows_of_knownloads);
ReducedK
```

The output then reads:

```
> ReducedK
        [,1]   [,2]   [,3]
[1,]  40000 -20000      0
[2,] -20000  40000 -20000
[3,]      0 -20000  40000
```

Step 5: Determine the unknown nodal displacements by using the sixth function:

```
UnknwonNodalDisp = LinearSpring_Element_NodalDisplacement
(ReducedK,ReducedForceVector);
UnknwonNodalDisp
```

The execution of the preceding block of code produced:

```
> UnknwonNodalDisp = LinearSpring_Element_NodalDisplacement
(ReducedK,ReducedForceVector);
> UnknwonNodalDisp
      [,1]
[1,] 0.125
[2,] 0.250
[3,] 0.125
```

Hence, the determined values of the displacements are:

$$\begin{Bmatrix} U_2 \\ U_3 \\ U_4 \end{Bmatrix} = \begin{Bmatrix} 0.125 \\ 0.250 \\ 0.125 \end{Bmatrix} m$$

For convenience, all nodal displacements are combined into a single vector and stored in the variable called **allglobalNodalDisp** as shown in the following:

```
allglobalNodalDisp=matrix(rep(0,Total_dof),byrow=T);
allglobalNodalDisp[vec_rows_of_
knownloads]=UnknwonNodalDisp
allglobalNodalDisp
```

Which if executed yields:

```
> allglobalNodalDisp
      [,1]
[1,] 0.000
[2,] 0.125
[3,] 0.250
[4,] 0.125
[5,] 0.000
```

We will use this variable in the succeeding code to obtain the reaction forces.

Step 6: With the nodal displacements found in Step 5, we obtain the element nodal forces using the function **LinearSpring_Element_Forces()**:

```
Element1_Forces = LinearSpring_Element_Forces(2,springsti
ffness,allglobalNodalDisp,1,2);
Element1_Forces
Element2_Forces = LinearSpring_Element_Forces(2,springsti
ffness,allglobalNodalDisp,2,3);
Element2_Forces
Element3_Forces = LinearSpring_Element_Forces(2,springsti
ffness,allglobalNodalDisp,3,4);
Element3_Forces
Element4_Forces = LinearSpring_Element_Forces(2,springsti
ffness,allglobalNodalDisp,4,5);
Element4_Forces
```

As observed, the function **LinearSpring_Element_Forces()** requires the vector of global nodal displacements (stored in **allglobalNodalDisp**) and the nodal labels of the respective element.

```
> Element1_Forces
      [,1]
[1,] -2500
[2,]  2500

> Element2_Forces = LinearSpring_Element_Forces(2,springs
tiffness,allglobalNodalDisp,2,3);
> Element2_Forces
      [,1]
[1,] -2500
[2,]  2500
>
> Element3_Forces = LinearSpring_Element_Forces(2,springs
tiffness,allglobalNodalDisp,3,4);
> Element3_Forces
      [,1]
[1,]  2500
[2,] -2500
>
> Element4_Forces = LinearSpring_Element_Forces(2,springs
tiffness,allglobalNodalDisp,4,5);
> Element4_Forces
      [,1]
[1,]  2500
[2,] -2500
```

It is apparent from the preceding listed results that spring elements 1 and 2 are in tension, whereas spring elements 3 and 4 are in compression.

Step 7: We determine the unknown global forces (including the reaction forces at nodes 1 and 5) by calling the function **LinearSpring_Global_Forces()**.

```
Globalforcevector = LinearSpring_Global_Forces(GlobalK,al
lglobalNodalDisp)
Globalforcevector
```

With the output being:

```
> Globalforcevector
      [,1]
[1,] -2500
[2,]     0
[3,]  5000
[4,]     0
[5,] -2500
```

The result shows that $F_1 = F_5 = -2500$ N. Finally, by considering the equilibrium equation of the system in x direction, one can confirm the accuracy of the determined reaction forces by noting that the equilibrium equation (for a one-dimensional bar) is satisfied as:

$$\sum F_x = -2500 + 5000 - 2500 = 0 \tag{3.15}$$

3.4.2 Example 3.2

For the spring assemblage shown in Figure 3.4, (a) formulate the global stiffness matrix; (b) apply the requisite boundary conditions to obtain the reduced stiffness matrix; (c) determine the nodal element forces in each spring element; and (d) find the reaction forces.

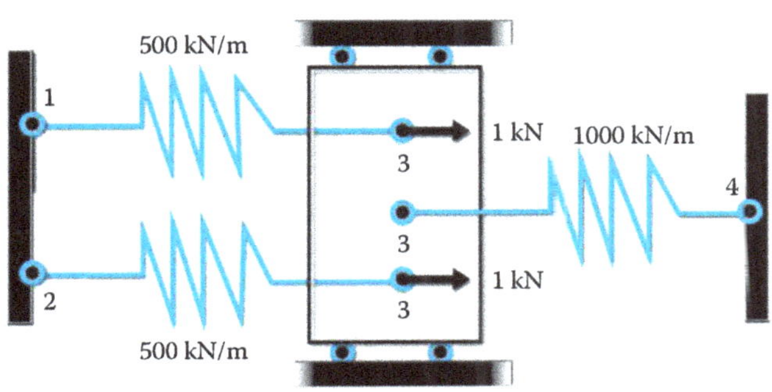

FIGURE 3.4
A system with parallel and serial combinations of springs.

TABLE 3.3

Element Property and Connectivity Table

Elements	Nodes (i, j)	Spring Stiffness
1	1,3	500 kN/m
2	2, 3	500 kN/m
3	3, 4	1000 kN/m

3.4.2.1 Solution

Based on the assemblage of Figure 3.4, we form the elements' connectivity in Table 3.3. In all, this system has four unique nodes with a total force 2 kN at node 3.

Step 1: With the stiffness stated in Table 3.3, use the **LinearSpring_Element_Matrix()** function to obtain the stiffness matrix for each of the spring elements.[1]

```
springstiffness1 = 500*10^3;
springstiffness2 = 500*10^3;
springstiffness3 = 1000*10^3;
k1 = LinearSpring_Element_Matrix(2,springstiffness1);k1
k2 = LinearSpring_Element_Matrix(2,springstiffness2);k2
k3 = LinearSpring_Element_Matrix(2,springstiffness3);k3
```

The output from running the above reads as:

```
> springstiffness1 = 500*10^3;
> springstiffness2 = 500*10^3;
> springstiffness3 = 1000*10^3;
> k1 = LinearSpring_Element_Matrix(2,springstiffness1);k1
        [,1]    [,2]
[1,]  500000 -500000
[2,] -500000  500000
> k2 = LinearSpring_Element_Matrix(2,springstiffness2);k2
        [,1]    [,2]
[1,]  500000 -500000
[2,] -500000  500000
> k3 = LinearSpring_Element_Matrix(2,springstiffness3);k3
         [,1]     [,2]
[1,]  1000000 -1000000
[2,] -1000000  1000000
```

[1] Note: If the RStudio program has been closed immediately after running the previous example, you should begin the current analysis by compiling the functions provided in Section 3.3.2 again.

Step 2: Determine the global degree of freedom of the spring assemblage and expand the matrix equation for each element to conform the size of the global degree of freedom.

This system has four unique nodes, resulting in a total degree of freedom that equals to four, stored in the variable **`Total_dof`** below.

```
n_nodes=4;Total_dof=n_nodes*1;
K1 = LinearSpring_ExpandedElement_Matrix(Total_
dof,k1,1,3);K1
K2 = LinearSpring_ExpandedElement_Matrix(Total_
dof,k2,2,3);K2
K3 = LinearSpring_ExpandedElement_Matrix(Total_
dof,k3,3,4);K3
```

The output of this is obtained as follows:

```
> K1 = LinearSpring_ExpandedElement_Matrix(Total_
dof,k1,1,3);K1
        [,1] [,2]    [,3] [,4]
[1,]  500000    0 -500000    0
[2,]       0    0       0    0
[3,] -500000    0  500000    0
[4,]       0    0       0    0
> K2 = LinearSpring_ExpandedElement_Matrix(Total_
dof,k2,2,3);K2
     [,1]    [,2]    [,3] [,4]
[1,]    0       0       0    0
[2,]    0  500000 -500000    0
[3,]    0 -500000  500000    0
[4,]    0       0       0    0
> K3 = LinearSpring_ExpandedElement_Matrix(Total_
dof,k3,3,4);K3
     [,1] [,2]     [,3]     [,4]
[1,]    0    0        0        0
[2,]    0    0        0        0
[3,]    0    0  1000000 -1000000
[4,]    0    0 -1000000  1000000
```

Step 3: Establish the global stiffness matrix by adding the expanded matrices of the elements as shown next.

```
GlobalK = K1+K2+K3;
GlobalK
```

The output of this gives the global stiffness matrix of the assemblage in Figure 3.4.

```
> GlobalK
        [,1]    [,2]     [,3]     [,4]
[1,]  500000       0  -500000        0
[2,]       0  500000  -500000        0
[3,] -500000 -500000  2000000 -1000000
[4,]       0       0 -1000000  1000000
```

Step 4: Identify the essential boundary condition(s) and apply these boundary condition(s) to the global matrix equation to obtain a reduced stiffness matrix and a reduced force vector.

The global matrix equation is given earlier as:

$$\begin{Bmatrix} F_1 \\ F_2 \\ F_3 \\ F_4 \end{Bmatrix} = 10^5 \begin{bmatrix} 5 & 0 & -5 & 0 \\ 0 & 5 & -5 & 0 \\ -5 & -5 & 20 & -10 \\ 0 & 0 & -10 & 10 \end{bmatrix} \begin{Bmatrix} U_1 \\ U_2 \\ U_3 \\ U_4 \end{Bmatrix} \tag{3.16}$$

From Figure 3.4, we identify the following boundary conditions:

$$U_1 = U_2 = U_4 = 0 \tag{3.17}$$

Further, the vector of known nodal forces (in *Newton*) can be identified as:

$$\{F_3\} = \{1000 + 1000\} \tag{3.18}$$

With the reduced force vector in Equation 3.18, we can then call the function **LinearSpring_Element_ReducedForceVector()** as shown in the following:

```
vec_values_of_knownloads=c(2000)
ReducedForceVector = LinearSpring_Element_
ReducedForceVector(vec_values_of_knownloads);
ReducedForceVector
```

Further, from Equation 3.18, it is noticed that the known nodal force is related to row 3. This information is useful in finding the reduced stiffness matrix using the function **LinearSpring_Element_ReducedStiffnessMatrix()** as shown in the following:

```
vec_rows_of_knownloads=c(3)
ReducedK = LinearSpring_Element_ReducedStiffnessMatrix
(GlobalK,vec_rows_of_knownloads);
ReducedK
```

The output then reads:

```
> ReducedK
[1] 2000000
```

Step 5: Determine the unknown nodal displacements by using the function **`LinearSpring_Element_NodalDisplacement()`**.

```
UnknwonNodalDisp = LinearSpring_Element_NodalDisplacement
(ReducedK,ReducedForceVector);
UnknwonNodalDisp
```

The output produces:

```
> UnknwonNodalDisp = LinearSpring_Element_NodalDisplacement
(ReducedK,ReducedForceVector);
> UnknwonNodalDisp
      [,1]
[1,] 0.001
```

As done in the previous example, we combine the determined displacement with the boundary condition:

```
allglobalNodalDisp=matrix(rep(0,Total_dof),byrow=T);
allglobalNodalDisp[vec_rows_of_knownloads]=UnknwonNodal
Disp
allglobalNodalDisp
```

Which then yields:

```
> allglobalNodalDisp
      [,1]
[1,] 0.000
[2,] 0.000
[3,] 0.001
[4,] 0.000
```

The determined values of the displacements are all in *meters*.

Step 6: With the nodal displacements found in Step 5, we obtain the nodal element forces (for the three elements) using the function **Linear Spring_Element_Forces()**.

```
Element1_Forces = LinearSpring_Element_Forces(2,springsti
ffness1,allglobalNodalDisp,1,3);
Element1_Forces

Element2_Forces = LinearSpring_Element_Forces(2,springsti
ffness2,allglobalNodalDisp,2,3);
Element2_Forces

Element3_Forces = LinearSpring_Element_Forces(2,springsti
ffness3,allglobalNodalDisp,3,4);
Element3_Forces
```

With the output being:

```
> Element1_Forces = LinearSpring_Element_Forces(2,springs
tiffness1,allglobalNodalDisp,1,3);
> Element1_Forces
     [,1]
[1,] -500
[2,]  500
>
> Element2_Forces = LinearSpring_Element_Forces(2,springs
tiffness2,allglobalNodalDisp,2,3);
> Element2_Forces
     [,1]
[1,] -500
[2,]  500
> Element3_Forces = LinearSpring_Element_Forces(2,springs
tiffness3,allglobalNodalDisp,3,4);
> Element3_Forces
      [,1]
[1,]  1000
[2,] -1000
```

Step 7: Obtain the unknown global forces (including the reaction forces at nodes 1 and 4).

For this, we call on the function **LinearSpring_Global_Forces()** as shown next:

```
Globalforcevector = LinearSpring_Global_Forces(GlobalK,al
lglobalNodalDisp)
Globalforcevector
```

Giving the reaction forces as:

```
> Globalforcevector = LinearSpring_Global_Forces(GlobalK,
allglobalNodalDisp)
> Globalforcevector
       [,1]
[1,]   -500
[2,]   -500
[3,]   2000
[4,]  -1000
```

Again, notice that the reaction forces add up to the applied force, and their sum (external forces plus support reaction forces) adds up to zero. Finally, although the implemented functions have been applied to only two examples, they can be applied to handle the finite element analyses of many variants of connected system of springs.

3.5 Exercises

1. For the spring assemblage shown in Figure 3.3, if the support at node 5 is removed and a force of 5 kN is applied at this node: (a) determine the unknown nodal displacements; (b) determine the nodal element forces in each spring element; and (c) determine the reaction forces.
2. For the spring assemblage shown in Figure 3.5: (a) formulate the global stiffness matrix; (b) apply the requisite boundary conditions to obtain the reduced stiffness matrix; (c) determine the nodal element forces in each spring element; and (d) find the reaction forces. The three springs have the same stiffness of 50 kN/m.

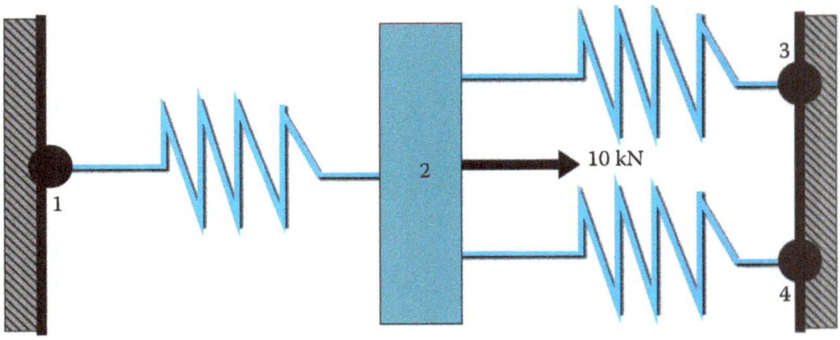

FIGURE 3.5
Exercise 3Q2.

References

1. S. S. Rao, F. F. Yap, *Mechanical Vibrations*, Prentice Hall: Upper Saddle River, NJ, 2011.
2. K. E. Kurrer, E. Ramm, *The History of the Theory of Structures: From Arch Analysis to Computational Mechanics*, John Wiley & Sons: Berlin, Germany, 2012.
3. D. L. Logan, *First Course in the Finite Element Method*, Thomson: Toronto, ON, 2007.
4. T. R. Chandrupatla, A. D. Belegundu, *Introduction to Finite Elements in Engineering*, Prentice Hall: Upper Saddle River, NJ, 1997.

4

Linear and Quadratic Bar Elements

4.1 Introduction

This chapter introduces the finite element models and associated computer implementations for straight elastic bars with constant cross-sectional areas. In classification, bars belong to the family of axially loaded members. In some sense, the behavior of bars is analogous to that of springs. Accordingly, the axial loading of a bar produces an associated deformation directed along its principal axis [1].

The bar element discussed in this chapter has applications for the analyses of axially loaded connected rods, struts, planar-tensioned cables, axially loaded pipes, thin-walled tubes. The treatment of bars in this chapter serves as a precursor to the analyses of trusses in Chapter 5. In Section 4.2, the finite element matrix equation for bars is derived using the energy approach (alternative approaches can be found in other References 2–5). The derivation is followed, in Section 4.3, by the presentation of the R functions for tackling problems involving complex systems of bars. Using the developed functions in Section 4.4, we presented examples to demonstrate their applications to the determination of reaction forces and nodal displacements in interconnected bar systems. The chapter ends with exercises (Section 4.5) and references.

4.2 Finite Element Equations for Bar Elements

4.2.1 The Two-Node Linear Bar Element

The derivation of the finite element matrix equation for the bar element hinges on the schematics shown in Figure 4.1.

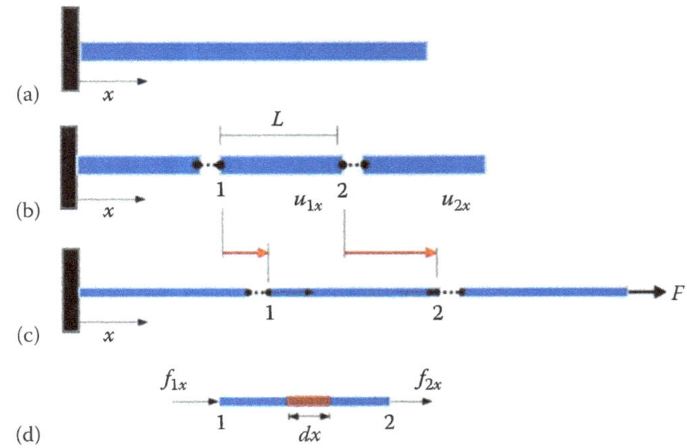

FIGURE 4.1
Straight elastic bars: (a) unloaded state, (b) segmented unloaded bar, (c) segmented loaded state, and (d) an isolated element of the bar.

For a start, consider the unloaded state, shown in Figure 4.1a, of a bar with cross-sectional area (A). In Figure 4.1b, we imposed a virtual segmentation on this unloaded bar, creating an element of length L possessing nodes 1 and 2. Under the influence of an axial force (F), the length and the cross-section of the bar change as illustrated in Figure 4.1c. Consequently, the displacements of nodes 1 and 2 of the bar element can be described as u_{1x} and u_{2x}, accompanying the reaction forces f_{1x} and f_{2x}.

Now, let us examine a slice of length dx of the isolated bar element as shown in Figure 4.1d. For this small slice, the differential strain energy associated with its deformation can be written as:

$$dU = \frac{1}{2}\sigma_x \varepsilon_x A dx \tag{4.1}$$

From Equation 4.1, the total strain energy of deformation across the entire element can be obtained by integrating over the length L as:

$$U = \frac{1}{2}\int_0^L \sigma_x \varepsilon_x A dx \tag{4.2}$$

For isotropic axially loaded elastic bars, a uniaxial state of stress exists, and the stress–strain relationship is:

$$\sigma_x = E\varepsilon_x \tag{4.3}$$

where E denotes the Young's modulus. Based on Equation 4.2, we can rewrite Equation 4.3 as:

$$U = \frac{1}{2}\int_0^L E\varepsilon_x^2 A\,dx \tag{4.4}$$

The strain–displacement relationship for the two-node bar element is of the following simple form:

$$\varepsilon_x = \frac{(u_{2x} - u_{1x})}{L} \tag{4.5}$$

By plugging Equation 4.5 into Equation 4.4, we obtain:

$$U = \frac{1}{2}\int_0^L E\left(\frac{u_{2x} - u_{1x}}{L}\right)^2 A\,dx \tag{4.6}$$

Integration of Equation 4.6 leads to:

$$U = \frac{EA}{2L}\left(u_{2x}^2 - 2u_{2x}u_{1x} + u_{1x}^2\right) \tag{4.7}$$

Meanwhile, the work done (W) by nodal forces f_{1x} and f_{2x} is obtained from the following relation:

$$W = f_{1x}u_{1x} + f_{2x}u_{2x} \tag{4.8}$$

With the aid of Equations 4.7 and 4.8, the finite element matrix equation for the two-node bar element depicted in Figure 4.2 is obtained by invoking the principle of minimum potential energy.

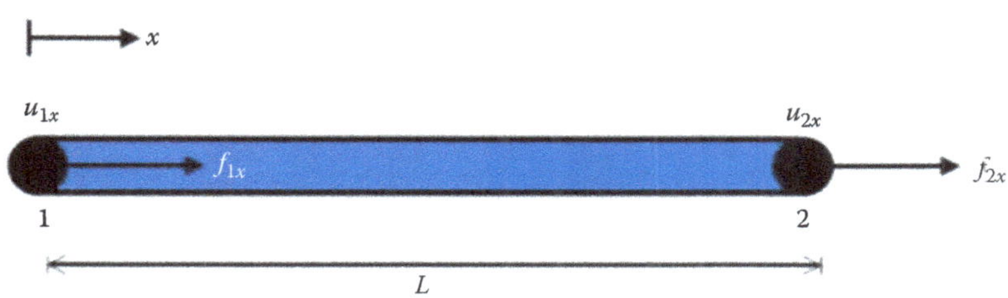

FIGURE 4.2
A one-dimensional linear bar element with two nodes.

Recalling the definition of the principle of minimum potential energy from Chapter 3, we write the total potential energy (Π_{bar}) of the bar element as:

$$\Pi_{\text{bar}} = U - W = \frac{EA}{2L}\left(u_{2x}^2 - 2u_{2x}u_{1x} + u_{1x}^2\right) - \left(f_{1x}u_{1x} + f_{2x}u_{2x}\right) \tag{4.9}$$

By minimizing Π_{bar} with respect to $u_k (k = 1,2)$, one obtains:

$$\frac{\partial \Pi_{\text{bar}}}{\partial u_{1x}} = \frac{EA}{L}(u_{1x} - u_{2x}) - f_{1x} \tag{4.10}$$

$$\frac{\partial \Pi_{\text{bar}}}{\partial u_{2x}} = \frac{EA}{L}(u_{2x} - u_{1x}) - f_{2x} \tag{4.11}$$

Rearranging and combining Equations 4.10 and 4.11 lead to the matrix equation for a bar element with concentrated nodal loads as:

$$\begin{Bmatrix} f_{1x} \\ f_{2x} \end{Bmatrix}^e = \frac{EA}{L}\begin{bmatrix} 1 & -1 \\ -1 & 1 \end{bmatrix}^e \begin{Bmatrix} u_{1x} \\ u_{2x} \end{Bmatrix}^e \text{ or } \boldsymbol{f}^e = \boldsymbol{k}^e \boldsymbol{u}^e \tag{4.12}$$

Equation 4.12 is similar to the finite element matrix equation presented in Chapter 3 for spring element, with the difference being that the stiffness matrix $\boldsymbol{k}^e$ of the bar element is characterized by the term $\frac{EA}{L}$, which describes the axial stiffness of an elastic bar.

4.2.2 The Three-Node Quadratic Bar Element

The two-node bar element presented in Section 4.2.1 is the most common element used to study the normal stress and deformation of axially loaded one-dimensional structures. However the quadratic bar element, which is a higher order element with three nodes as shown in Figure 4.3, also finds

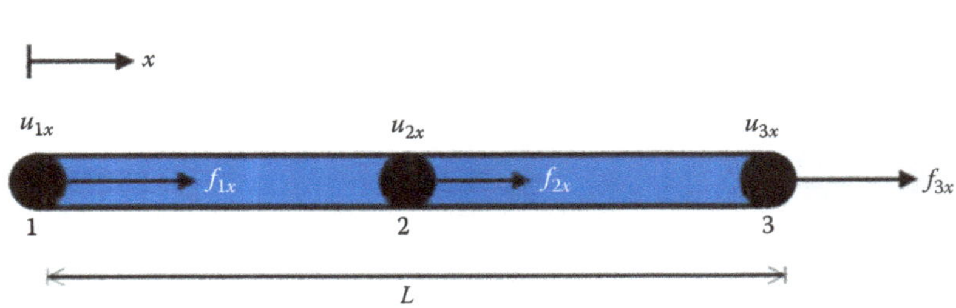

FIGURE 4.3
A one-dimensional quadratic bar element with three nodes.

application in some practical situations. For brevity sake, a formal derivation of the expressions for this element is not pursued here. Instead, we will employ the derived expressions for this element and provide its implementation using the R programming language later in this chapter.

The finite element matrix equation for the quadratic bar element is [6]:

$$\begin{Bmatrix} f_{1x} \\ f_{2x} \\ f_{3x} \end{Bmatrix} = \frac{EA}{3L}\begin{bmatrix} 7 & -8 & 1 \\ -8 & 16 & -8 \\ 1 & -8 & 7 \end{bmatrix}\begin{Bmatrix} u_{1x} \\ u_{2x} \\ u_{3x} \end{Bmatrix} \tag{4.13}$$

where f_{1x}, f_{2x}, and f_{3x} characterize the bar nodal forces, whereas u_{1x}, u_{2x}, and u_{3x} are the three nodal displacements that accompanied the nodal forces.

4.3 Computer Implementation

An aim of the computer implementations is to ease the process of dealing with problems that involve system of connected bars. In such situations, the global matrix equation relates the vector of global nodal forces $\{F\}$, the global stiffness matrix $[K]$, and the vector of global nodal displacements $\{U\}$ as:

$$\{F\} = [K]\{U\} \tag{4.14}$$

where K would be formed by making use of either Equation 4.12 or 4.13 depending on if we are dealing with two-node linear bar elements or three-node quadratic bar elements. The solution of Equation 4.14 generally follows a systematic procedure.

4.3.1 Systematic Procedure for Solving Static Problems of Longitudinally Connected Elastic Bars

For problems that involve analyses of connected bars, the following steps are recommended.

4.3.1.1 Precomputation Phase

- Discretize the system into a required number of bar elements.
- Label the nodes of the discretized structure, with no two elements having the same set of nodal labels.

4.3.1.2 Computation Phase

1. Write the finite element matrix equation for each bar element using Equation 4.12.
2. Determine the global degree of freedom of the assembly of discretized bars and expand the matrix equation for each element (from Step 1) to conform the size of the global degree of freedom.
3. Form the global matrix equation by summing up the expanded matrices of the elements.
4. Identify the essential boundary condition(s) and apply the boundary condition(s) to the global matrix equation. Programmatically, this step involves two actions:
   a. Obtain the reduced force vector.
   b. Use the identified boundary condition to obtain the reduced stiffness matrix.
5. Determine the unknown nodal displacements of the bars by using the reduced matrix equation. Using the determined nodal displacements, one may then obtain other information such as the nodal element forces and the unknown global forces, which will include the reaction forces(s).

4.3.2 Implemented R Functions for the Linear Bar Element

```
#**********************************************#
# R functions for the 2-node linear bar element #
# KB Mustapha                                  #
#**********************************************#

Bar_Element_Matrix = function(DOF=2,stiffbar)
{
  ematrix=stiffbar*matrix(c(1,-1,-1,1),nrow=DOF,byrow=T);
  return (ematrix)
}
```

```
Bar_ExpandedElement_Matrix = function(TDOF,eMatrix,i,j)
{
  bigMatrix = matrix(vector(l = TDOF * TDOF),nrow =
  TDOF,byrow = T);
  bigMatrix[c(i,j),c(i,j)]= eMatrix;
  return (bigMatrix)
}

Bar_Element_ReducedStiffnessMatrix = function(bigKmatrix,known
forcenodes)
{
  reducedsmatrix=bigKmatrix[c(knownforcenodes),(knownforcenodes)];
  return(reducedsmatrix)
}

Bar_Element_ReducedForceVector = function(forcevector)
{
  reducedforcevector = matrix(forcevector,ncol = 1)
  return(reducedforcevector)
}

Bar_Element_NodalDisplacement = function(reducedmatrix,reduced
forcevec)
{
  return(solve(reducedmatrix,reducedforcevec))
}
Bar_Element_Forces = function(DOF=2,stiffbar,totalnodaldisp,
i,j)
{
  nodaldisplacment=totalnodaldisp[c(i,j)]
  stiffnessmatrix=stiffbar*matrix(c(1,-1,-1,1),nrow=DOF,
  byrow=T)
  nodal_forces=stiffnessmatrix %*% nodaldisplacment
  return(nodal_forces)
}

Bar_Global_Forces = function(bigKmatrix,totalnodaldisp)
{
  global_forces = bigKmatrix %*% totalnodaldisp
  return(round(global_forces))
}
```

Tables 4.1 and 4.2 describe each of the listed functions along with the parameters they accept.

TABLE 4.1

Functions for Preprocessing Phase

Description of Functions
Bar_Element_Matrix(DOF=2,stiffbar) This function generates the stiffness matrix for a two-node bar element. It accepts two parameters: **DOF**, which stands for degree of freedom of the bar (predefined as 2); and **stiffbar**, which represents the bar's stiffness (*AE/L*).
Bar_ExpandedElement_Matrix(TDOF,eMatrix,i,j) This function forms the expanded stiffness matrix for a two-node bar. It accepts four parameters. The first two parameters are **TDOF** (total degree of freedom of the connected system) and **eMatrix** (which represents the unexpanded stiffness matrix of the element). The third and fourth parameters relate to the element's nodal numbers.
Bar_Element_ReducedStiffnessMatrix(bigKmatrix,knownforcenodes) This function generates a reduced stiffness matrix for a bar assemblage. Its parameters are **bigKmatrix** (the global stiffness matrix) and **knownforcenodes** (a vector specifying the rows of known nodal loads).
Bar_Element_ReducedForceVector(forcevector) This function generates a reduced force vector. It accepts the parameter **forcevector,** (a vector of known nodal loads) which it then transforms to a column matrix.

TABLE 4.2

Functions for the Solution Phase

Description of Functions
Bar_Element_NodalDisplacement(reducedmatrix, reducedforcevec) Given the parameters **reducedmatrix** (reduced stiffness matrix) and **reducedforcevec** (reduced force vector), this function solves for the unknown nodal displacements.
Bar_Element_Forces(DOF=2,stiffbar,totalnodaldisp, i,j) This function determines the nodal forces in a bar element. The required parameters are **DOF**, **stiffbar** (the axial stiffness), **totalnodaldisp** (a vector of global nodal displacements), and **i, j** (nodal labels of the element).
Bar_Global_Forces(bigKmatrix,totalnodaldisp) This function calculates the global nodal forces (and global reaction forces) in a system of connected bars. Its required parameters are **bigKmatrix** (the global stiffness matrix) and **totalnodaldisp** (a vector of the global nodal displacement).

4.3.3 Implemented R Functions for the Quadratic Bar Element

```
#**********************************************#
# R functions for the 3-node quadratic         #
#   bar element                                #
#   KB Mustapha                                #
#**********************************************#
```

```
QuadraticBar_Element_Matrix = function(DOF = 3,stiffbar)
{
  ematrix =(1/3)*stiffbar*matrix(c(7,-8,1,-8,16,-8,1,-8,7),
  nrow = DOF,byrow =T);
  return(ematrix)
}

QuadraticBar_ExpandedElement_Matrix = function(TDOF,eMatrix,i,j,k)
{
  bigMatrix = matrix(vector(l = TDOF * TDOF),nrow = TDOF,
  byrow = T);
  bigMatrix[c(i,j,k),c(i,j,k)]= eMatrix;
  return (bigMatrix)
}

QuadraticBar_Element_ReducedStiffnessMatrix = function(big
Kmatrix,knownforcenodes)
{
  reducedstiffnessmatrix = bigKmatrix[c(knownforcenodes),(known
  forcenodes)]
  return(reducedstiffnessmatrix)
}

QuadraticBar_Element_ReducedForceVector = function(forcevector)
{
  reducedforcevector = matrix(forcevector,ncol = 1)
  return(reducedforcevector)
}

QuadraticBar_Element_NodalDisplacement = function(reducedmatrix,
reducedforcevec)
{
  return(solve(reducedmatrix,reducedforcevec))
}

QuadraticBar_Element_Forces = function(DOF,stiff,totalnodaldisp,
i,j,k)
{
  nodaldisplacment=totalnodaldisp[c(i,j,k)]
  stiffnessmatrix =(1/3)*stiff*matrix(c(7,-8,1,-8,16,-8,1,
  -8,7),nrow = DOF,byrow = T)
  nodal_forces = stiffnessmatrix %*% nodaldisplacment
  return(nodal_forces)
}

QuadraticBar_Global_Forces = function(bigKmatrix,totalnodaldisp)
{
  global_forces = bigKmatrix %*% totalnodaldisp
  return(round(global_forces))
}
```

The descriptions of the functions contained in the preceding block of code bear similarity with those provided in Tables 4.1 and 4.2. In the following, we shall demonstrate the applications of the functions with three examples in Section 4.4.

4.4 Examples

4.4.1 Example 4.1

A bar assembly is loaded as shown in Figure 4.4. Establish the global stiffness matrix and determine the unknown nodal displacements. $E = 201 \times 10^9\ \mathrm{N/m^2}$, $A = 1 \times 10^{-4}\ \mathrm{m^2}$, and $L = 1\ \mathrm{m}$.

4.4.1.1 Solution

Based on the discretized system provided, we form the connectivity table revealed in Table 4.3. For each element, the axial stiffness is calculated by using the expression *AE/L*.

In the following steps 1–3, we will tackle the first part of the question (i.e., to establish the global stiffness matrix), whereas the second part of the question (finding the nodal displacements) is addressed in Steps 4–6.

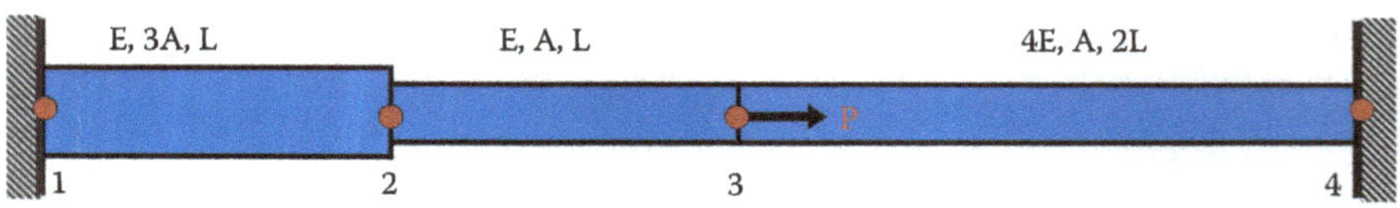

FIGURE 4.4
A longitudinally connected three-element bar system.

TABLE 4.3
Element's Connectivity and Stiffness

Elements	Nodes (i, j)	Axial Stiffness
1	1,2	630 MN/m
2	2,3	210 MN/m
3	3,4	420 MN/m

Step 1: Write the stiffness matrix for each element using the function **`Bar_Element_Matrix()`**.

However, **`Bar_Element_Matrix()`** requires information about the properties of the elements of the structure. Hence, we start with the declaration of the elements properties.

```
E=210*10^9; A=1*10^(-4); L=1;p=10000;
Axialstiffness1=(E*3*A)/L;
Axialstiffness2=E*A/L;
Axialstiffness3=(4*E*A)/(2*L);
```

The first line of the preceding block of code declares the following: (a) values of the Young's modulus (E), (b) area of the cross-section of the element (A), and (c) length of the element (L). The next three lines calculate the axial stiffness (AE/L) for each element.

Using the computed axial stiffness for the members of the structure, we will now obtain the stiffness matrix for each element as shown in the following:

```
k1=Bar_Element_Matrix(2,Axialstiffness1);k1
k2=Bar_Element_Matrix(2,Axialstiffness2);k2
k3=Bar_Element_Matrix(2,Axialstiffness3);k3
```

The output of which produces:

```
> k1=Bar_Element_Matrix(2,Axialstiffness1);k1
          [,1]      [,2]
[1,]  63000000 -63000000
[2,] -63000000  63000000
> k2=Bar_Element_Matrix(2,Axialstiffness2);k2
          [,1]      [,2]
[1,]  21000000 -21000000
[2,] -21000000  21000000
> k3=Bar_Element_Matrix(2,Axialstiffness3);k3
          [,1]      [,2]
[1,]  42000000 -42000000
[2,] -42000000  42000000
```

Step 2: Determine the global degree of freedom of the bar assemblage and expand the matrix equation for each element (from Step 1).

The two-node bar element has one degree of freedom per node. Given that there exist four nodes for the system, its total degree of freedom will be 4. In the following code snippet, the variable **Total_dof** holds this value, and it is one of the parameters required by the function **Bar_ExpandedElement_Matrix()**.

```
n_nodes=4;Total_dof=n_nodes*1;
K1=Bar_ExpandedElement_Matrix(Total_dof,k1,1,2);K1
K2=Bar_ExpandedElement_Matrix(Total_dof,k2,2,3);K2
K3=Bar_ExpandedElement_Matrix(Total_dof,k3,3,4);K3
```

The output reveals the expanded matrix for each element as:

```
> K1=Bar_ExpandedElement_Matrix(Total_dof,k1,1,2);K1
          [,1]      [,2] [,3] [,4]
[1,]  63000000 -63000000    0    0
[2,] -63000000  63000000    0    0
[3,]         0         0    0    0
[4,]         0         0    0    0
> K2=Bar_ExpandedElement_Matrix(Total_dof,k2,2,3);K2
     [,1]      [,2]      [,3] [,4]
[1,]    0         0         0    0
[2,]    0  21000000 -21000000    0
[3,]    0 -21000000  21000000    0
[4,]    0         0         0    0
> K3=Bar_ExpandedElement_Matrix(Total_dof,k3,3,4);K3
     [,1] [,2]      [,3]      [,4]
[1,]    0    0         0         0
[2,]    0    0         0         0
[3,]    0    0  42000000 -42000000
[4,]    0    0 -42000000  42000000
```

Step 3: Establish the global stiffness matrix by adding the expanded matrices of the elements as shown in the following:

```
GlobalK=K1+K2+K3;GlobalK
```

Which produces:

```
> GlobalK=K1+K2+K3;GlobalK
          [,1]      [,2]      [,3]      [,4]
[1,]  63000000 -63000000         0         0
[2,] -63000000  84000000 -21000000         0
[3,]         0 -21000000  63000000 -42000000
[4,]         0         0 -42000000  42000000
```

As shown, the global stiffness matrix for the system is thus:

$$[K] = 10^5 \begin{bmatrix} 630 & -630 & 0 & 0 \\ -630 & 840 & -210 & 0 \\ 0 & -210 & 630 & -420 \\ 0 & 0 & -420 & 420 \end{bmatrix} \tag{4.15}$$

Step 4: Identify the essential boundary condition(s) and apply the boundary condition(s) to the global matrix equation.

In Step 3, we obtained the global stiffness matrix, which is now used to write the global matrix equation as:

$$\begin{Bmatrix} F_1 \\ F_2 \\ F_3 \\ F_4 \end{Bmatrix} = 10^5 \begin{bmatrix} 630 & -630 & 0 & 0 \\ -630 & 840 & -210 & 0 \\ 0 & -210 & 630 & -420 \\ 0 & 0 & -420 & 420 \end{bmatrix} \begin{Bmatrix} U_1 \\ U_2 \\ U_3 \\ U_4 \end{Bmatrix} \tag{4.16}$$

Apparently, the assemblage is fixed at nodes 1 and 4 (see Figure 4.4), thus the boundary conditions are:

$$U_1 = U_4 = 0 \tag{4.17}$$

Further, the vector of known nodal loads (in *Newton*) can be identified as:

$$\begin{Bmatrix} F_2 \\ F_3 \end{Bmatrix} = \begin{Bmatrix} 0 \\ 10000 \end{Bmatrix} \tag{4.18}$$

Considering Equation 4.18, the reduced force vector can be established via the function **`Bar_Element_ReducedForceVector()`**.

```
vec_values_of_knownloads=c(0,10000)
ReducedForceVector=Bar_Element_ReducedForceVector(vec_
values_of_knownloads);
ReducedForceVector
```

Further, armed with the knowledge that the known forces belong to nodes 2 and 3 as shown in Equation 4.18, the reduced stiffness matrix can be obtained as:

```
vec_rows_of_knownloads=c(2,3)
ReducedK=Bar_Element_ReducedStiffnessMatrix(GlobalK,
vec_rows_of_knownloads);
ReducedK
```

The output reads:

```
> ReducedK
          [,1]      [,2]
[1,]  84000000 -21000000
[2,] -21000000  63000000
```

From which we see that the reduced matrix is:

$$10^5\begin{bmatrix} 840 & -210 \\ -210 & 630 \end{bmatrix} \tag{4.19}$$

Step 5: Determine the unknown nodal displacements by using the function **Bar_Element_NodalDisplacement()**:

```
UnknwonNodalDisp=Bar_Element_NodalDisplacement(ReducedK,
ReducedForceVector);
UnknwonNodalDisp
```

Which produces the unknown nodal displacements as:

```
> UnknwonNodalDisp
             [,1]
[1,] 0.00004329004
[2,] 0.00017316017
```

From the above, we have: $U_2 = 4.329\times10^{-5}$ m and $U_3 = 1.7316\times10^{-4}$ m.

An analytical expression for this problem specifies the values of U_2 and U_3 as [1]:

$$\begin{Bmatrix} U_2 \\ U_3 \end{Bmatrix} = \frac{PL}{11AE}\begin{Bmatrix} 1 \\ 4 \end{Bmatrix} \tag{4.20}$$

Evaluating Equation 4.20 using the fact that $E = 201 \times 10^9\ \text{N/m}^2$, $A = 1 \times 10^{-4}\ \text{m}^2, P = 10\ \text{kN}$, and $L = 1$ m, we have:

$$\begin{Bmatrix} U_2 \\ U_3 \end{Bmatrix} = \frac{PL}{11AE} \begin{Bmatrix} 1 \\ 4 \end{Bmatrix} = 10^{-4} \begin{Bmatrix} 0.4329 \\ 1.7316 \end{Bmatrix} \text{m} \tag{4.21}$$

which coincides with the answer that we have just got.

4.4.2 Example 4.2

A four-element bar assemblage is loaded as shown in Figure 4.5 [7]. Given that $E = 201$ GPa and $A = 3 \times 10^{-4}\ \text{m}^2$:

- Establish the global stiffness matrix.
- Determine the unknown nodal displacements.
- Determine the reaction forces.

4.4.2.1 Solution

Let us designate the element of the assemblage as shown in Table 4.4, which also contains the axial stiffness for each element (calculated from AE/L).

The rest of the solution procedure follows.

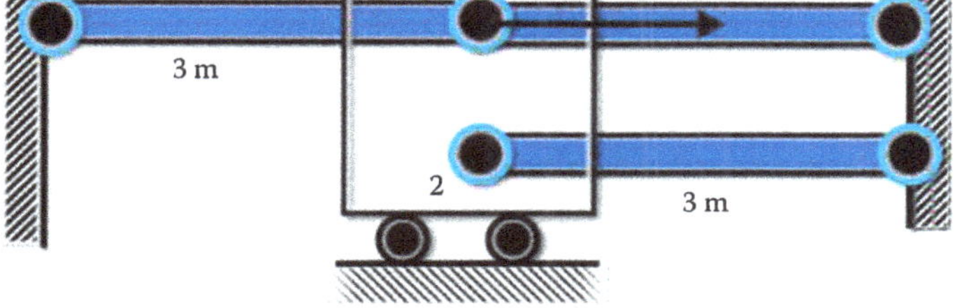

FIGURE 4.5
Parallel and serial combinations of bars.

TABLE 4.4
Elements' Connectivity and Properties for Figure 4.5

Elements	Nodes (i, j)	Axial Stiffness
1	1,2	21 MN/m
2	2,3	21 MN/m
3	2,4	21 MN/m
4	2,5	21 MN/m

Step 1: We write the element matrix using the information about the axial stiffness.

```
Axialstiffness=(210*(10^9)*(3*10^(-4)))/3;Axialstiffness
k1=Bar_Element_Matrix(2,Axialstiffness);k1
k2=Bar_Element_Matrix(2,Axialstiffness);k2
k3=Bar_Element_Matrix(2,Axialstiffness);k3
k4=Bar_Element_Matrix(2,Axialstiffness);k4
```

The output yields:

```
> Axialstiffness=(210*(10^9)*(3*10^(-4)))/3;Axialstiffness
[1] 21000000
> k1=Bar_Element_Matrix(2,Axialstiffness);k1
          [,1]      [,2]
[1,]  21000000 -21000000
[2,] -21000000  21000000
> k2=Bar_Element_Matrix(2,Axialstiffness);k2
          [,1]      [,2]
[1,]  21000000 -21000000
[2,] -21000000  21000000
> k3=Bar_Element_Matrix(2,Axialstiffness);k3
          [,1]      [,2]
[1,]  21000000 -21000000
[2,] -21000000  21000000
> k4=Bar_Element_Matrix(2,Axialstiffness);k4
          [,1]      [,2]
[1,]  21000000 -21000000
[2,] -21000000  21000000
```

Step 2: Determine the global degree of freedom of the bar assemblage and expand the element stiffness matrix.

```
n_nodes=5;Total_dof=n_nodes*1;
K1=Bar_ExpandedElement_Matrix(Total_dof,k1,1,2);K1
K2=Bar_ExpandedElement_Matrix(Total_dof,k2,2,3);K2
K3=Bar_ExpandedElement_Matrix(Total_dof,k3,2,4);K3
K4=Bar_ExpandedElement_Matrix(Total_dof,k4,2,5);K4
```

Step 3: Establish the global stiffness matrix by adding the expanded matrices.

```
GlobalK=K1+K2+K3+K4;
GlobalK
```

The output of this gives the global stiffness matrix of the assemblage in Figure 3.4.

```
> GlobalK
          [,1]      [,2]      [,3]      [,4]      [,5]
[1,]  21000000 -21000000         0         0         0
[2,] -21000000  84000000 -21000000 -21000000 -21000000
[3,]         0 -21000000  21000000         0         0
[4,]         0 -21000000         0  21000000         0
[5,]         0 -21000000         0         0  21000000
```

Hence, the global stiffness matrix (as required by the problem statement) goes thus:

$$[K]=10^7\begin{bmatrix} 2.1 & -2.1 & 0 & 0 & 0 \\ -2.1 & 8.4 & -2.1 & -2.1 & -2.1 \\ 0 & -2.1 & 2.1 & 0 & 0 \\ 0 & -2.1 & 0 & 2.1 & 0 \\ 0 & -2.1 & 0 & 0 & 2.1 \end{bmatrix} \tag{4.22}$$

Step 4: Identify the essential boundary condition(s) and obtain the reduced stiffness matrix and reduced force vector.

With the global stiffness matrix obtained in Step 3, we can write the global matrix equation for the system.

$$\begin{Bmatrix} F_1 \\ F_2 \\ F_3 \\ F_4 \\ F_5 \end{Bmatrix} = 10^7 \begin{bmatrix} 2.1 & -2.1 & 0 & 0 & 0 \\ -2.1 & 8.4 & -2.1 & -2.1 & -2.1 \\ 0 & -2.1 & 2.1 & 0 & 0 \\ 0 & -2.1 & 0 & 2.1 & 0 \\ 0 & -2.1 & 0 & 0 & 2.1 \end{bmatrix} \begin{Bmatrix} U_1 \\ U_2 \\ U_3 \\ U_4 \\ U_5 \end{Bmatrix} \tag{4.23}$$

Further, knowing that the assemblage is fixed at nodes 1, 3, 4, and 5, we establish the boundary conditions as:

$$U_1 = U_3 = U_4 = U_5 = 0 \tag{4.24}$$

Besides, as only node 2 is loaded, the vector of known nodal forces (in *Newton*) can be identified as:

$$\{F_2\} = \{30000\} \tag{4.25}$$

As can be noticed in Equation 4.25, the known force belongs to row 2. This information is used with the function **Bar_Element_ReducedForceVector()** as:

```
vec_rows_of_knownloads=c(2);
ReducedK=Bar_Element_ReducedStiffnessMatrix(GlobalK,
vec_rows_of_knownloads);
ReducedK
```

The reduced force vector is also obtained as:

```
vec_values_of_knownloads=c(30000)
ReducedForceVector=Bar_Element_ReducedForceVector(vec_
values_of_knownloads);
ReducedForceVector
```

Step 5: Determine the unknown nodal displacement with the function **Bar_Element_NodalDisplacement()**.

```
UnknwonNodalDisp=Bar_Element_NodalDisplacement(ReducedK,
ReducedForceVector);
UnknwonNodalDisp
```

```
>
UnknwonNodalDisp=Bar_Element_NodalDisplacement(ReducedK,
ReducedForceVector);
> UnknwonNodalDisp
             [,1]
[1,] 0.0003571429
```

The unknown nodal displacement is thus:

$$U_2 = 3.571 \times 10^{-4}\,\text{m} \tag{4.26}$$

With this, we can write the vector of nodal displacements for the system as:

$$\begin{Bmatrix} U_1 \\ U_2 \\ U_3 \\ U_4 \\ U_5 \end{Bmatrix} = \begin{Bmatrix} 0 \\ 3.571 \times 10^{-4} \\ 0 \\ 0 \\ 0 \end{Bmatrix} \tag{4.27}$$

Equation 4.27 is reproduced in the variable **allglobalNodalDisp()**, which is shown in the following:

```
allglobalNodalDisp=matrix(rep(0,Total_dof),byrow=T);
allglobalNodalDisp[vec_rows_of_knownloads]=UnknwonNodalDisp
allglobalNodalDisp
```

Finally, we determine the global nodal forces by using the function **Bar_Global_Forces()**.

```
globalforces=Bar_Global_Forces(GlobalK,allglobalNodalDisp);
globalforces
```

The output from which reads:

```
> globalforces=Bar_Global_Forces(GlobalK,allglobalNodalDisp);
> globalforces
      [,1]
[1,] -7500
[2,] 30000
[3,] -7500
[4,] -7500
[5,] -7500
```

The output from this gives the values of the global nodal forces as:

$$\begin{Bmatrix} F_{1x} \\ F_{2x} \\ F_{3x} \\ F_{4x} \\ F_{5x} \end{Bmatrix} = \begin{Bmatrix} -7500 \\ 30000 \\ -7500 \\ -7500 \\ -7500 \end{Bmatrix} \text{N} \tag{4.28}$$

4.4.3 Example 4.3

The structure shown in Figure 4.6 shows a fixed-free longitudinally connected bar with five nodes. If the bar has a Young's modulus of $E = 210$ GPa and a cross-sectional area of $A = 3\times10^{-3}\text{m}^2$, use the three-node bar element to establish

- The global stiffness matrix.
- The unknown nodal displacements.
- The global nodal forces.

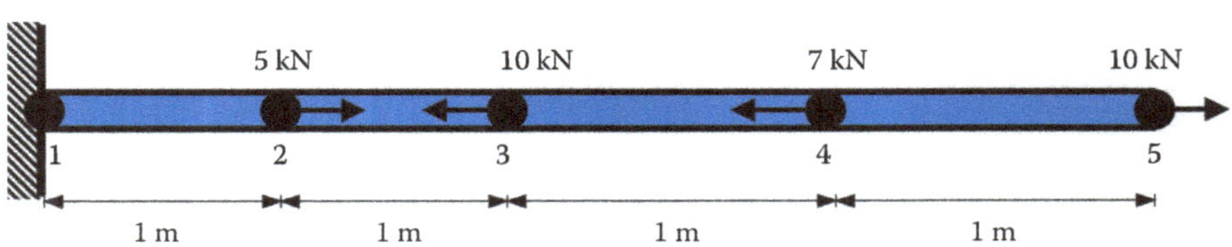

FIGURE 4.6
A fixed-free longitudinally connected bars with five nodes.

4.4.3.1 Solution

For each of the steps that follow, we will be using the functions for the quadratic bar element presented in Section 4.3.3. With respect to the present problem, let us designate the element of the assemblage as shown in Table 4.5 and then employ the implemented functions to deal with each step of the solution procedure that follows.

Step 1: We write the elements matrices using the information about the axial stiffness.

```
Axialstiffness=(0.003*210*10^9)/2;Axialstiffness
k1=QuadraticBar_Element_Matrix(3,Axialstiffness);k1
k2=QuadraticBar_Element_Matrix(3,Axialstiffness);k2
```

Which yields:

```
> k1=QuadraticBar_Element_Matrix(3,Axialstiffness);k1
           [,1]       [,2]       [,3]
[1,]  735000000 -840000000  105000000
[2,] -840000000 1680000000 -840000000
[3,]  105000000 -840000000  735000000
> k2=QuadraticBar_Element_Matrix(3,Axialstiffness);k2
           [,1]       [,2]       [,3]
[1,]  735000000 -840000000  105000000
[2,] -840000000 1680000000 -840000000
[3,]  105000000 -840000000  735000000
```

Step 2: Determine the global degree of freedom of the bar assemblage and expand the element matrix.

```
n_nodes=5;Total_dof=n_nodes*1;
K1=QuadraticBar_ExpandedElement_Matrix(Total_dof, k1,1,2,3);K1
K2=QuadraticBar_ExpandedElement_Matrix(Total_dof, k1,3,4,5);K2
```

TABLE 4.5

Elements' Connectivity and Properties for Figure 4.6

Elements	Nodes (i, j, m)	Axial Stiffness
1	1,2,3	315 MN/m
2	3,4,5	315 MN/m

Step 3: Establish the global stiffness matrix by summing up the expanded matrices as follows.

```
GlobalK=K1+K2;GlobalK
```

Producing:

```
> GlobalK
            [,1]       [,2]       [,3]       [,4]      [,5]
[1,]   735000000 -840000000  105000000          0         0
[2,]  -840000000 1680000000 -840000000          0         0
[3,]   105000000 -840000000 1470000000 -840000000 105000000
[4,]           0          0 -840000000 1680000000 -840000000
[5,]           0          0  105000000 -840000000 735000000
```

Step 4: Identify the essential boundary condition(s) and obtain the reduced stiffness matrix and reduced force vector. Using the global stiffness matrix obtained in Step 3, we can write the global matrix equation.

$$\begin{Bmatrix} F_1 \\ F_2 \\ F_3 \\ F_4 \\ F_5 \end{Bmatrix} = 10^8 \begin{bmatrix} 7.35 & -8.4 & 1.05 & 0 & 0 \\ -8.4 & 16.8 & -8.4 & 0 & 0 \\ 1.05 & -8.4 & 14.7 & -8.4 & 1.05 \\ 0 & 0 & -8.4 & 16.8 & -8.4 \\ 0 & 0 & 1.05 & -8.4 & 7.35 \end{bmatrix} \begin{Bmatrix} U_1 \\ U_2 \\ U_3 \\ U_4 \\ U_5 \end{Bmatrix} \tag{4.29}$$

The structure is fixed at node 1, which means the boundary condition is:

$$U_1 = 0 \tag{4.30}$$

Further, the vector of known nodal forces (in *Newton*) can be identified as:

$$\begin{Bmatrix} F_{2x} \\ F_{3x} \\ F_{4x} \\ F_{5x} \end{Bmatrix} = \begin{Bmatrix} 5000 \\ -10000 \\ -7000 \\ 10000 \end{Bmatrix} \text{N} \tag{4.31}$$

With the knowledge of the rows and values of the applied nodal forces, we can obtain the reduced stiffness matrix and reduced force vector as:

```
vec_rows_of_knownloads=c(2,3,4,5)
vec_values_of_knownloads=c(5000,-10000,-7000,10000)
ReducedK=QudraticBar_Element_ReducedStiffnessMatrix
(GlobalK,vec_rows_of_knownloads);
ReducedForceVector=QuadraticBar_Element_ReducedForce
Vector(vec_values_of_knownloads);
```

Step 5: Determine the unknown nodal displacement.

From the preceding, it is apparent that the nodal displacements can be written compactly as:

```
UnknwonNodalDisp=QuadraticBar_Element_NodalDisplacement
(ReducedK,ReducedForceVector);
```

```
> UnknwonNodalDisp
                   [,1]
[1,] -0.000004166667
[2,] -0.000014285714
[3,] -0.000008134921
[4,]  0.000006349206
```

$$\begin{Bmatrix} U_{2x} \\ U_{3x} \\ U_{4x} \\ U_{5x} \end{Bmatrix} = 10^{-4} \begin{Bmatrix} -4.167 \\ -0.142 \\ -0.0813 \\ 0.0635 \end{Bmatrix} \text{m} \tag{4.32}$$

Step 6: With the nodal displacements found in Equation 4.32, we obtain the global nodal forces.

```
allglobalNodalDisp=matrix(rep(0,Total_dof),byrow=T);
allglobalNodalDisp[vec_rows_of_knownloads]=UnknwonNodalDisp
allglobalNodalDisp
globalforces=QuadraticBar_Global_Forces(GlobalK,allglobal
NodalDisp);globalforces
```

Producing:

```
> globalforces=QuadraticBar_Global_Forces(GlobalK,allglobal
NodalDisp);globalforces
        [,1]
[1,]    2000
[2,]    5000
[3,]  -10000
[4,]   -7000
[5,]   10000
```

The output from this gives the values of the global nodal forces as:

$$\begin{Bmatrix} F_{1x} \\ F_{2x} \\ F_{3x} \\ F_{4x} \\ F_{5x} \end{Bmatrix} = \begin{Bmatrix} 2000 \\ 5000 \\ -10000 \\ -7000 \\ 10000 \end{Bmatrix} \text{N} \tag{4.33}$$

It can be observed that the sum of the determined reaction forces along with the externally applied forces obeys the axial equilibrium equation.

4.5 Exercises

1. An assembly consisting of a system of connected bars made from copper alloy with $E = 101$ GPa is depicted in Figure 4.7. Each member of the assembly has diameter $D = 30$ mm and length as shown in the figure.
   a. Formulate the global stiffness matrix for the assembly.
   b. Determine the displacement of node 3 (which supports a total force of 100 kN).
   c. The reaction forces at nodes 1, 2, and 4.
2. Employ two quadratic bar elements to determine the unknown nodal displacements and the reaction force for the system in Figure 4.8. Let $E = 210$ GPa and $A = 4 \times 10^{-4}$.

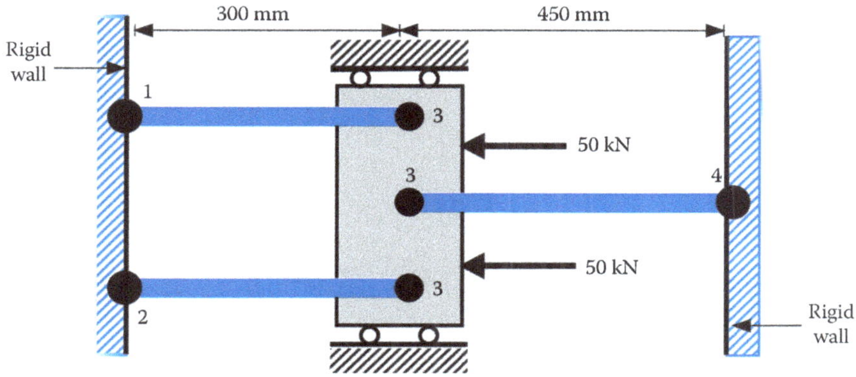

FIGURE 4.7
Exercise 4Q1.

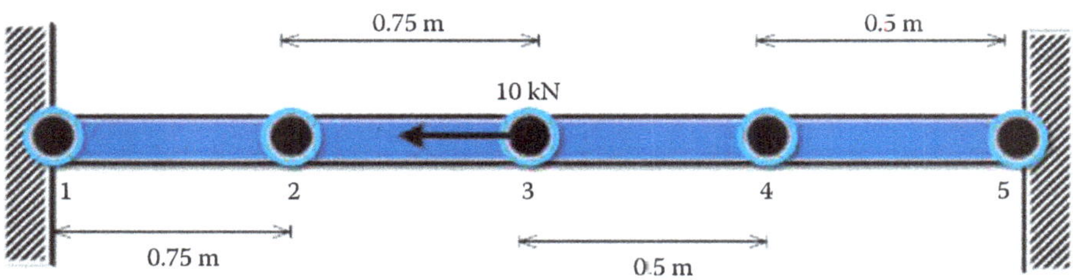

FIGURE 4.8
Exercise 4Q2.

References

1. J. F. Doyle, *Static and Dynamic Analysis of Structures: with An Emphasis on Mechanics and Computer Matrix Methods*, Springer, Dordrecht, the Netherlands, 1991.
2. T. R. Chandrupatla, A. D. Belegundu, *Introduction to Finite Elements in Engineering*, Prentice Hall, Upper Saddle River, NJ, 1997.
3. S. S. Rao, *The Finite Element Method in Engineering*, Elsevier, Burlington, NJ, 2010.
4. K. S. Surana, J. N. Reddy, *The Finite Element Method for Boundary Value Problems: Mathematics and Computations*, CRC Press LLC, Boca Raton, FL, 2016.
5. O. C. Zienkiewicz, R. L. Taylor, J. Z. Zhu, *The Finite Element Method: Its Basis and Fundamentals*, Elsevier Science, UK, 2005.
6. A. Öchsner, *Computational Statics and Dynamics: An Introduction Based on the Finite Element Method*, Springer, Singapore, 2016.
7. D. L. Logan, *First Course in the Finite Element Method*, Thomson, Canada, Ontario, ON, 2007.

5

Plane and Space Truss Elements

5.1 Introduction

In Chapter 4 we studied the static analyses of horizontally positioned bars using two-node and three-node bar elements. In doing so, we established the stiffness matrices for the horizontally positioned bar elements and discussed the fact that these elements have a single degree of freedom per node.

In this chapter, we aim to consider plane truss and space truss elements. These elements have higher nodal degrees of freedom and are suitable for the treatment of bars with arbitrary orientations in two- and three-dimensional spaces. Such arbitrarily oriented bars are found in the formation of many interesting structures derived from trusses. A truss may be aptly described "as a structure composed of pin-jointed members arranged to form one or more triangles" [1]. Various types of truss are used in practice, for instance, Pratt truss, Warren truss, W-truss, and V-truss [2]. Trusses find applications in many mechanical and civil engineering systems. They feature heavily as construction elements [3] and are also widely used in telecommunication towers, load-carrying cranes, railway bridges, roofs of residential and commercial buildings, aircraft, and space structures [4]. The analyses of trusses provide information about the (a) internal forces developed within the constituent members, (b) displacements of the joints under the influence of a coplanar system of forces, and (c) axial stresses experienced by the members.

The derivations of the finite element equations for the plane and space truss elements appear in many of the excellent texts on FEM, for instance, [5–9]. Nevertheless, for completeness, we briefly explored these equations in Sections 5.2 and 5.3 for the plane and space truss elements, respectively. In Section 5.4 we provide relevant implemented R functions for the analyses of these two elements. Demonstrations of the implemented functions are shown in Section 5.5.

5.2 Finite Element Equations for the Plane Truss Element

5.2.1 Displacement Function

In this section, we will extend the modeling of the horizontally positioned bar presented in Chapter 4. A unique consideration in the treatment of a truss, which differentiates it from that in Chapter 4, is the fact that the elements of a truss have orientations. It is these orientations that call for the establishment of two coordinate systems: the first being the local coordinate system (which aligns with the element) and the second is the global coordinate system (which is the reference system for the whole truss). We will base the derivation of the finite element matrix equation for the plane truss element upon Figure 5.1a, which depicts an oriented prismatic bar of area A (in the unloaded state), Young's modulus E, and length L_e. Further, we symbolize the global coordinate system, as shown in Figure 5.1a, by the XY system, whereas xy represents the local coordinate system (with x directed along the length of the element and θ being the orientation of x with X).

Figure 5.1b is a schematic of the undeformed and the deformed bars, originally collinear (only separated to allow for clearer geometric analyses). Going by Figure 5.1b, let the nodal displacements, u_{1x} and u_{2x}, be experienced by the bar under the influence of the nodal forces (f_{1x}, f_{2x}). With these axial loads, we assume the element undergoes pure axial displacements with negligible transverse displacements. Consequently, we may prescribe a linear function for the variation of the displacement along its length in the form:

$$u = A_1 + A_2 x \tag{5.1}$$

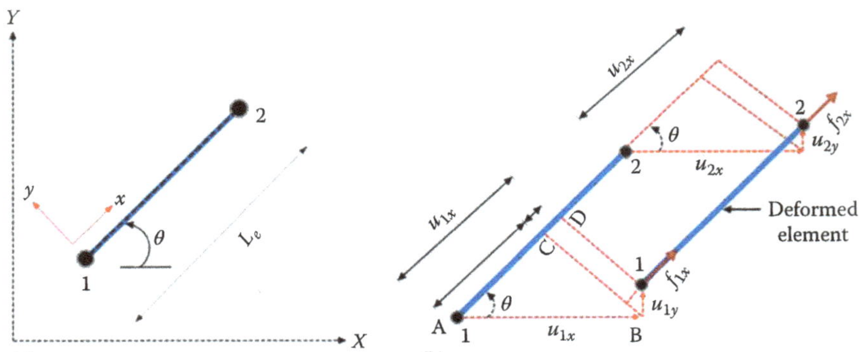

FIGURE 5.1
(a) The local (x, y) and global (X, Y) coordinates for a two-dimensional truss element and (b) the undeformed and deformed truss elements.

where A_1 and A_2 are unknown coefficients and x is within the range $0 \le x \le L_e$. In Equation 5.1, the total number of coefficients A_i is typically chosen to be equal to the total number of degrees of freedom associated with the element (the current element has 2 degrees of freedom in the local coordinate).

We solve the unknown coefficients by imposing boundary conditions on Equation 5.1 as follows:

- At node 1: $x = 0$, $u = u_{1x}$
- At node 2, where $x = L_e$, $u = u_{2x}$

Substituting the boundary conditions into Equation 5.1 leads to:

$$u_{1x} = A_1 \tag{5.2}$$

$$u_{2x} = A_1 + A_2 L_e \tag{5.3}$$

Equations 5.2 and 5.3 can be rewritten compactly as:

$$\underbrace{\begin{Bmatrix} u_{1x} \\ u_{2x} \end{Bmatrix}}_{\{u_e\}} = \underbrace{\begin{bmatrix} 1 & 0 \\ 1 & L_e \end{bmatrix}}_{[H]} \underbrace{\begin{Bmatrix} A_1 \\ A_2 \end{Bmatrix}}_{\{A\}} \tag{5.4}$$

For compactness, we may write Equation 5.4 in the form:

$$\{u_e\} = [H]\{A\} \tag{5.5}$$

where $[H]$ will be referred to as a coordinate matrix and $\{A\}$ is the vector of the unknown coefficients. Using Equation 5.5, one may write:

$$\{A\} = [H]^{-1}\{u_e\} \tag{5.6}$$

where

$$\{u_e\} = \begin{Bmatrix} u_{1x} \\ u_{2x} \end{Bmatrix} \tag{5.7}$$

$$[H]^{-1} = \begin{bmatrix} 1 & 0 \\ -\dfrac{1}{L_e} & \dfrac{1}{L_e} \end{bmatrix} \tag{5.8}$$

$$\{A\} = \begin{Bmatrix} A_1 \\ A_2 \end{Bmatrix} \tag{5.9}$$

Equation 5.6 gives the values of the unknown constants in terms of the geometric parameter and the local nodal displacement values (u_{1x} and u_{2x}). For generalization purpose, it is desired to decompose the local nodal displacement values into the global coordinate system. To do this, we consider triangles *BAC* and *CBD* in Figure 5.1b, through which we may write the vector u_{1x} as:

$$u_{1x} = u_{1X}\cos\theta + u_{1Y}\sin\theta \tag{5.10}$$

A similar line of analysis (i.e., if one considers the triangles at node 2) leads to:

$$u_{2x} = u_{2X}\cos\theta + u_{2Y}\sin\theta \tag{5.11}$$

Using Equations 5.10 and 5.11, we rewrite Equation 5.7 as:

$$\{u_e\} = \begin{Bmatrix} u_{1x} \\ u_{2x} \end{Bmatrix} = \begin{bmatrix} \cos\theta & \sin\theta & 0 & 0 \\ 0 & 0 & \cos\theta & \sin\theta \end{bmatrix} \begin{Bmatrix} u_{1X} \\ u_{1Y} \\ u_{2X} \\ u_{2Y} \end{Bmatrix} \tag{5.12}$$

We will employ the new expression for $\{u_e\}$ established in Equation 5.12 in Equation 5.6 to obtain:

$$\begin{Bmatrix} A_1 \\ A_2 \end{Bmatrix} = \underline{\begin{bmatrix} 1 & 0 \\ -\dfrac{1}{L_e} & \dfrac{1}{L_e} \end{bmatrix} \begin{bmatrix} \cos\theta & \sin\theta & 0 & 0 \\ 0 & 0 & \cos\theta & \sin\theta \end{bmatrix}} \begin{Bmatrix} u_{1X} \\ u_{1Y} \\ u_{2X} \\ u_{2Y} \end{Bmatrix} \tag{5.13}$$

Multiplying the underlined terms leads to:

$$\begin{Bmatrix} A_1 \\ A_2 \end{Bmatrix} = \begin{bmatrix} \cos\theta & \sin\theta & 0 & 0 \\ -\dfrac{\cos\theta}{L_e} & -\dfrac{\sin\theta}{L_e} & \dfrac{\cos\theta}{L_e} & \dfrac{\sin\theta}{L_e} \end{bmatrix} \begin{Bmatrix} u_{1X} \\ u_{1Y} \\ u_{2X} \\ u_{2Y} \end{Bmatrix} \tag{5.14}$$

Clearly, Equation 5.14 allows us to define A_1 and A_2 in terms of the orientation angle θ and the global nodal displacement variables. Substituting the values of A_1 and A_2 into Equation 5.1 results in:

$$u = \begin{bmatrix} \cos\theta\left(1-\dfrac{x}{L}\right) & \sin\theta\left(1-\dfrac{x}{L}\right)\left(\dfrac{x}{L}\cos\theta\right) & \left(\dfrac{x}{L}\sin\theta\right) \end{bmatrix} \begin{Bmatrix} u_{1X} \\ u_{1Y} \\ u_{2X} \\ u_{2Y} \end{Bmatrix} \tag{5.15}$$

where the shape functions $[N]$ in the local coordinate can be extracted from Equation 5.15 as:

$$[N]=\left[\left(1-\frac{x}{L}\right)\quad \frac{x}{L}\right] \tag{5.16}$$

5.2.2 Strain–Displacement Relationship and Constitutive Law

The expression of the general state of strain (ε) in a 3D element with displacements u, v, w along x, y, and z directions is:

$$\varepsilon=\begin{bmatrix}\varepsilon_{xx} & \varepsilon_{xy} & \varepsilon_{xz}\\ \varepsilon_{yx} & \varepsilon_{yy} & \varepsilon_{yz}\\ \varepsilon_{zx} & \varepsilon_{zy} & \varepsilon_{zz}\end{bmatrix}=\begin{bmatrix}\dfrac{\partial u}{\partial x} & \dfrac{1}{2}\left(\dfrac{\partial u}{\partial y}+\dfrac{\partial v}{\partial x}\right) & \dfrac{1}{2}\left(\dfrac{\partial u}{\partial z}+\dfrac{\partial w}{\partial x}\right)\\ \dfrac{1}{2}\left(\dfrac{\partial u}{\partial y}+\dfrac{\partial v}{\partial x}\right) & \dfrac{\partial v}{\partial y} & \dfrac{1}{2}\left(\dfrac{\partial v}{\partial z}+\dfrac{\partial w}{\partial y}\right)\\ \dfrac{1}{2}\left(\dfrac{\partial u}{\partial z}+\dfrac{\partial w}{\partial x}\right) & \dfrac{1}{2}\left(\dfrac{\partial v}{\partial z}+\dfrac{\partial w}{\partial y}\right) & \dfrac{\partial w}{\partial z}\end{bmatrix} \tag{5.17}$$

For a bar in a uniaxial deformation, the only non-zero component of the strain is ε_{xx}, which according to Equation 5.17, is defined as:

$$\varepsilon_{xx}=\frac{\partial u}{\partial x} \tag{5.18}$$

Bearing in mind Equation 5.1, we can evaluate Equation 5.18 to have

$$\varepsilon_{xx}=\frac{\partial}{\partial x}(A_1+A_2x)=A_2 \tag{5.19}$$

From Equation 5.14, we know the form of A_2, and we can thus rewrite Equation 5.19 as:

$$\varepsilon_{xx}=\underline{\frac{1}{L_e}\left[-\cos\theta \quad -\sin\theta \quad \cos\theta \quad \sin\theta\right]}\begin{Bmatrix}u_{1X}\\ u_{1Y}\\ u_{2X}\\ u_{2Y}\end{Bmatrix} \tag{5.20}$$

For simplification, we drop the subscript associated with the strain. Further, let us denote the underlined term as:

$$[B]=\frac{1}{L_e}\left[-\cos\theta \quad -\sin\theta \quad \cos\theta \quad \sin\theta\right] \tag{5.21}$$

Then, in compact form, Equation 5.20 becomes:

$$\{\varepsilon\} = [B]\{u_e\} \tag{5.22}$$

We will now employ the constitutive rule that maps the stress field to the strain field. From the generalized Hooke's law, we have:

$$\{\sigma\} = [D]\{\varepsilon\} \tag{5.23}$$

where $[D]$ is called the *material property matrix*. For a uniaxial loading of an isotropic homogeneous bar, $[D] = E$ (the Young's modulus). Therefore:

$$\{\sigma\} = E\{\varepsilon\} \tag{5.24}$$

Substituting the expression for $\{\varepsilon\}$ from Equation 5.22 into 5.24, one arrives at:

$$\{\sigma\} = E[B]\{u_e\} \tag{5.25}$$

5.2.3 Minimization of the Total Potential Energy for the Plane Truss Element

The total potential energy (P_t) for the bar is:

$$P_t = U - W \tag{5.26}$$

As shown in Chapter 4 (Section 4.2), if we consider a small slice of the element of length dx, we can write the strain energy (U) and the work done by external forces (W) as:

$$U = \frac{1}{2}\int_0^{L_e} \{\sigma\}^T \{\varepsilon\}(A_e dx) \tag{5.27}$$

$$W = \{u_e\}^T \{f_e\} \tag{5.28}$$

We substitute Equations 5.25 and 5.22 into Equation 5.27 to obtain:

$$U = \frac{1}{2}\int_0^{L_e} \{\sigma\}^T \{\varepsilon\}(A_e dx) = \frac{EA_e}{2}\int_0^{L_e} \left([B]\{u_e\}\right)^T \left([B]\{u_e\}]\right) dx \tag{5.29}$$

Minimizing P_t with respect to u_e leads to:

$$\frac{\partial}{\partial\{u_e\}^T} P_t = 0 = \left(A_e E L_e [B]^T [B]\right)\{u_e\} - \{f_e\} \tag{5.30}$$

And consequently, we obtain the desired relation as:

$$\left(A_e E L_e [B]^T [B]\right)\{u_e\} - \{f_e\} = 0 \tag{5.31}$$

By rearranging Equation 5.31, we obtain the finite element matrix equation for a 2D plane truss element as:

$$\{f_e\} = \left(A_e E L_e [B]^T [B]\right)\{u_e\} = [k_e]\{u_e\} \tag{5.32}$$

In full form,

$$\{f_e\} = \begin{Bmatrix} f_{1X} \\ f_{1Y} \\ f_{2X} \\ f_{2Y} \end{Bmatrix}; \qquad \{u_e\} = \begin{Bmatrix} u_{1X} \\ u_{1Y} \\ u_{2X} \\ u_{2Y} \end{Bmatrix} \tag{5.33}$$

We make use of the definition of $[B]$ from Equation 5.21 to obtain the stiffness matrix $[k_e]$ of the element as:

$$\frac{A_e E}{L_e}\begin{bmatrix} \cos^2\theta & \cos\theta\sin\theta & -\cos^2\theta & -\cos\theta\sin\theta \\ \cos\theta\sin\theta & \sin^2\theta & -\cos\theta\sin\theta & -\sin^2\theta \\ -\cos^2\theta & -\cos\theta\sin\theta & \cos^2\theta & \cos\theta\sin\theta \\ -\cos\theta\sin\theta & -\sin^2\theta & \cos\theta\sin\theta & \sin^2\theta \end{bmatrix} \tag{5.34}$$

5.3 Finite Element Equations for the Space Truss Element

5.3.1 Displacement Function

When a bar is arbitrarily oriented in three-dimensional space, the finite element matrix equation becomes more involved. To obtain the matrix equation for the space truss, we will start from the relation that describes the transformation of the local nodal displacements into the global coordinate system. That is [5]:

$$u_{1x} = u_{1X}\cos\theta_X + u_{1X}\cos\theta_Y + u_{1Z}\cos\theta_Z \tag{5.35}$$

$$u_{2x} = u_{2X}\cos\theta_X + u_{2Y}\cos\theta_Y + u_{2Z}\cos\theta_Z \tag{5.36}$$

where Equations 5.35 and 5.36 involve the direction cosine angles θ_X, θ_Y, and θ_Z and the global displacement components u_X, u_Y, and u_Z. Equations 5.35 and 5.36 are rewritten in matrix form as:

$$\{u_e\} = \begin{Bmatrix} u_{1x} \\ u_{2x} \end{Bmatrix} = \begin{bmatrix} \cos\theta_X & \cos\theta_Y & \cos\theta_Z & 0 & 0 & 0 \\ 0 & 0 & 0 & \cos\theta_X & \cos\theta_Y & \cos\theta_Z \end{bmatrix} \begin{Bmatrix} u_{1X} \\ u_{1Y} \\ u_{1Z} \\ u_{2X} \\ u_{2Y} \\ u_{2Z} \end{Bmatrix} \tag{5.37}$$

Making use of the new expression for $\{u_e\}$ in Equation 5.6 leads to:

$$\begin{Bmatrix} A_1 \\ A_2 \end{Bmatrix} = \underline{\begin{bmatrix} 1 & 0 \\ -\dfrac{1}{L_e} & \dfrac{1}{L_e} \end{bmatrix} \begin{bmatrix} \cos\theta_X & \cos\theta_Y & \cos\theta_Z & 0 & 0 & 0 \\ 0 & 0 & 0 & \cos\theta_X & \cos\theta_Y & \cos\theta_Z \end{bmatrix}} \begin{Bmatrix} u_{1X} \\ u_{1Y} \\ u_{1Z} \\ u_{2X} \\ u_{2Y} \\ u_{2Z} \end{Bmatrix} \tag{5.38}$$

Upon multiplication of the underlined terms, Equation 5.38 evaluates to:

$$\begin{Bmatrix} A_1 \\ A_2 \end{Bmatrix} = \begin{bmatrix} \cos\theta_X & \cos\theta_Y & \cos\theta_Z & 0 & 0 & 0 \\ -\dfrac{\cos\theta_X}{L_e} & -\dfrac{\cos\theta_Y}{L_e} & -\dfrac{\cos\theta_Z}{L_e} & \dfrac{\cos\theta_X}{L_e} & \dfrac{\cos\theta_Y}{L_e} & \dfrac{\cos\theta_Z}{L_e} \end{bmatrix} \begin{Bmatrix} u_{1X} \\ u_{1Y} \\ u_{1Z} \\ u_{2X} \\ u_{2Y} \\ u_{2Z} \end{Bmatrix} \tag{5.39}$$

The axial strain for the space truss element thus becomes:

$$\varepsilon_{xx} = \begin{bmatrix} -\dfrac{\cos\theta_X}{L_e} & -\dfrac{\cos\theta_Y}{L_e} & -\dfrac{\cos\theta_Z}{L_e} & \dfrac{\cos\theta_X}{L_e} & \dfrac{\cos\theta_Y}{L_e} & \dfrac{\cos\theta_Z}{L_e} \end{bmatrix} \begin{Bmatrix} u_{1X} \\ u_{1Y} \\ u_{1Z} \\ u_{2X} \\ u_{2Y} \\ u_{2Z} \end{Bmatrix} \tag{5.40}$$

We denote the vector of trigonometric functions as $[B]$:

$$[B] = \begin{bmatrix} -\dfrac{\cos\theta_X}{L_e} & -\dfrac{\cos\theta_Y}{L_e} & -\dfrac{\cos\theta_Z}{L_e} & \dfrac{\cos\theta_X}{L_e} & \dfrac{\cos\theta_Y}{L_e} & \dfrac{\cos\theta_Z}{L_e} \end{bmatrix} \tag{5.41}$$

We then have the strain–displacement and stress–strain matrix relationships as:

$$\{\varepsilon\} = [B]\{u_e\} \tag{5.42}$$

$$\{\sigma\} = E\{\varepsilon\} \tag{5.43}$$

Or:

$$\{\sigma\} = E[B]\{u_e\} \tag{5.44}$$

5.3.2 Minimization of the Total Potential Energy for the Space Truss Element

Following the procedure of Section 5.2.3, we minimize the total potential energy of the space truss element with respect to u_e, leading to:

$$\frac{\partial P_t}{\partial \{u_e\}^T} = \left(A_e E L_e [B]^T [B]\right)\{u_e\} - \{f_e\} = 0 \tag{5.45}$$

Accordingly, we obtain the desired relation for the finite element matrix equation for a space truss element as:

$$\{f_e\} = \underbrace{\left(A_e E L_e [B]^T [B]\right)}_{[k_e]}\{u_e\} \tag{5.46}$$

The full forms of the load and displacement vectors as well as the transformed stiffness matrix of the system are:

$$\{f_e\} = \begin{Bmatrix} f_{1X} \\ f_{1Y} \\ f_{1Z} \\ f_{2X} \\ f_{2Y} \\ f_{2Z} \end{Bmatrix}; \qquad \{u_e\}\begin{Bmatrix} u_{1X} \\ u_{1Y} \\ u_{1Z} \\ u_{2X} \\ u_{2Y} \\ u_{2Z} \end{Bmatrix} \tag{5.47}$$

$$k_e = A_e E L_e [B]^T [B] = \frac{A_e E}{L_e}\begin{bmatrix} C_x^2 & C_xC_y & C_xC_y & -C_x^2 & -C_xC_y & -C_xC_z \\ C_xC_y & C_y^2 & C_yC_z & -C_xC_y & -C_y^2 & -C_yC_z \\ C_xC_z & C_yC_z & C_z^2 & -C_xC_z & -C_yC_z & -C_x^2 \\ -C_x^2 & -C_xC_y & -C_xC_z & C_x^2 & C_xC_y & C_xC_z \\ -C_xC_y & -C_y^2 & -C_yC_z & C_xC_y & C_y^2 & C_yC_z \\ -C_xC_z & -C_yC_z & -C_z^2 & C_xC_z & C_yC_z & C_z^2 \end{bmatrix} \tag{5.48}$$

Where

$$C_x = \cos\theta_X;\; C_y = \cos\theta_Y;\; C_z = \cos\theta_Z \tag{5.49}$$

Geometrically, if the coordinates of the nodes of the elements in space are x_1, y_1, z_1 and x_2, y_2, z_2, respectively, then we can estimate the cosine angles and the length of the element L_e as follows:

$$\cos\theta_X = \frac{x_2 - x_1}{L_e};\; \cos\theta_Y = \frac{y_2 - y_1}{L_e};\; \cos\theta_Z = \frac{z_2 - z_1}{L_e} \tag{5.50}$$

$$L_e = \left[\left(x_2 - x_1\right)^2 + \left(y_2 - y_1\right)^2 + \left(z_2 - z_1\right)^2\right]^{\frac{1}{2}} \tag{5.51}$$

5.4 Computer Implementation

The computer implementation of this section will ease the process of analyzing problems that involve truss structures having bars with different orientation angles. Of course, in such situations, one must form the global matrix equation that relates the vector of global nodal forces $\{\boldsymbol{F}\}$, the global stiffness matrix $[\boldsymbol{K}]$, and the vector of global nodal displacements $\{\boldsymbol{U}\}$ and then apply the appropriate boundary condition to solve:

$$\{\boldsymbol{F}\} = [\boldsymbol{K}]\{\boldsymbol{U}\} \tag{5.52}$$

For the plane truss, the equation for each member of the truss is formed from:

$$\begin{Bmatrix} f_{1X} \\ f_{1Y} \\ f_{2X} \\ f_{2Y} \end{Bmatrix} = \frac{A_e E}{L_e}\begin{bmatrix} \cos^2\theta & \cos\theta\sin\theta & -\cos^2\theta & -\cos\theta\sin\theta \\ \cos\theta\sin\theta & \sin^2\theta & -\cos\theta\sin\theta & -\sin^2\theta \\ -\cos^2\theta & -\cos\theta\sin\theta & \cos^2\theta & \cos\theta\sin\theta \\ -\cos\theta\sin\theta & -\sin^2\theta & \cos\theta\sin\theta & \sin^2\theta \end{bmatrix} \tag{5.53}$$

And for the space truss, the equation for each member is formed from:

$$\begin{Bmatrix} f_{1X} \\ f_{1Y} \\ f_{1Z} \\ f_{2X} \\ f_{2Y} \\ f_{2Z} \end{Bmatrix} = \frac{A_e E}{L_e}\begin{bmatrix} C_x^2 & C_xC_y & C_xC_y & -C_x^2 & -C_xC_y & -C_xC_z \\ C_xC_y & C_y^2 & C_yC_z & -C_xC_y & -C_y^2 & -C_yC_z \\ C_xC_z & C_yC_z & C_z^2 & -C_xC_z & -C_yC_z & -C_x^2 \\ -C_x^2 & -C_xC_y & -C_xC_z & C_x^2 & C_xC_y & C_xC_z \\ -C_xC_y & -C_y^2 & -C_yC_z & C_xC_y & C_y^2 & C_yC_z \\ -C_xC_z & -C_yC_z & -C_z^2 & C_xC_z & C_yC_z & C_z^2 \end{bmatrix}\begin{Bmatrix} u_{1X} \\ u_{1Y} \\ u_{1Z} \\ u_{2X} \\ u_{2Y} \\ u_{2Z} \end{Bmatrix} \tag{5.54}$$

5.4.1 Systematic Procedure for Analyses of Truss Structures

Solution of Equation 5.52 follows a systematic procedure. The following steps are recommended to be carried out before running the codes for the analyses of plane and space trusses.

5.4.1.1 *Precomputation Phase*

- Establish the global coordinate system. For a plane truss, it makes sense to have the origin located at the joint for which the coordinates for all the other joints are positive.
- Discretize the truss into individual-oriented bars.
- Label the nodes and elements (this helps to form the connectivity table).

5.4.1.2 *Computation Phase*

1. Specify and/or determine the geometric lengths and the orientation angles (using the nodal coordinates) if they are not explicitly provided.
2. Establish the finite element matrix equation for each oriented bar element (using Equation 5.53 for a plane truss and Equation 5.54 for the space truss). Specifically, we only need to form the stiffness matrix for each element.
3. Determine the global degree of freedom of the truss structure and carry out the expansion of the matrices (from Step 1) to conform to the size of the global degree of freedom.
4. Form the global stiffness by summing up the expanded matrices and use the global stiffness matrix to form the global matrix equation.
5. Identify the essential boundary condition(s) and apply the boundary conditions to the global matrix equation to obtain a reduced matrix equation. Programmatically, this step involves two actions:
   a. Establishing the reduced force vector
   b. Using the identified boundary condition to obtain the reduced stiffness matrix
6. Determine the unknown global nodal displacements by using the reduced matrix equation. From this, other variables such as local force vector, global force vector, and stress values can be found.
   a. Obtain the global nodal element forces using the global nodal displacement (from Step 6) and the element stiffness matrix.
   b. Obtain the stress developed in each member and the local forces (if desired).

5.4.2 Implemented R Functions for the Plane Truss Elements

```
#*********************************************************#
# R functions for the plane truss element                #
# KB Mustapha                                             #
#*********************************************************#

PlaneTruss_LengthOrientation = function(vec_nodalcoordinates)
{
  ysquared=(vec_nodalcoordinates[4]-vec_nodalcoordinates[2])^2
  xsquared=(vec_nodalcoordinates[3]-vec_nodalcoordinates[1])^2
   length_of_element=sqrt(ysquared+xsquared)
   ydiff=(vec_nodalcoordinates[4]-vec_nodalcoordinates[2])
   xdiff=(vec_nodalcoordinates[3]-vec_nodalcoordinates[1])
   orientation_of_element=(180/pi)*atan(ydiff/xdiff)
   return(round(c(length_of_element,orientation_of_element),3))

}

PlaneTruss_GeometricLengths = function(num_ele,vec_xnodal
coordinates,vec_ynodalcoordinates)
{
   length_of_nodalcoordinates=length(vec_xnodalcoordinates)
   numberofelement=length_of_nodalcoordinates/2
   length_of_element=vector(mode="numeric",length=numberofelement)
   counter=1;
   for (j in seq(1,length_of_nodalcoordinates-1,by=2)){
     ysquared=(vec_ynodalcoordinates[j+1]-
vec_ynodalcoordinates[j])^2
     xsquared=(vec_xnodalcoordinates[j+1]-
vec_xnodalcoordinates[j])^2
     length_of_element[counter]=sqrt(ysquared+xsquared)
     counter=counter+1

   }

   return(round(length_of_element))

}


PlaneTruss_Orientations = function(num_ele,vec_xnodal
coordinates,vec_ynodalcoordinates)
{
```

```
  length_of_nodalcoordinates=length(vec_xnodalcoordinates)
  numberofelement=length_of_nodalcoordinates/2
  orientation_of_element=vector(mode="numeric",length=
number ofelement)
  counter=1
  for (j in seq(1,length_of_nodalcoordinates-1,by=2)){
    ydiff=(vec_ynodalcoordinates[j+1]-vec_ynodal
coordinates[j])
    xdiff=(vec_xnodalcoordinates[j+1]-vec_xnodal
coordinates[j])
    orientation_of_element[counter]=(180/pi)*atan(ydiff/xdiff)
    counter=counter+1
  }

  return(round(orientation_of_element,0))

}


PlaneTruss_Element_Matrix = function(DOF = 4,stifftruss,theta)
{
  m=cos(theta*pi/180);
  n=sin(theta*pi/180);
  stiffnesscoefficients=c(m^2,m*n,-m^2,-m*n,m*n,n^2,-m*n,
-n^2,-m^2,-m*n,m^2,m*n,-m*n,-n^2,m*n,n^2)
  ematrix=stifftruss*matrix(stiffnesscoefficients,
nrow=DOF,byrow=T)
}

PlaneTruss_ExpandedElement_Matrix = function(TDOF,eMatrix,i,j)
{

  r1=(i-1)+i;r2=(i-1)+(i+1);r3=(j-2)+(j+1);r4=(j-2)+(j+2);
  bigMatrix=matrix(vector(l=TDOF*TDOF),nrow=TDOF,byrow=T);
  bigMatrix[c(r1,r2,r3,r4),c(r1,r2,r3,r4)]=eMatrix;
  return (bigMatrix)
}

PlaneTruss_ReducedStiffnessMatrix = function(bigKmatrix,
knownforcenodes)
{
  reducedstiffnessmatrix = bigKmatrix[c(knownforcenodes),
(knownforcenodes)]
  return(reducedstiffnessmatrix)
}
```

```
PlaneTruss_ReducedForceVector = function(forcevector)
{
  reducedforcevector = matrix(forcevector,ncol = 1)
  return(reducedforcevector)
}

PlaneTruss_NodalDisplacement = function(reducedmatrix,
vec_reducedforce)
{
  return(solve(reducedmatrix,vec_reducedforce))
}

PlaneTruss_Global_Forces = function(bigKmatrix,vec_
globalnodaldisp)
{
  global_forces = bigKmatrix %*% vec_globalnodaldisp
  return(round(global_forces))
}

PlaneTruss_Local_Forces = function(stifftruss,theta,
vec_globalnodaldisp,i,j)
{
  m=cos(theta*pi/180);
  n=sin(theta*pi/180);
  r1=(i-1)+i;r2=(i-1)+(i+1);r3=(j-2)+(j+1);r4=(j-2)+(j+2);
  elementdisp=vec_globalnodaldisp[c(r1,r2,r3,r4)];
  localstiffnessmatrix=matrix(c(1,-1,-1,1), byrow=T,nrow=2);
  transformationmatrix=matrix(c(m,n,0,0,0,0,m,n),byrow=T,
nrow=2);
  local_forces = (stifftruss)*localstiffnessmatrix
%*%transformationmatrix %*%elementdisp

  return(round(local_forces))
}

PlaneTruss_AxialStress = function(YoungMod,elem_
len,theta,vec_globalnodaldisp,i,j)
{
  m=cos(theta*pi/180);
  n=sin(theta*pi/180);
  r1=(i-1)+i;r2=(i-1)+(i+1);r3=(j-2)+(j+1);r4=(j-2)+(j+2);
  elementdisp=vec_globalnodaldisp[c(r1,r2,r3,r4)];
  transformationmatrix=matrix(c(-m,-n,m,n),nrow=1,byrow=T);
  local_stress=(YoungMod/elem_len)*transformationmatrix%*
%elementdisp;
  return(local_stress)
}
```

We describe each of the listed functions along with the parameters they accept in Tables 5.1 through 5.3.

TABLE 5.1

Functions for Evaluating Geometric Properties (Plane Truss Element)

Description of Functions
`PlaneTruss_LengthOrientation(vec_nodalcoordinates)` This function returns a vector of the geometric length and orientation of a given element. It accepts the parameter **`vec_nodalcoordinates`**, which is a vector of nodal coordinates of the element (in the form (x_i, y_i, x_j, y_j)).
`PlaneTruss_GeometricLengths(num_ele,vec_xnodalcoordinates, vec_ynodalcoordinates)` This function returns a vector of the geometric lengths for all elements. It accepts the parameters: **`num_ele`**, which denotes the number of elements; **`vec_xnodalcoordinates`**, a vector of the x nodal coordinates of all elements (in the form $x_i, x_j, \ldots x_n$); and **`vec_ynodalcoordinates`**, a vector of the y nodal coordinates of all elements (in the form $y_i, y_j, \ldots y_n$).
`PlaneTruss_Orientations(num_ele,vec_xnodalcoordinates,vec_ynodalcoordinates)` This function returns a vector of the orientation angles for all elements. It accepts the same set of parameters as those of the function in the previous row.

TABLE 5.2

Functions for the Formulation of Elements' Matrices and Vectors (Plane Truss Element)

Description of Functions
`PlaneTruss_Element_Matrix (DOF = 4,stifftruss, theta)` This function generates and returns the stiffness matrix for a plane truss element. It accepts three parameters: **`DOF`**, which stands for degree of freedom of the truss (predefined as 4); **`stifftruss`**, which represents a member's axial stiffness (i.e., AE/L); and **`theta`**, which denotes a member's orientation.
`PlaneTruss_ExpandedElement_Matrix(TDOF,eMatrix,i,j)` This function forms the expanded stiffness matrix. It accepts four parameters. The first two parameters are: **`TDOF`**, which stands for the total degree of freedom of the whole truss structures and **`eMatrix`**, which represents the unexpanded stiffness matrix of a member of the truss structure. The third and fourth parameters relate to the member's nodal numbers.
`PlaneTruss_ReducedStiffnessMatrix(bigKmatrix,knownforcenodes)` This function generates a reduced stiffness matrix for the system. Its parameters are **`bigKmatrix`** (the global stiffness matrix) and **`knownforcenodes`** (a vector specifying the rows of known forces).
`PlaneTruss_ReducedForceVector(forcevector)` This function establishes the reduced force vector needed to find the unknown nodal displacements. It accepts the parameter **`forcevector`** (a vector of known forces) and transforms this vector to a column matrix.

TABLE 5.3

Functions for the Solution Phase in the Analysis of a Plane Truss

Description of Functions
PlaneTruss_NodalDisplacement(reducedmatrix,reducedforcevec) Given the parameters **reducedmatrix** (reduced stiffness matrix) and **reducedforcevec** (reduced force vector), this function solves for the unknown nodal displacements.
PlaneTruss_Global_Forces (bigKmatrix,vec_globalnodaldisp) This function calculates the global forces (including reaction forces). Its parameters are **bigKmatrix** (the global stiffness matrix) and **vec_globalnodaldisp** (a vector of the global nodal displacement).
PlaneTruss_Local_Forces(stifftruss,theta,vec_globalnodaldisp,i,j) This function calculates the local forces in a member of the truss. Its parameters are **stifftruss**, which represents an element's axial stiffness (AE/L); **theta**, the element's orientation; **vec_globalnodaldisp** (a vector of the global nodal displacement); and the member's nodal labels **i,j**.
PlaneTruss_AxialStress(YoungMod,elem_len,theta,vec_globalnodaldisp,i,j) This function calculates stress developed in a member. Its parameters are **YoungMod**, which represents the member's Young's modulus; **elem_len**, which denotes the element's geometric length, **theta**, which denotes the elements orientation; **vec_globalnodaldisp** (a vector of the global nodal displacement); and the member's nodal labels **i,j**.

5.4.3 Implemented R Functions for the Space Truss Elements

```
#*********************************************#
# R functions for the space truss element      #
# KB Mustapha                                  #
#*********************************************#

SpaceTruss_LengthOrientations = function(vec_nodalcoordinates)
{
  xsquared=(vec_nodalcoordinates[4]-vec_nodalcoordinates[1])^2
  ysquared=(vec_nodalcoordinates[5]-vec_nodalcoordinates[2])^2
  zsquared=(vec_nodalcoordinates[6]-vec_nodalcoordinates[3])^2
  length_of_element=sqrt(xsquared+ysquared+zsquared)

  cx=(1/length_of_element)*(vec_nodalcoordinates[4]-vec_nodalcoordinates[1])
  cy=(1/length_of_element)*(vec_nodalcoordinates[5]-vec_nodalcoordinates[2])
  cz=(1/length_of_element)*(vec_nodalcoordinates[6]-vec_nodalcoordinates[3])
  orientation_of_element=(180/pi)*acos(c(cx,cy,cz))
  return(c(length_of_element,orientation_of_element))

}

SpaceTruss_Element_Matrix = function(DOF=6,stifftruss,thetas){
  cx=cos(thetas[1]*pi/180);
  cy=cos(thetas[2]*pi/180);
  cz=cos(thetas[3]*pi/180);

  submatrix=stifftruss*matrix(c(cx^2,cx*cy,cx*cz,cy*cx,cy^2,cy*cz,
cz*cx,cz*cy,cz^2),nrow=3,byrow=T);
  eMatrix=matrix(vector(l=DOF*DOF),nrow=DOF,ncol=DOF);
```

```
    eMatrix[1:3,1:3]=submatrix;
    eMatrix[4:6,1:3]=-submatrix;
    eMatrix[1:3,4:6]=-submatrix;
    eMatrix[4:6,4:6]=submatrix;
    return(round(eMatrix,2))
  }

  SpaceTruss_ExpandedElement_Matrix = function(TDOF,eMatrix,i,j)
  {

    r1=2*(i-1)+i;
    r2=2*(i-1)+(i+1);
    r3=2*(i-1)+(i+2);
    r4=(j-1)+(j+2)+(j-3);
    r5=(j-1)+(j+3)+(j-3);
    r6=(j-1)+(j+4)+(j-3);
    bigMatrix=matrix(vector(l=TDOF*TDOF),nrow=TDOF,byrow=T);
    bigMatrix[c(r1,r2,r3,r4,r5,r6),c(r1,r2,r3,r4,r5,r6)]=eMatrix;
    return (bigMatrix)

  }

  SpaceTruss_ReducedStiffnessMatrix = function(bigKmatrix,knownforcenodes)
  {
    reducedstiffnessmatrix = bigKmatrix[c(knownforcenodes),(knownforcenodes)]
    return(reducedstiffnessmatrix)
  }

  SpaceTruss_ReducedForceVector = function(forcevector)
  {
    reducedforcevector=matrix(forcevector,ncol = 1)
    return(reducedforcevector)
  }

  SpaceTruss_NodalDisplacement = function(reducedmatrix,vec_reducedforce)
  {
    return(solve(reducedmatrix,vec_reducedforce))
  }

  SpaceTruss_Global_Forces = function(bigKmatrix,vec_globalnodaldisp)
  {
    global_forces = bigKmatrix %*% vec_globalnodaldisp
    return(round(global_forces))
  }

  SpaceTruss_Local_Forces = function(YoungMod,Area,thetas,vec_globalnodaldisp)
  {
    cx=cos(thetas[1]*pi/180);
    cy=cos(thetas[2]*pi/180);
    cz=cos(thetas[3]*pi/180);
    localstiffnessmatrix=matrix(c(1,-1,-1,1), byrow=T,nrow=2);

    transformationmatrix=matrix(c(cx,cy,cz,0,0,0,0,0,0,cx,cy,cz).byrow=T,
  nrow=2);
    local_forces = (YoungMod/Area)*localstiffnessmatrix
  %*%transformationmatrix %*%
      vec_globalnodaldisp
    return(round(local_forces))
  }
```

```
Spacetruss_AxialStress = function(YoungMod,elem_len,thetas,vec_
globalnodaldisp)
{
  cx=cos(thetas[1]*pi/180);
  cy=cos(thetas[2]*pi/180);
  cz=cos(thetas[3]*pi/180);

  disp_vec=matrix(vec_globalnodaldisp,nrow=length(vec_globalnodaldisp),
byrow=T,ncol = 1);
  transformationmatrix=matrix(c(-cx,-cy,-cz,cx,cy,cz),nrow=1,byrow=T);
  local_stress=(YoungMod/elem_len)*transformationmatrix%*%disp_vec;
  return(local_stress)
}
```

The descriptions of the usage and arguments for the listed functions are provided in Tables 5.4 through 5.6.

Three examples are selected to illustrate the use of the implemented functions for both plane truss and space truss elements in the section that follows.

TABLE 5.4

Functions for Evaluating Geometric Properties (Space Truss Element)

Description of Functions
`SpaceTruss_LengthOrientations (vec_nodalcoordinates)` This function returns a vector of the geometric length (L) and the orientation angles ($\theta_X, \theta_Y, \theta_Z$) of a given element. It accepts the parameter **`vec_nodalcoordinates`**, which is a vector of nodal coordinates of the element in the form $(x_i, y_i, z_i, x_j, y_j, z_j)$.

TABLE 5.5

Functions for the Formulation of Element's Matrices and Vectors (Space Truss Element)

Description of Functions
`SpaceTruss_Element_Matrix (DOF = 6,stifftruss,thetas)` This function generates and returns the stiffness matrix for a space truss element. It accepts three parameters: **`DOF`**, which stands for degree of freedom of the truss (predefined as 6); **`stifftruss`**, which represents a member's axial stiffness (AE/L); and **`thetas`**, which denotes a member's orientation angles (i.e., θ_X, θ_Y, θ_Z).
`SpaceTruss_ExpandedElement_Matrix (TDOF,eMatrix,i,j)` This function forms the expanded stiffness matrix, and its parameters are of the same form as those of the plane truss elements.
`SpaceTruss_ReducedStiffnessMatrix(bigKmatrix,knownforcenodes)` This function generates a reduced stiffness matrix for the system, and its parameters are of the same form as those of the plane truss elements.
`SpaceTruss_ReducedForceVector(forcevector)` This function generates a reduced force vector needed to find the unknown nodal displacements. It accepts the parameter **`forcevector`** (a vector of known forces) and transforms this vector to a column matrix.

TABLE 5.6

Functions for the Solution Phase in the Analyses of a Space Truss Structure

Description of Functions
`SpaceTruss_NodalDisplacement(reducedmatrix,reducedforcevec)` Given the parameters **`reducedmatrix`** (reduced stiffness matrix) and **`reducedforcevec`** (reduced force vector), this function solves for the unknown nodal displacements.
`SpaceTruss_Global_Forces (bigKmatrix,vec_globalnodaldisp)` This function calculates the global forces (and global reaction forces) in a space truss structure. Its parameters are **`bigKmatrix`** (the global stiffness matrix) and **`vec_globalnodaldisp`** (a vector of the global nodal displacement).
`SpaceTruss_Local_Forces(YoungMod,Area,thetas,vec_globalnodaldisp)` This function calculates the local forces in a member of the space truss. Its parameters are **`YoungMod`**, which denotes the Young's modulus; **`Area`**, which denotes the area; **`thetas`**, which denotes a member's orientation angles (i.e., θ_X, θ_Y, θ_Z); and **`vec_globalnodaldisp`** (a vector of the global nodal displacement).
`SpaceTruss_AxialStress(YoungMod,elem_len,thetas,vec_globalnodaldisp)` This function returns the stress developed in a member of a space truss. Its parameters are **`YoungMod`**, which represents the member's Young's modulus; **`elem_len`**, which denotes the element's geometric length, **`thetas`**, which denotes the element's orientations; and **`vec_globalnodaldisp`** (a vector of the global nodal displacement).

5.5 Examples

5.5.1 Example 5.1

Consider the plane truss shown in Figure 5.2 with members made from A36 structural steel and each having a cross-sectional area of $10 \times 10^{-5}\,\text{m}^2$. Determine the displacements of the nodes and the forces in each member.

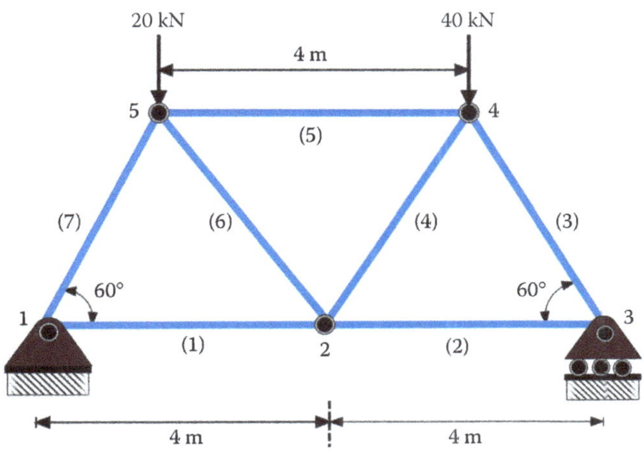

FIGURE 5.2
Plane truss with seven elements.

TABLE 5.7

Connectivity Table from the Discretized System

Elements	Nodes (i, j)	Coordinates (x_i, y_i, x_j, y_j)
1	1,2	(0,0,4,0)
2	2,3	(4,0,8,0)
3	3,4	(8,0,6,3.46)
4	2,4	(4,0,6,3.46)
5	4,5	(6,3.46,2,3.46)
6	2,5	(4,0,2,3.46)
7	1,5	(0,0,2,3.46)

5.5.1.1 Solution

We will now employ the steps outlined in Section 5.4.1 and the R functions provided in Section 5.4.2 to solve the problem. The solution approach is to treat each member of the truss as a plane truss element, and for convenience, the joints of the truss have been labeled accordingly. The connectivity of the elements is shown in Table 5.7. The A36 structural steel has Young's modulus of 200 GPa.

Step 1: Obtain/specify the geometric and material properties of each elements.

The necessary geometric properties include the length and orientation angle for each element. Three functions have been implemented to determine the length and orientation angles of members of a plane truss:

- **`PlaneTruss_LengthOrientation()`**
- **`PlaneTruss_GeometricLengths()`**
- **`PlaneTruss_Orientations()`**

For conciseness, we use the last two functions to obtain the lengths and orientations for all elements at once as shown in the following block of code:

```
YoungModulus=200*10^9;A=1*10^(-4);
Axialstiffness=(YoungModulus*A)/lengths;Axialstiffness
```

```
n_ele=7;
nodalxcoord=c(0,4,4,8,8,6,4,6,6,2,4,2,0,2)
nodalycoord=c(0,0,0,0,0,3.46,0,3.46,3.46,3.46,0,3.46,0,3.46)
lengths=PlaneTruss_GeometricLengths(n_ele,nodalxcoord,nod
alycoord);lengths
orientations=PlaneTruss_Orientations(n_ele,nodalxcoord,no
dalycoord);orientations
```

Running the lines of code mentioned earlier produces the values of the axial stiffness, geometric length, and orientation angle for all the elements.

```
> Axialstiffness=(YoungModulus*A)/lengths;Axialstiffness
[1] 5000000 5000000 5000000 5000000 5000000 5000000 5000000
> n_ele=7;
> nodalxcoord=c(0,4,4,8,8,6,4,6,6,2,4,2,0,2)
> nodalycoord=c(0,0,0,0,0,3.46,0,3.46,3.46,3.46,0,3.46,0,3.46)
>
lengths=PlaneTruss_GeometricLengths(n_ele,nodalxcoord,nod
alycoord);lengths
[1] 4 4 4 4 4 4 4
> orientations=PlaneTruss_Orientations(Nelement,nodalxcoo
rd,nodalycoord);orientations
[1]   0   0 -60  60   0 -60  60
```

Step 2: Write the stiffness matrix for each element using the function **PlaneTruss_Element_Matrix()**.

```
k1=PlaneTruss_Element_Matrix(4,Axialstiffness[1],orientations[1]);k1
k2=PlaneTruss_Element_Matrix(4,Axialstiffness[2],orientations[2]);k2
k3=PlaneTruss_Element_Matrix(4,Axialstiffness[3],orientations[3]);k3
k4=PlaneTruss_Element_Matrix(4,Axialstiffness[4],orientations[4]);k4
k5=PlaneTruss_Element_Matrix(4,Axialstiffness[5],orientations[5]);k5
k6=PlaneTruss_Element_Matrix(4,Axialstiffness[6],orientations[6]);k6
k7=PlaneTruss_Element_Matrix(4,Axialstiffness[7],orientations[7]);k7
```

The computed stiffness matrices are shown in the following block:

```
> k1=PlaneTruss_Element_Matrix(4,Axialstiffness[1],
orientations[1]);k1
        [,1] [,2]     [,3] [,4]
[1,]  5000000    0 -5000000    0
[2,]        0    0        0    0
[3,] -5000000    0  5000000    0
[4,]        0    0        0    0
> k2=PlaneTruss_Element_Matrix(4,Axialstiffness[2],
orientations[2]);k2
        [,1] [,2]     [,3] [,4]
[1,]  5000000    0 -5000000    0
[2,]        0    0        0    0
[3,] -5000000    0  5000000    0
[4,]        0    0        0    0
> k3=PlaneTruss_Element_Matrix(4,Axialstiffness[3],
orientations[3]);k3
        [,1]     [,2]     [,3]     [,4]
[1,]  1250000 -2165064 -1250000  2165064
[2,] -2165064  3750000  2165064 -3750000
[3,] -1250000  2165064  1250000 -2165064
[4,]  2165064 -3750000 -2165064  3750000
> k4=PlaneTruss_Element_Matrix(4,Axialstiffness[4],
orientations[4]);k4
        [,1]     [,2]     [,3]     [,4]
[1,]  1250000  2165064 -1250000 -2165064
[2,]  2165064  3750000 -2165064 -3750000
[3,] -1250000 -2165064  1250000  2165064
[4,] -2165064 -3750000  2165064  3750000
> k5=PlaneTruss_Element_Matrix(4,Axialstiffness[5],
orientations[5]);k5
        [,1] [,2]     [,3] [,4]
[1,]  5000000    0 -5000000    0
[2,]        0    0        0    0
[3,] -5000000    0  5000000    0
[4,]        0    0        0    0
> k6=PlaneTruss_Element_Matrix(4,Axialstiffness[6],
orientations[6]);k6
        [,1]     [,2]     [,3]     [,4]
[1,]  1250000 -2165064 -1250000  2165064
[2,] -2165064  3750000  2165064 -3750000
[3,] -1250000  2165064  1250000 -2165064
[4,]  2165064 -3750000 -2165064  3750000
> k7=PlaneTruss_Element_Matrix(4,Axialstiffness[7],
orientations[7]);k7
        [,1]     [,2]     [,3]     [,4]
[1,]  1250000  2165064 -1250000 -2165064
[2,]  2165064  3750000 -2165064 -3750000
[3,] -1250000 -2165064  1250000  2165064
[4,] -2165064 -3750000  2165064  3750000
```

Step 3: Determine the global degree of freedom of the truss and expand the matrix equation for each element (from Step 2) to conform to the global degree of freedom.

The expansion of the element matrices is done with the function **`PlaneTruss_ExpandedElement_Matrix()`**. The total degree of freedom of the structure is calculated and contained in the variable **`Total_dof`**.

```
n_nodes=5;Total_dof=n_nodes*2;
K1Expanded=PlaneTruss_ExpandedElement_Matrix(Total_dof,
k1,1,2);
K2Expanded=PlaneTruss_ExpandedElement_Matrix(Total_dof,
k2,2,3);
K3Expanded=PlaneTruss_ExpandedElement_Matrix(Total_dof,
k3,3,4);
K4Expanded=PlaneTruss_ExpandedElement_Matrix(Total_dof,
k4,2,4);
K5Expanded=PlaneTruss_ExpandedElement_Matrix(Total_dof,
k5,4,5);
K6Expanded=PlaneTruss_ExpandedElement_Matrix(Total_dof,
k6,2,5);
K7Expanded=PlaneTruss_ExpandedElement_Matrix(Total_dof,
k7,1,5);
```

Step 4: Establish the global stiffness matrix by adding the expanded matrices of the elements as shown in the following:

```
GlobalStiffnessMatrix=K1Expanded+K2Expanded+K3Expanded+
                      K4Expanded+K5Expanded+K6Expanded+K7Expanded;
GlobalStiffnessMatrix
```

The output of this gives the elements of the global stiffness matrix $[K]$.

```
> GlobalStiffnessMatrix
           [,1]     [,2]     [,3]     [,4]     [,5]     [,6]     [,7]     [,8]     [,9]    [,10]
 [1,]  6250000  2165064 -5000000        0        0        0        0        0 -1250000 -2165064
 [2,]  2165064  3750000        0        0        0        0        0        0 -2165064 -3750000
 [3,] -5000000        0 12500000        0 -5000000        0 -1250000 -2165064 -1250000  2165064
 [4,]        0        0        0  7500000        0        0 -2165064 -3750000  2165064 -3750000
 [5,]        0        0 -5000000        0  6250000 -2165064 -1250000  2165064        0        0
 [6,]        0        0        0        0 -2165064  3750000  2165064 -3750000        0        0
 [7,]        0        0 -1250000 -2165064 -1250000  2165064  7500000        0 -5000000        0
 [8,]        0        0 -2165064 -3750000  2165064 -3750000        0  7500000        0        0
 [9,] -1250000 -2165064 -1250000  2165064        0        0 -5000000        0  7500000        0
[10,] -2165064 -3750000  2165064 -3750000        0        0        0        0        0  7500000
```

The 10×10 global stiffness matrix equation is reproduced in the following for clarity:

$$[K] = 10^4 \begin{bmatrix} 625 & 216.51 & -500 & 0 & 0 & 0 & 0 & 0 & -125 & -216.51 \\ 216.51 & 375 & 0 & 0 & 0 & 0 & 0 & 0 & 216.51 & -375 \\ -500 & 0 & 1250 & 0 & -500 & 0 & -125 & -216.51 & -125 & 216.51 \\ 0 & 0 & 0 & 750 & 0 & 0 & -216.51 & -375 & 216.51 & -375 \\ 0 & 0 & -500 & 0 & 625 & -216.51 & -125 & 216.51 & 0 & 0 \\ 0 & 0 & 0 & 0 & -216.51 & 375 & 216.51 & -375 & 0 & 0 \\ 0 & 0 & -125 & -216.51 & -375 & 216.50 & 750 & 0 & -500 & 0 \\ 0 & 0 & -216.51 & -375 & -216.51 & -375 & 0 & 750 & 0 & 0 \\ -125 & -216.51 & -125 & 216.51 & 0 & 0 & -500 & 0 & 750 & 0 \\ -216.51 & -375 & 216.51 & -375 & 0 & 0 & 0 & 0 & 0 & 750 \end{bmatrix}$$

Using the global stiffness matrix, we may now form the global matrix equation for the system as:

$$\begin{Bmatrix} F_{1X} \\ F_{1Y} \\ F_{2X} \\ F_{2Y} \\ F_{3X} \\ F_{3Y} \\ F_{4X} \\ F_{4Y} \\ F_{5X} \\ F_{5Y} \end{Bmatrix} = [K] \begin{Bmatrix} U_{1X} \\ U_{1Y} \\ U_{2X} \\ U_{2Y} \\ U_{3X} \\ U_{3Y} \\ U_{4X} \\ U_{4Y} \\ U_{5X} \\ U_{5Y} \end{Bmatrix}$$

Step 5: Identify the essential boundary condition(s) and apply the boundary condition(s) to modify the global matrix equation.

To identify the boundary conditions, we look at the constraint applied to the system and analyze the displacements. For instance, we notice that the truss is fixed at node 1 (translating to zero displacements at node 1), and it has a roller support at node 3 that restrains its movement along the Y axis but allows it to move along the X axis. Based on this, the identified displacement boundary conditions are:

$$U_{1X} = U_{1Y} = U_{3Y} = 0 \tag{5.55}$$

From the system, the following seven known nodal forces (in *Newton*) can also be identified as:

$$F_{2X} = F_{2Y} = F_{3X} = F_{4X} = F_{5X} = 0; \quad F_{4Y} = 40{,}000 \text{ N}; \quad F_{5Y} = 20{,}000 \text{ N} \tag{5.56}$$

Using the rows of the identified known force values, one obtains the reduced stiffness matrix via the function **PlaneTruss_ReducedStiffness Matrix()** as shown in the following:

```
vec_rows_of_knownloads=c(3,4,5,7,8,9,10);
KglobalReduced=PlaneTruss_ReducedStiffnessMatrix(GlobalSt
iffnessMatrix,vec_rows_of_knownloads);
KglobalReduced
```

Which produces:

```
> KglobalReduced
         [,1]     [,2]     [,3]     [,4]     [,5]     [,6]     [,7]
[1,] 12500000        0 -5000000 -1250000 -2165064 -1250000  2165064
[2,]        0  7500000        0 -2165064 -3750000  2165064 -3750000
[3,] -5000000        0  6250000 -1250000  2165064        0        0
[4,] -1250000 -2165064 -1250000  7500000        0 -5000000        0
[5,] -2165064 -3750000  2165064        0  7500000        0        0
[6,] -1250000  2165064        0 -5000000        0  7500000        0
[7,]  2165064 -3750000        0        0        0        0  7500000
```

The reduced force vector is also established as:

```
 vec_values_of_knownloads=c(0,0,0,0,-40000,0,-20000);
ReducedForceVector=PlaneTr
uss_ReducedForceVector(vec_values_of_knownloads);
ReducedForceVector
```

The output of which is:

```
> ReducedForceVector
       [,1]
[1,]      0
[2,]      0
[3,]      0
[4,]      0
[5,] -40000
[6,]      0
[7,] -20000
```

Combining the reduced stiffness matrix and the values of the known nodal force leads to the reduced matrix equation. That is:

$$10^3\begin{Bmatrix} 0 \\ 0 \\ 0 \\ 0 \\ -40 \\ 0 \\ -20 \end{Bmatrix} = 10^4\begin{bmatrix} 1250 & 0 & -500 & -125 & -216.51 & -125 & 216.51 \\ 0 & 750 & 0 & -216.51 & -375 & 216.51 & -375 \\ -500 & 0 & 625 & -125 & 216.51 & 0 & 0 \\ -125 & -216.51 & -125 & 750 & 0 & -500 & 0 \\ -216.51 & -375 & 216.51 & 0 & 750 & 0 & 0 \\ -125 & 216.51 & 0 & -500 & 0 & 750 & 0 \\ 216.51 & -375 & 0 & 0 & 0 & 0 & 750 \end{bmatrix}\begin{Bmatrix} U_{2X} \\ U_{2Y} \\ U_{3X} \\ U_{4X} \\ U_{4Y} \\ U_{5X} \\ U_{5Y} \end{Bmatrix} \quad (5.57)$$

Step 6: Determine the unknown nodal displacement by using the function **PlaneTruss_NodalDisplacement()**.

```
UnknwonNodalDisp=PlaneTruss_NodalDisplacement(KglobalRedu
ced,ReducedForceVector)
round(UnknwonNodalDisp,4)
```

From which we find the unknown nodal displacements, rounded to the fourth decimal place in meters, as:

```
> round(UnknwonNodalDisp,4)
        [,1]
[1,]  0.0029
[2,] -0.0120
[3,]  0.0069
[4,]  0.0014
[5,] -0.0125
[6,]  0.0049
[7,] -0.0095
```

The above shows that:

$$U_{2X}=0.0029;\ \ U_{2Y}=-0.0120;\ \ U_{3X}=0.0069;\ \ U_{4X}=0.0014;$$
$$U_{4Y}=-0.0125;\ \ U_{5X}=0.0049;\ \ U_{5Y}=-0.0095$$

For convenience, we now combine the determined displacements and the boundary condition (from Equation 5.55).

```
allglobalNodalDisp=matrix(rep(0,Total_dof),byrow=T);
allglobalNodalDisp[vec_rows_of_knownloads]=UnknwonNodalDisp
round(allglobalNodalDisp,4)
```

This allows us to see all the displacements of the joints of the truss in one vector as shown in the following:

```
> round(allglobalNodalDisp,4)
         [,1]
 [1,]  0.0000
 [2,]  0.0000
 [3,]  0.0029
 [4,] -0.0120
 [5,]  0.0069
 [6,]  0.0000
 [7,]  0.0014
 [8,] -0.0125
 [9,]  0.0049
[10,] -0.0095
```

With the complete nodal displacements found, we obtain the element forces via the function **PlaneTruss_Local_Forces()**.

As explained in Table 5.3, this function accepts five parameters: (1) axial stiffness, (2) orientation, (3) vector of global nodal displacements, and (4) a member's nodal labels. Note that we have already stored the global nodal displacements in the variable **allglobalNodalDisp** in Step 5. Besides, each member's nodal labels can be identified from the connectivity table. With these variables, the local forces are computed as illustrated in the following block of code:

```
Element1_force=PlaneTruss_Local_Forces(Axialstiffness[1],
orientations[1],allglobalNodalDisp,1,2);
Element1_force
Element2_force=PlaneTruss_Local_Forces(Axialstiffness[2],
orientations[2],allglobalNodalDisp,2,3);
Element2_force
Element3_force=PlaneTruss_Local_Forces(Axialstiffness[3],
orientations[3],allglobalNodalDisp,3,4);
Element3_force
Element4_force=PlaneTruss_Local_Forces(Axialstiffness[4],
orientations[4],allglobalNodalDisp,2,4);
Element4_force
Element5_force=PlaneTruss_Local_Forces(Axialstiffness[5],
orientations[5],allglobalNodalDisp,4,5);
Element5_force
Element6_force=PlaneTruss_Local_Forces(Axialstiffness[6],
orientations[6],allglobalNodalDisp,2,5);
Element6_force
Element7_force=PlaneTruss_Local_Forces(Axialstiffness[7],
orientations[7],allglobalNodalDisp,1,5);
Element7_force
```

Running the previous returns the local force developed within each element of the truss as:

```
Element1_force=PlaneTruss_Local_Forces(Axialstiffness[1],
orientations[1],allglobalNodalDisp,1,2);
> Element1_force
        [,1]
[1,] -14434
[2,]  14434
> Element2_force=PlaneTruss_Local_Forces(Axialstiffness[2],
orientations[2],allglobalNodalDisp,2,3);
> Element2_force
        [,1]
[1,] -20207
[2,]  20207
> Element3_force=PlaneTruss_Local_Forces(Axialstiffness[3],
orientations[3],allglobalNodalDisp,3,4);
> Element3_force
        [,1]
[1,] -40415
[2,]  40415
> Element4_force=PlaneTruss_Local_Forces(Axialstiffness[4],
orientations[4],allglobalNodalDisp,2,4);
> Element4_force
       [,1]
[1,]  5774
[2,] -5774
> Element5_force=PlaneTruss_Local_Forces(Axialstiffness[5],
orientations[5],allglobalNodalDisp,4,5);
> Element5_force
        [,1]
[1,] -17321
[2,]  17321
> Element6_force=PlaneTruss_Local_Forces(Axialstiffness[6],
orientations[6],allglobalNodalDisp,2,5);
> Element6_force
       [,1]
[1,]  5774
[2,] -5774
> Element7_force=PlaneTruss_Local_Forces(Axialstiffness[7],
orientations[7],allglobalNodalDisp,1,5);
> Element7_force
        [,1]
[1,]  28868
[2,] -28868
```

5.5.2 Example 5.2

The three-member plane truss shown in Figure 5.3 is made from a material with $E = 210$ GPa. Assuming that each member of the truss has a cross-sectional area of $300\,mm^2$, determine the displacements of the nodes and the axial stresses in the members.

5.5.2.1 Solution

We will again use the steps outlined in Section 5.4.1 and the code snippet provided in Section 5.4.2 to solve the problem. The elements' connectivity for this problem is shown in Table 5.8.

Step 1: Obtain the geometric lengths and orientation angles of each elements by making calls to the following R functions:

- **`PlaneTruss_GeometricLengths()`**
- **`PlaneTruss_Orientations()`**

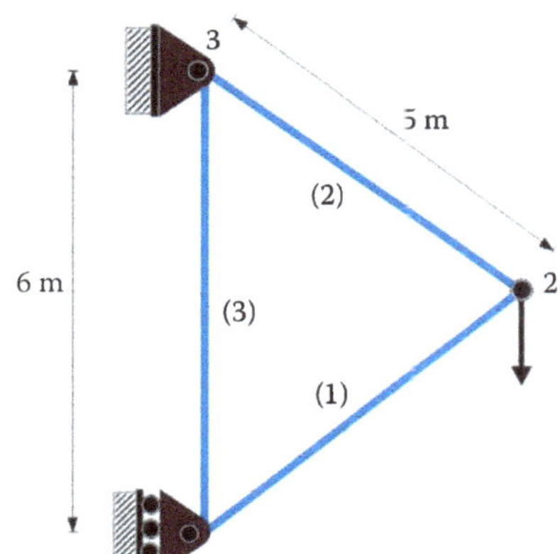

FIGURE 5.3
A three-member plane truss.

TABLE 5.8
Connectivity Table from the Discretized System

Elements	Nodes (i, j)	Coordinates (x_i, y_i, x_j, y_j)
1	1,2	(0,0,4,3)
2	2,3	(4,3,0,6)
3	1,3	(0,0,0,6)

```
n_ele=3;
nodalxcoord=c(0,4,4,0,0,0)
nodalycoord=c(0,3,3,6,0,6)
lengths=PlaneTruss_GeometricLengths(n_ele,nodalxcoord,nod
alycoord);
orientations=PlaneTruss_Orientations(n_ele,nodalxcoord,no
dalycoord);
lengths
orientations
```

Which produces:

```
> lengths
[1] 5 5 6
> orientations
[1]  37 -37  90
```

The produced output shows that the geometric lengths for elements 1, 2, and 3 are $L_1 = L_2 = 5\text{m}$, and $L_3 = 6\text{m}$, respectively. The orientations are also found as $\theta_1 = 37°, \theta_2 = -37°, \theta_3 = 90$.

Notice that the variable **lengths** and **orientations** in the preceding code block are vectors. Using the vectorized capability of the R language, we will obtain the axial stiffness of the three elements at once by using the variable **lengths** combined with the Young's modulus and cross-sectional area A of the elements as shown in the following block of code:

```
YoungModulus=210*10^9;A=3*10^(-4);
Axialstiffness=(YoungModulus*A)/lengths;
Axialstiffness
```

The output of which gives the axial stiffness for the three elements as:

```
> Axialstiffness
[1] 12600000 12600000 10500000
```

Step 2: Write the stiffness matrix for each element using **PlaneTruss_ Element_Matrix()**.

```
k1=PlaneTruss_Element_Matrix(4,Axialstiffness[1],orientations[1]);
round(k1)
k2=PlaneTruss_Element_Matrix(4,Axialstiffness[2],orientations[2]);
round(k2)
k3=PlaneTruss_Element_Matrix(4,Axialstiffness[3],orientations[3]);
round(k3)
```

It is worth pointing out that we have used the R in-built function **round()** here (and in many instances within this text) to prevent the printing of unnecessary trailing zeros or decimal points.

Running the earlier block of code yields the computed stiffness matrices as:

```
> k1=PlaneTruss_Element_Matrix(4,Axialstiffness[1],
orientations[1]);round(k1)
         [,1]     [,2]     [,3]     [,4]
[1,]  8036515  6055949 -8036515 -6055949
[2,]  6055949  4563485 -6055949 -4563485
[3,] -8036515 -6055949  8036515  6055949
[4,] -6055949 -4563485  6055949  4563485
> k2=PlaneTruss_Element_Matrix(4,Axialstiffness[2],
orientations[2]);round(k2)
         [,1]     [,2]     [,3]     [,4]
[1,]  8036515 -6055949 -8036515  6055949
[2,] -6055949  4563485  6055949 -4563485
[3,] -8036515  6055949  8036515 -6055949
[4,]  6055949 -4563485 -6055949  4563485
> k3=PlaneTruss_Element_Matrix(4,Axialstiffness[3],
orientations[3]);round(k3)
     [,1]      [,2] [,3]      [,4]
[1,]    0         0    0         0
[2,]    0  10500000    0 -10500000
[3,]    0         0    0         0
[4,]    0 -10500000    0  10500000
```

Step 3: Determine the global degree of freedom of the truss and expand the matrix equation for each element (from Step 2) to conform to the global degree of freedom.

```
n_nodes=3;Total_dof=n_nodes*2;
K1Expanded=PlaneTruss_ExpandedElement_Matrix(Total_dof,k1,1,2);
K2Expanded=PlaneTruss_ExpandedElement_Matrix(Total_dof,k2,2,3);
K3Expanded=PlaneTruss_ExpandedElement_Matrix(Total_dof,k3,1,3);
```

Step 4: Establish the global stiffness matrix by adding the expanded matrices of the elements as shown in the following:

```
GlobalStiffnessMatrix=K1Expanded+K2Expanded+K3Expanded
round(GlobalStiffnessMatrix)
```

The output of this gives the elements of the global stiffness matrix **K**.

```
> round(GlobalStiffnessMatrix)
         [,1]      [,2]     [,3]     [,4]     [,5]      [,6]
[1,]  8036515   6055949 -8036515 -6055949        0         0
[2,]  6055949  15063485 -6055949 -4563485        0 -10500000
[3,] -8036515  -6055949 16073031        0 -8036515   6055949
[4,] -6055949  -4563485        0  9126969  6055949  -4563485
[5,]        0         0 -8036515  6055949  8036515  -6055949
[6,]        0 -10500000  6055949 -4563485 -6055949  15063485
```

The 6×6 global stiffness matrix $[K]$ is reproduced in the following for clarity purpose:

$$[K] = 10^4 \begin{bmatrix} 803.6515 & 605.5949 & -803.6515 & -605.5949 & 0 & 0 \\ 605.5949 & 1506.3485 & -605.5949 & -456.3585 & 0 & -1050 \\ -803.6515 & -605.5949 & 1607.3031 & 0 & -803.6515 & 605.5949 \\ -605.5949 & -456.3485 & 0 & 912.6969 & 605.5949 & -456.3485 \\ 0 & 0 & -803.6515 & 605.5949 & 803.6515 & -605.5949 \\ 0 & -1050 & 605.5949 & -456.3485 & -605.5949 & 1506.3485 \end{bmatrix}$$

Consequently, the global matrix equation for the system takes the form:

$$\begin{Bmatrix} F_{1X} \\ F_{1Y} \\ F_{2X} \\ F_{2Y} \\ F_{3X} \\ F_{3Y} \end{Bmatrix} = 10^4 \begin{bmatrix} 803.6515 & 605.5949 & -803.6515 & -605.5949 & 0 & 0 \\ 605.5949 & 1506.3485 & -605.5949 & -456.3585 & 0 & -1050 \\ -803.6515 & -605.5949 & 1607.3031 & 0 & -803.6515 & 605.5949 \\ -605.5949 & -456.3485 & 0 & 912.6969 & 605.5949 & -456.3485 \\ 0 & 0 & -803.6515 & 605.5949 & 803.6515 & -605.5949 \\ 0 & -1050 & 605.5949 & -456.3485 & -605.5949 & 1506.3485 \end{bmatrix} \begin{Bmatrix} U_{1X} \\ U_{1Y} \\ U_{2X} \\ U_{2Y} \\ U_{3X} \\ U_{3Y} \end{Bmatrix} \tag{5.58}$$

Step 5: Identify the essential boundary condition(s) and apply the boundary condition(s) to modify the global matrix equation.

The boundary conditions are derived from the knowledge that the truss is: (i) fixed at node 1 (translating to zero displacements) and (ii) constrained by a roller support at node 3 (which allows it to move along Y, but restrains its movement along X). Consequently, the identified displacement boundary conditions are:

$$U_{1X} = U_{1Y} = U_{3X} = 0 \tag{5.59}$$

Further, the known nodal forces (in *Newton*) are identified as:

$$F_{1Y} = 0 = F_{2X} = 0; \quad F_{2Y} = 80{,}000\,\text{N} \tag{5.60}$$

The rows of the forces in Equation 5.60 can be identified from Equation 5.58 as 2, 3, and 4. Using the rows of the identified known forces, one obtains the reduced stiffness matrix via **PlaneTruss_ReducedStiffnessMatrix()** as:

```
vec_rows_of_knownloads=c(2,3,4);
KglobalReduced=PlaneTruss_ReducedStiffnessMatrix(GlobalSt
iffnessMatrix,vec_rows_of_knownloads);
KglobalReduced
```

Which yields:

```
> KglobalReduced
         [,1]     [,2]     [,3]
[1,] 15063485 -6055949 -4563485
[2,] -6055949 16073031        0
[3,] -4563485        0  9126969
```

In the next line of code, we will use **PlaneTruss_ReducedForceVector()** to transform the reduced force vector into a column matrix.

```
vec_values_of_knownloads=c(0,0,-80000);
ReducedForceVector=PlaneTruss_ReducedForceVector
(vec_values_of_knownloads);
ReducedForceVector
```

Combining the reduced stiffness matrix and the values of the known nodal force leads to the reduced matrix equation.

$$10^3 \begin{Bmatrix} 0 \\ 0 \\ -80 \end{Bmatrix} = \begin{bmatrix} 15063485 & -6055949 & -4563485 \\ -6055949 & 16073031 & 0 \\ -4563485 & 0 & 9126969 \end{bmatrix} \begin{Bmatrix} U_{1X} \\ U_{2X} \\ U_{2Y} \end{Bmatrix} \quad (5.61)$$

Step 6: Determine the unknown nodal displacement.

```
UnknwonNodalDisp=PlaneTruss_NodalDisplacement(Kglobal
Reduced,ReducedForceVector)
UnknwonNodalDisp
```

The determined displacements in *meters* are obtained as:

```
> UnknwonNodalDisp=PlaneTruss_NodalDisplacement(Kglobal
Reduced,ReducedForceVector)
> UnknwonNodalDisp
             [,1]
[1,] -0.003809524
[2,] -0.001435341
[3,] -0.010669994
```

The previous content shows that, to the fourth decimal place, the nodal displacements are:

$$U_{1X} = -0.0038\text{ m};\ U_{2X} = -0.0014\text{ m};\ U_{2Y} = -0.0107\text{ m}$$

For convenience, we now combine the determined displacements and the boundary condition (from Equation 5.59).

```
allglobalNodalDisp=matrix(rep(0,Total_dof),byrow=T);
allglobalNodalDisp[vec_rows_of_knownloads]=UnknwonNodalDisp
round(allglobalNodalDisp,4)
```

Thereby producing the following:

```
> allglobalNodalDisp[vec_rows_of_knownloads]=UnknwonNodalDisp
> round(allglobalNodalDisp,4)
        [,1]
[1,]  0.0000
[2,] -0.0038
[3,] -0.0014
[4,] -0.0107
[5,]  0.0000
[6,]  0.0000
```

Step 7: With the complete nodal displacements found, we obtain the stress in each element via the function **PlaneTruss_AxialStress()**.

It is recalled that the function **PlaneTruss_AxialStress()** accepts six arguments relating to a specific member of a truss. The first, second, and third arguments are the Young's modulus, geometric length, and orientation angle of a specific element. The fourth parameter is the vector of global nodal displacements stored in the variable **allglobalNodalDisp**, whereas the fifth and sixth refer to the nodal labels (given in the connectivity table).

```
Element1_Stress=PlaneTruss_AxialStress(YoungModulus,lengths
[1],orientations[1],allglobalNodalDisp,1,2);
Element1_Stress
Element2_Stress=PlaneTruss_AxialStress(YoungModulus,lengths
[2],orientations[2],allglobalNodalDisp,2,3);
Element2_Stress
Element3_Stress=PlaneTruss_AxialStress(YoungModulus,lengths
[3],orientations[3],allglobalNodalDisp,1,3);
Element3_Stress
```

Running the earlier function returns the axial stress (in Pa) developed within each element of the truss.

```
> Element1_Stress
          [,1]
[1,] -221552019
> Element2_Stress
          [,1]
[1,] -221552019
> Element3_Stress
         [,1]
[1,] 133333333
```

It becomes apparent from the earlier that the stress in elements 1, 2, and 3 is approximately –221.55 MPa, 221.55 MPa and 133.33 MPa, respectively.

5.5.3 Example 5.3

For the space truss shown in Figure 5.4, determine the displacements at node 1 and the reaction forces at nodes 2, 3, and 4. The areas of elements 1, 2, and 3 are $10\times10^{-4}\text{m}^2$, $20\times10^{-4}\text{m}^2$, and $30\times10^{-4}\text{m}^2$, respectively. For all elements $E = 210$ GPa [5].

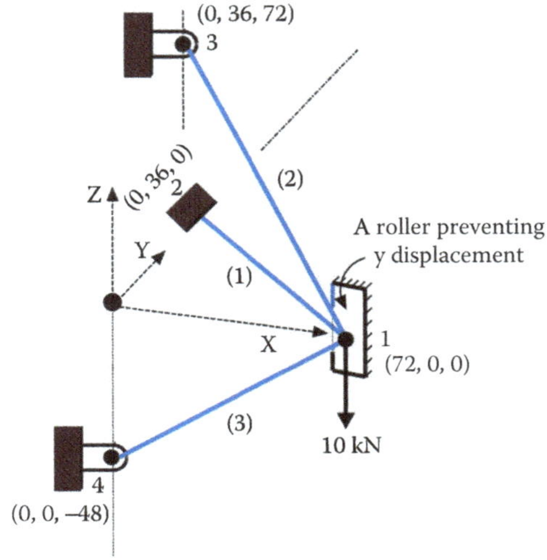

FIGURE 5.4
A three-element space truss.

5.5.3.1 Solution

For this problem, we will employ the implemented R functions for the space truss (provided in Section 5.4.3). For convenience, the structure has been discretized, and its node and the elements are labeled. The connectivity of the elements is shown in Table 5.9.

TABLE 5.9
Connectivity Table from the Discretized System in Figure 5.4

Elements	Nodes (*i*,*j*)	Coordinates ($x_i, y_i, z_i, x_j, y_j, z_j$)
1	1,2	(72,0,0,0,36,0)
2	1,3	(72,0,0,0,36,72)
3	1,4	(72,0,0,0,0,–48)

Step 1: Obtain the geometric lengths and orientation angles of each elements.

This is done by making calls to the R function **SpaceTruss_ LengthOrientations()**, which requires the nodal coordinates as input. Its output is a vector that contains four values: $L_e, \theta_{eX}, \theta_{eY}, \theta_{eZ}$, where e denotes the respective elements.

```
L1=SpaceTruss_LengthOrientations(c(72,0,0,0,36,0))[1];L1
L2=SpaceTruss_LengthOrientations(c(72,0,0,0,36,72))[1];L2
L3=SpaceTruss_LengthOrientations(c(72,0,0,0,0,-48))[1];L3

thetas1=SpaceTruss_LengthOrientations(c(72,0,0,0,36,0))
[2:4];thetas1
thetas2=SpaceTruss_LengthOrientations(c(72,0,0,0,36,72))
[2:4];thetas2
thetas3=SpaceTruss_LengthOrientations(c(72,0,0,0,0,-48))
[2:4];thetas3
```

In the first three lines of the previous block of code, we compute the length of each element by supplying **SpaceTruss_LengthOrientations()** with an element's nodal coordinate. However, **SpaceTruss_Length Orientations()** returns four numbers whenever it is called: the first is the length; and the second, third, and fourth represent the three orientation angles. Hence, length and angles must be extracted using the indexing number(s) in square brackets as done previously as length is the first element of the returned vector whenever **SpaceTruss_LengthOrientations()** is called. This explains why [1] is used to extract the geometric length and why [2:4] is used to extract the three orientation angles (again recalled, as explained in Chapter 2, that 2 : 4 is a sequence that means 2,3,4).

```
> L1=SpaceTruss_LengthOrientations(c(72,0,0,0,36,0))
[1];L1
[1] 80.49845
> L2=SpaceTruss_LengthOrientations(c(72,0,0,0,36,72))
[1];L2
[1] 108
> L3=SpaceTruss_LengthOrientations(c(72,0,0,0,0,-48))
[1];L3
[1] 86.53323
>
> thetas1=SpaceTruss_LengthOrientations(c(72,0,0,0,36,0))
[2:4];thetas1
[1] 153.43495  63.43495  90.
> thetas2=SpaceTruss_LengthOrientations(c(72,0,0,0,36,72))
[2:4];thetas2
[1] 131.81031  70.52878  48.18969
> thetas3=SpaceTruss_LengthOrientations(c(72,0,0,0,0,-48))
[2:4];thetas3
[1] 146.3099  90.0000 123.6901
```

Based on the preceding code snippet, the geometric lengths for elements 1, 2, and 3 are determined as $L_1 = 80.49$ m; $L_2 = 108$ m, and $L_3 = 86.53$ m, respectively. Their orientations are also found. For instance, for element 1: $\theta_{1X} = 153.43°$, $\theta_{1Y} = 63.43°$; $\theta_{1Z} = 90°$.

With the geometric length of the elements found, we will obtain the axial stiffness (AE/L) of all the elements.

```
E1=210*10^9;A1=10*10^(-4);
E2=210*10^9;A2=20*10^(-4);
E3=210*10^9;A3=30*10^(-4);
stiffness1=E1*A1/L1;stiffness1
stiffness2=E2*A2/L2;stiffness2
stiffness3=E3*A3/L3;stiffness3
```

The output of which reads:

```
> stiffness1=E1*A1/L1;stiffness1
[1] 2608746
> stiffness2=E2*A2/L2;stiffness2
[1] 3888889
> stiffness3=E3*A3/L3;stiffness3
[1] 7280440
```

Step 2: Write the stiffness matrix for each element using the function **SpaceTruss_Element_Matrix()**.

```
k1=SpaceTruss_Element_Matrix(6,stiffness1,thetas1);k1
k2=SpaceTruss_Element_Matrix(6,stiffness2,thetas2);k2
k3=SpaceTruss_Element_Matrix(6,stiffness3,thetas3);k3
```

The stiffness matrices are shown as:

```
> k1=SpaceTruss_Element_Matrix(6,stiffness1,thetas1):k1
          [,1]       [,2] [,3]      [,4]       [,5] [,6]
[1,]   2086997 -1043498.4    0 -2086997  1043498.4    0
[2,] -1043498   521749.2    0  1043498  -521749.2    0
[3,]        0        0.0    0        0        0.0    0
[4,] -2086997  1043498.4    0  2086997 -1043498.4    0
[5,]  1043498  -521749.2    0 -1043498   521749.2    0
[6,]        0        0.0    0        0        0.0    0
> k2=SpaceTruss_Element_Matrix(6,stiffness2,thetas2);k2
          [,1]       [,2]       [,3]       [,4]      [,5]       [,6]
[1,]  1728395.1 -864197.5 -1728395.1 -1728395.1  864197.5  1728395.1
[2,]  -864197.5  432098.8   864197.5   864197.5 -432098.8  -864197.5
[3,] -1728395.1  864197.5  1728395.1  1728395.1 -864197.5 -1728395.1
[4,] -1728395.1  864197.5  1728395.1  1728395.1 -864197.5 -1728395.1
[5,]   864197.5 -432098.8  -864197.5  -864197.5  432098.8   864197.5
[6,]  1728395.1 -864197.5 -1728395.1 -1728395.1  864197.5  1728395.1
> k3=SpaceTruss_Element_Matrix(6,stiffness3,thetas3);k3
          [,1] [,2]     [,3]     [,4] [,5]     [,6]
[1,]   5040305    0  3360203 -5040305    0 -3360203
[2,]         0    0        0        0    0        0
[3,]   3360203    0  2240135 -3360203    0 -2240135
[4,]  -5040305    0 -3360203  5040305    0  3360203
[5,]         0    0        0        0    0        0
[6,]  -3360203    0 -2240135  3360203    0  2240135
```

Step 3: Determine the global degree of freedom of the truss and expand the matrix equation for each element.

```
n_nodes=4;Total_dof=n_nodes*3;
K1=SpaceTruss_ExpandedElement_Matrix(12,k1,1,2);
K2=SpaceTruss_ExpandedElement_Matrix(12,k2,1,3);
K3=SpaceTruss_ExpandedElement_Matrix(12,k3,1,4);
```

Step 4: Establish the global stiffness matrix by adding the expanded matrices of the elements as shown in the following:

```
Kglobal=K1+K2+K3;
Kglobal
```

The output of this gives the elements of the global stiffness matrix $[K]$ as:

```
> Kglobal=K1+K2+K3;Kglobal
            [,1]       [,2]       [,3]     [,4]       [,5] [,6]       [,7]      [,8]       [,9]   [,10] [,11]    [,12]
 [1,]  8855696.5 -1907695.9  1631808.0 -2086997  1043498.4    0 -1728395.1  864197.5  1728395.1 -5040305     0 -3360203
 [2,] -1907695.9   953848.0   864197.5  1043498  -521749.2    0   864197.5 -432098.8  -864197.5        0     0        0
 [3,]  1631808.0   864197.5  3968530.5        0        0.0    0  1728395.1 -864197.5 -1728395.1 -3360203     0 -2240135
 [4,] -2086996.8  1043498.4        0.0  2086997 -1043498.4    0        0.0       0.0        0.0        0     0        0
 [5,]  1043498.4  -521749.2        0.0 -1043498   521749.2    0        0.0       0.0        0.0        0     0        0
 [6,]        0.0        0.0        0.0        0        0.0    0        0.0       0.0        0.0        0     0        0
 [7,] -1728395.1   864197.5  1728395.1        0        0.0    0  1728395.1 -864197.5 -1728395.1        0     0        0
 [8,]   864197.5  -432098.8  -864197.5        0        0.0    0  -864197.5  432098.8   864197.5        0     0        0
 [9,]  1728395.1  -864197.5 -1728395.1        0        0.0    0 -1728395.1  864197.5  1728395.1        0     0        0
[10,] -5040304.7        0.0 -3360203.1        0        0.0    0        0.0       0.0        0.0  5040305     0  3360203
[11,]        0.0        0.0        0.0        0        0.0    0        0.0       0.0        0.0        0     0        0
[12,] -3360203.1        0.0 -2240135.4        0        0.0    0        0.0       0.0        0.0  3360203     0  2240135
```

Consequently, the global matrix equation for the system takes the following form:

$$\begin{Bmatrix} F_{1X} \\ F_{1Y} \\ F_{1Z} \\ F_{2X} \\ F_{2Y} \\ F_{2Z} \\ F_{3X} \\ F_{3Y} \\ F_{3Z} \end{Bmatrix} = [K] \begin{Bmatrix} U_{1X} \\ U_{1Y} \\ U_{1Z} \\ U_{2X} \\ U_{2Y} \\ U_{2Z} \\ U_{3X} \\ U_{3Y} \\ U_{3Z} \end{Bmatrix} \tag{5.62}$$

Where

$$[K] = 10^5 \begin{bmatrix} 88.557 & -19.077 & 16.318 & -20.870 & 10.435 & 0 & -17.284 & 8.642 & 17.284 & -50.403 & 0 & -33.602 \\ -19.077 & 9.538 & 8.642 & 10.435 & -5.217 & 0 & 8.642 & -4.321 & -8.642 & 0 & 0 & 0 \\ 16.318 & 8.642 & 39.685 & 0 & 0 & 0 & 17.284 & -8.642 & -17.284 & -33.602 & 0 & -22.401 \\ -20.870 & 10.435 & 0 & 20.870 & -10.435 & 0 & 0 & 0 & 0 & 0 & 0 & 0 \\ 10.435 & -5.217 & 0 & -10.435 & 5.217 & 0 & 0 & 0 & 0 & 0 & 0 & 0 \\ 0 & 0 & 0 & 0 & 0 & 0 & 0 & 0 & 0 & 0 & 0 & 0 \\ -17.284 & 8.642 & 17.284 & 0 & 0 & 0 & 17.284 & -8.642 & -17.284 & 0 & 0 & 0 \\ 8.642 & -4321 & -8.642 & 0 & 0 & 0 & -8.642 & 4.321 & 8.642 & 0 & 0 & 0 \\ 17.284 & -8.642 & -17.284 & 0 & 0 & 0 & -17.484 & 8.642 & 17.284 & 0 & 0 & 0 \\ -50.403 & 0 & -33.602 & 0 & 0 & 0 & 0 & 0 & 0 & 50.403 & 0 & 33.602 \\ 0 & 0 & 0 & 0 & 0 & 0 & 0 & 0 & 0 & 0 & 0 & 0 \\ -33.602 & 0 & -22.401 & 0 & 0 & 0 & 0 & 0 & 0 & 33.602 & 0 & 22.401 \end{bmatrix}$$

Step 5: Identify the essential boundary condition(s) and apply the boundary condition(s) to modify the global matrix equation.

The boundary conditions are derived from the knowledge that the space truss is fixed at nodes 2, 3, and 4 (translating to zero displacements at these nodes). It is also constrained by a roller support at node 1 (which allows it to move along X and Z, but restrains its movement along Y). Accordingly, the identified displacement boundary conditions will be:

$$U_{1Y} = 0; U_{2X} = U_{2Y} = U_{2Z} = 0; U_{3X} = U_{3Y} = U_{3Z} = 0; U_{4X} = U_{4Y} = U_{4Z} = 0 \tag{5.63}$$

By realizing that we have 10 nodal displacement boundary conditions, we conclude that there must exist two known nodal force values. The known nodal forces (in *Newton*) are identified as:

$$F_{1X} = 0; F_{1Z} = 10{,}000 \text{ N} \tag{5.64}$$

The rows of the forces in Equation 5.64 can be identified in the matrix in Equation 5.62 as 1 and 3, respectively. Using the set of rows of the identified known force values, one obtains the reduced stiffness matrix via the function **SpaceTruss_ReducedStiffnessMatrix()**. The following code is used for this purpose.

```
vec_rows_of_knownloads=c(1,3);
KglobalReduced=SpaceTruss_ReducedStiffnessMatrix(Kglobal,
vec_rows_of_knownloads);
KglobalReduced
```

```
> KglobalReduced
        [,1]    [,2]
[1,] 8855697 1631808
[2,] 1631808 3968530
```

In other words, the reduced stiffness matrix is:

$$[K_{reduced}] = \begin{bmatrix} 8855697 & 1631808 \\ 1631808 & 3968530 \end{bmatrix}$$

```
vec_values_of_knownloads=c(0,-10000);
ReducedForceVector=SpaceTruss_ReducedForceVector
(vec_values_of_knownloads);
ReducedForceVector
```

In the following block of code, we use the function **SpaceTruss_ReducedForceVector()** to establish the reduced force vector:

Combining the reduced stiffness matrix and the values of the known nodal force leads to the reduced matrix equation.

$$10^3 \begin{Bmatrix} 0 \\ -10 \end{Bmatrix} = \begin{bmatrix} 8855697 & 1631808 \\ 1631808 & 3968530 \end{bmatrix} \begin{Bmatrix} U_{1X} \\ U_{1Z} \end{Bmatrix} \tag{5.65}$$

Step 6: Determine the unknown nodal displacements and the nodal global forces by using:

- **SpaceTruss_NodalDisplacement()**
- **SpaceTruss_Global_Forces()**

```
UnknwonNodalDisp=SpaceTruss_NodalDisplacement(KglobalRedu
ced,ReducedForceVector)
UnknwonNodalDisp
```

Running the previous lines of code returns the nodal displacements in meters as:

```
> UnknwonNodalDisp
                [,1]
[1,]   0.0005023838
[2,]  -0.0027263981
```

Which shows that, to the third decimal place, the displacements are:

$$U_{1X} = -0.502 \text{ mm}; U_{1Z} = -2.726 \text{ mm} \tag{5.66}$$

For convenience, we now combine the determined displacements and the boundary condition (from Equation 5.63).

```
allglobalNodalDisp=matrix(rep(0,Total_dof),byrow=T);
allglobalNodalDisp[vec_rows_of_knownloads]=UnknwonNodalDisp
round(allglobalNodalDisp,4)
```

Having determined the nodal displacements, we may now find the global forces (through which the reaction forces can be determined). We will use the function **SpaceTruss_Global_Forces()** for this purpose. It is recalled that this function requires the global stiffness matrix (represented by the variable **Kglobal** determined in Step 4) and the vector of global nodal displacements (contained in the variable **allglobalNodalDisp**).

```
NodalGlobalForces=SpaceTruss_Global_Forces(Kglobal,allglo
balNodalDisp);
NodalGlobalForces
```

Running the preceding code returns the global forces as:

```
> NodalGlobalForces
       [,1]
 [1,]     0
 [2,] -3315
 [3,] -10000
 [4,] -1048
 [5,]   524
 [6,]     0
 [7,] -5581
 [8,]  2790
 [9,]  5581
[10,]  6629
[11,]     0
[12,]  4419
```

Based on the preceding output, the reaction forces at the nodes are as follows.

$$F_{1Y} = -3315\,N;\ F_{2X} = -1048\,N;\ F_{2Y} = 524\,N;\ F_{2Z} = 0\,N;\ F_{3X} = -5581\,N;$$

$$F_{3Y} = 2790\,N;\ F_{3Z} = 5581\,N;\ F_{4X} = 6629\,N;\ F_{4Y} = 0\,N;\ F_{4Z} = 4419\,N$$

5.6 Exercises

1. For the planar truss shown in Figure 5.5, solve the following: (i) the horizontal and vertical displacements at node 3 and (ii) determine the forces in each element. Let $E = 70$ GPa and $A = 4\times10^{-4}$m for all elements.
2. For the truss shown in Figure 5.6, solve the horizontal and vertical displacements at nodes 2 and 3. Also determine the forces in each element. Let $E = 200$ GPa and $A = 6\times10^{-4}$ m for all elements.
3. For the space truss shown in Figure 5.7, determine the nodal displacements and the stresses developed in each member. Let $E = 210$ GPa and $A = 10\times10^{-4}$ m for all elements [10].

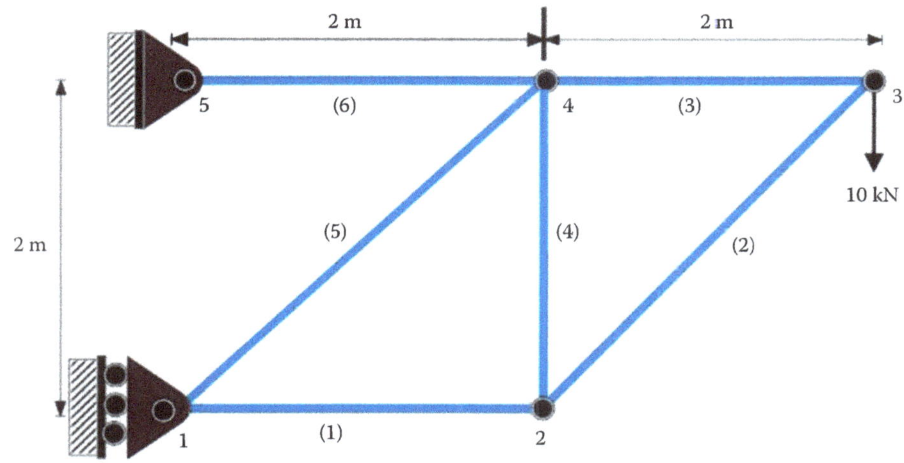

FIGURE 5.5
Exercise 5Q1.

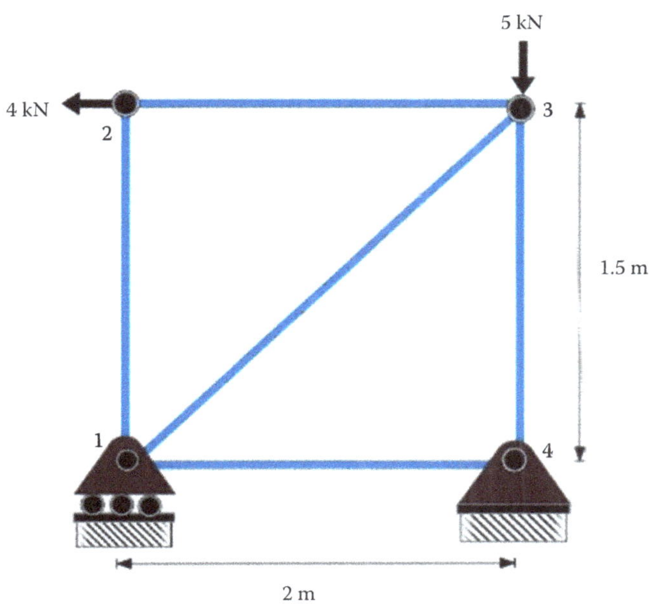

FIGURE 5.6
Exercise 5Q2.

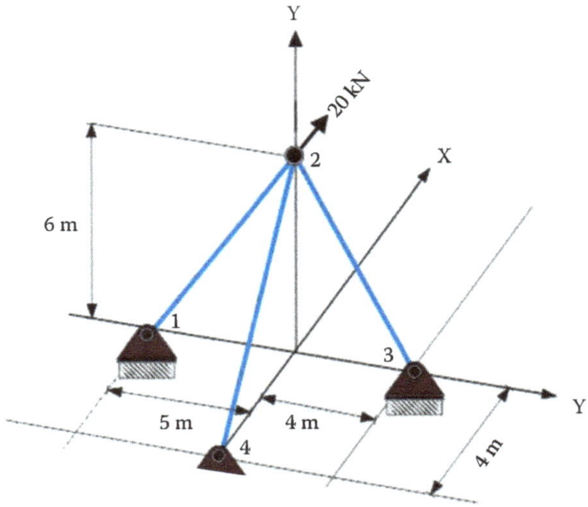

FIGURE 5.7
Exercise 5Q3.

References

1. J. F. Doyle, *Nonlinear Analysis of Thin-Walled Structures: Statics, Dynamics, and Stability*, Springer Science & Business Media, New York, 2013.
2. P. Marti, Stress resultants and state diagrams. In *Theory of Structures*, Wiley-VCH Verlag GmbH & Co. KGaA, Berlin, Germany, 2013, pp. 159–175.
3. J. Ambrose, *Simplified Mechanics and Strength of Materials*, vol. 34, John Wiley & Sons, New York, 2011.
4. F. A. Leckie, D. J. Bello, *Strength and Stiffness of Engineering Systems*, Springer Science & Business Media, New York, 2009.
5. D. L. Logan, *First Course in the Finite Element Method*, Thomson, Toronto, ON, 2007.
6. S. S. Rao, *The Finite Element Method in Engineering*, Elsevier Science, Burlington, NJ, 2010.
7. T. J. Hughes, *The Finite Element Method: Linear Static and Dynamic Finite Element Analysis*, Courier Corporation, New York, 2012.
8. M. A. Bhatti, *Fundamental Finite Element Analysis and Applications: with Mathematica and Matlab Computations*, Wiley, Hoboken, NJ, 2005.
9. A. Khennane, *Introduction to Finite Element Analysis using MATLAB® and Abaqus*, CRC Press, Boca Raton, FL, 2013.
10. P. I. Kattan, *MATLAB Guide to Finite Elements: An Interactive Approach*, Springer, Berlin, Germany, 2010.

6

Shear Rigid and Shear Elastic Beam Elements

6.1 Introduction

This chapter considers the formulation of beam elements suitable for the static analyses of flexural members. We will present the finite element formulations for the Euler–Bernoulli (EB) and Timoshenko beams (TB). The former represents a beam model that neglects shear-deformation effect (hence the name shear-rigid beam), whereas the latter is a higher order beam model that accounts for the effect of shear deformation.

Beams, just like bars treated in Chapter 4, are slender structural components, the motions and deformations of which can be described in terms of a single spatial variable. They are predominantly designed to support transverse loads (acting perpendicular to the member's principal axis), and they can resist shear force and bending moments. As one of the most familiar structural elements, beams function as supporting members in many engineering applications (e.g., as propellers for aircrafts and boats, as models for the idealization of turbine blades and support structures in automobile, bridges, and numerous machines). The main interest in the analyses of beams is to evaluate the internal forces (shear forces and bending moments), the distribution of the deflection, and the stresses experienced by the beam under a system externally applied load.

In the remainder of the chapter, we will walk through the finite element models for the EB and TB elements in Sections 6.2 and 6.3, respectively. Section 6.4 contains the implemented R functions related to these two elements. Examples of the applications of the implemented functions are detailed in Section 6.5.

6.2 Finite Element Equations for the Euler–Bernoulli Beam Element

6.2.1 Displacement and Shape Functions

Figure 6.1a displays a beam with generic transverse loads: concentrated moment M_z, concentrated force P, distributed moment $M(x)$, and distributed force $q(x)$. The beam is endowed with a cross-sectional area depicted in Figure 6.1b. Our aim is to obtain a simple beam element that can be used to determine the displacements, and stresses under such loadings.

Let us designate the unloaded segment AB (in Figure 6.1a) as a two-node EB beam element with two degrees of freedom per node as illustrated in Figure 6.1c. Furthermore, let v_i and θ_i represent the vertical displacement and slope at node i, whereas v_j and θ_j are the vertical displacement and slope at node j. For this element, the slope at any point along the beam equals the first derivative of the deflection curve. For reference, in addition to these degrees of freedom, Figure 6.1c shows the adopted sign convention for positive bending moments (counterclockwise) and shear forces (in the direction of positive y).

Consider that the beam undergoes a moderate deflection under the effects of the nodal loads (shear force and bending moments), then we may prescribe a cubic function for the variation of the transverse deflection along the length of the element in the form [1,2]:

$$v = A_1 + A_2 x + A_3 x^2 + A_4 x^3 \tag{6.1}$$

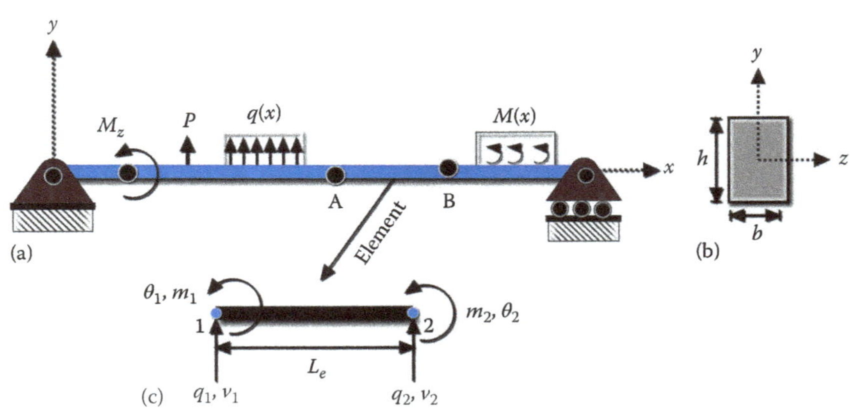

FIGURE 6.1
(a) A beam with generic transverse loads, (b) cross-sectional area, and (c) a beam segment to be used as an element.

Or rewritten in a matrix form:

$$v = \begin{bmatrix} 1 & x & x^2 & x^3 \end{bmatrix} \begin{Bmatrix} A_1 \\ A_2 \\ A_3 \\ A_4 \end{Bmatrix} \tag{6.2}$$

By using Equations 6.1 and 6.2, we obtain the slope ($\theta = dv/dx$) at any point along the neutral axis of the beam as

$$\theta = A_2 + 2A_3 x + 3A_4 x^2 \tag{6.3}$$

where A_k ($k = 1, 2, 3, 4$) are unknown coefficients, and x is within the range $0 \le x \le L_e$. As reflected in Chapter 5, we have chosen the number of unknown coefficients A_k to be equal to the total number of degrees of freedom associated with the element.

Next, we solve for the unknown coefficients by imposing boundary conditions on Equations 6.1 and 6.2 as follows:

- At node 1, $x = 0$: $v = v_1$
- At node 1, $x = 0$: $\theta = \theta_1$
- At node 2, $x = L_e$: $v = v_2$
- At node 2, $x = L_e$: $\theta = \theta_2$

By substituting these boundary conditions into Equations 6.1 through 6.3, one obtains a set of equations that can be written in the matrix form as

$$\underbrace{\begin{Bmatrix} v_1 \\ \theta_1 \\ v_2 \\ \theta_2 \end{Bmatrix}}_{\{u_e\}} = \underbrace{\begin{bmatrix} 1 & 0 & 0 & 0 \\ 0 & 1 & 0 & 0 \\ 1 & L_e & L_e^2 & L_e^3 \\ 0 & 1 & 2L_e & 3L_e^2 \end{bmatrix}}_{[H]} \underbrace{\begin{Bmatrix} A_1 \\ A_2 \\ A_3 \\ A_4 \end{Bmatrix}}_{\{A\}} \tag{6.4}$$

where $\{u_e\}$ is a vector of the nodal displacements and slopes, $[H]$ is the coordinate matrix, and $\{A\}$ is the vector of the unknown coefficients. From Equation 6.4, the unknown coefficients are determined as

$$\{A\} = [H]^{-1} \{u_e\} \tag{6.5}$$

where

$$[H]^{-1} = \begin{bmatrix} 1 & 0 & 0 & 0 \\ 0 & 1 & 0 & 0 \\ -\frac{3}{L_e^2} & -\frac{2}{L_e} & \frac{3}{L_e^2} & -\frac{1}{L_e} \\ \frac{2}{L_e^3} & \frac{1}{L_e^2} & -\frac{2}{L_e^3} & \frac{1}{L_e^2} \end{bmatrix} \tag{6.6}$$

Having found the unknowns constants A, we now rewrite Equations 6.1 and 6.2 as

$$v = \begin{bmatrix} 1 & x & x^2 & x^3 \end{bmatrix} \begin{bmatrix} 1 & 0 & 0 & 0 \\ 0 & 1 & 0 & 0 \\ -\frac{3}{L_e^2} & -\frac{2}{L_e} & \frac{3}{L_e^2} & -\frac{1}{L_e} \\ \frac{2}{L_e^3} & \frac{1}{L_e^2} & -\frac{2}{L_e^3} & \frac{1}{L_e^2} \end{bmatrix} \begin{Bmatrix} v_1 \\ \theta_1 \\ v_2 \\ \theta_2 \end{Bmatrix} \tag{6.7}$$

Carrying out the multiplication of the first two matrix terms in Equation 6.7, one obtains

$$v = \underbrace{\begin{bmatrix} \left(1 - \frac{3x^2}{L_e^2} + \frac{2x^3}{L_e^3}\right) & \left(x - \frac{2x^2}{L_e} + \frac{x^3}{L_e^2}\right) & \left(\frac{3x^2}{L_e^2} - \frac{2x^3}{L_e^3}\right) & \left(-\frac{x^2}{L_e} + \frac{x^3}{L_e^2}\right) \end{bmatrix}}_{[N]} \underbrace{\begin{Bmatrix} v_1 \\ \theta_1 \\ v_2 \\ \theta_2 \end{Bmatrix}}_{\{u_e\}} \tag{6.8}$$

where $[N]$ is a vector of shape functions.

6.2.2 Strain–Displacement Relationship and Constitutive Law

Under the assumption of the EB beam theory, the non-zero component of the strain tensor is the axial normal strain (ε_{xx}), which quantifies the extension of the beam's fibers when deflection occurs. By considering the kinematic of the beam's deformation, ε_{xx} is defined as

$$\varepsilon_{xx} = y\frac{\partial^2 v}{\partial x^2} \tag{6.9}$$

With Equation 6.8, we may recast Equation 6.9 as

$$\varepsilon_{xx} = y\frac{\partial^2[N]}{\partial x^2}\{u_e\} \tag{6.10}$$

$$\varepsilon_{xx} = y\underbrace{\left[\left(-\frac{6}{L_e^2}+\frac{12x}{L_e^3}\right)\ \left(-\frac{4}{L_e}+\frac{6x}{L_e^2}\right)\ \left(\frac{6}{L_e^2}-\frac{12x}{L_e^3}\right)\ \left(-\frac{2}{L_e}+\frac{6x}{L_e^2}\right)\right]}_{[B]}\begin{Bmatrix} v_1 \\ \theta_1 \\ v_2 \\ \theta_2 \end{Bmatrix} \tag{6.11}$$

Or:

$$\varepsilon_{xx} = y[B]\{u_e\} \tag{6.12}$$

For a one-dimensional isotropic elastic beam, the stress–strain relationship is

$$\{\sigma\} = E\{\varepsilon\} \tag{6.13}$$

where E is the Young's modulus. Substituting the expression for the strain from Equation 6.12 into Equation 6.13, one arrives at

$$\{\sigma\} = Ey[B]\{u_e\} \tag{6.14}$$

6.2.3 Minimization of the Total Potential Energy for the Euler–Bernoulli Beam Element

We recall the expression for the total potential energy (P_t) as

$$P_t = U - W \tag{6.15}$$

Furthermore, we may write the strain energy (U) and the work done by the system of external forces (W) on the element shown in Figure 6.1c as

$$U = \frac{1}{2}\int_0^{L_e}\int_A \{\sigma\}^T\{\varepsilon\}\,dx\,dA \tag{6.16}$$

$$W = q_1v_1 + m_1\theta_1 + q_2v_2 + m_2\theta_2 = \{u_e\}^T\{f_e\} \tag{6.17}$$

It follows from Equations 6.11 through 6.14 that

$$U = \frac{1}{2}\int_0^{L_e}\int_A \{\sigma\}^T\{\varepsilon\}\,dx\,dA = \frac{EI}{2}\int_0^{L_e}\left([B]\{u_e\}\right)^T\left([B]\{u_e\}\right)dx \tag{6.18}$$

where we have used the fact that $I = \int y^2 \, dA$ in Equation 6.18. We substitute Equations 6.17 and 6.18 into Equation 6.15 and then minimize P_t with respect to u_e to obtain

$$\frac{\partial}{\partial \{u_e\}^T} P_t = 0 = EI \int_0^{L_e} \left([B]^T [B] \right) \{u_e\} - \{f_e\} \tag{6.19}$$

Consequently, we obtain the desired relation as

$$\left(EI \int_0^{L_e} \left([B]^T [B] \right) dx \right) \{u_e\} - \{f_e\} = 0 \tag{6.20}$$

By rearranging Equation 6.20, we obtain the finite element matrix equation for the EB beam element as

$$\{f_e\} = \left(EI \int_0^{L_e} \left([B]^T [B] \right) dx \right) \{u_e\} = [k_e]\{u_e\} \tag{6.21}$$

In full form, the stiffness matrix $[k_e]$, which results from the multiplication of the matrix $[B]$ and its transpose, reads as

$$[k_e] = \frac{EI}{L_e^3} \begin{bmatrix} 12 & 6L_e & -12 & 6L_e \\ 6L_e & 4L_e^2 & -6L_e & 2L_e^2 \\ -12 & -6L_e & 12 & -6L_e \\ 6L_e & 2L_e^2 & -6L_e & 4L_e^2 \end{bmatrix} \tag{6.22}$$

Accordingly, the complete finite element equation for the EB beam element is as follows:

$$\begin{Bmatrix} q_1 \\ m_1 \\ q_2 \\ m_2 \end{Bmatrix} = \frac{EI}{L_e^3} \begin{bmatrix} 12 & 6L_e & -12 & 6L_e \\ 6L_e & 4L_e^2 & -6L_e & 2L_e^2 \\ -12 & -6L_e & 12 & -6L_e \\ 6L_e & 2L_e^2 & -6L_e & 4L_e^2 \end{bmatrix} \begin{Bmatrix} v_1 \\ \theta_1 \\ v_2 \\ \theta_2 \end{Bmatrix} \tag{6.23}$$

6.3 Finite Element Equations for the Shear-Deformable Beams

The EB beam finite element derived in the preceding section does not account for shear-deformation effect. This element is mostly suitable for idealizing the mechanical response of shallow beams ($L/h > 10$) under transverse loads.

TABLE 6.1

Highlight of the Differences between the Euler–Bernoulli and Timoshenko Beam Elements

	Euler–Bernoulli Beam Element	Timoshenko Beam Element
Nodal deformation variables	• Vertical displacement (v) • Slope of deflected curve (θ) due to bending moments where $\theta = \frac{dv}{dx}$	• Vertical displacement (v) • Rotation of cross-section (ϕ) due to bending where $\phi = \frac{dv}{dx} - \gamma$
Strain components	• Normal strain: $\left(\varepsilon_{xx} = -\frac{yd^2v}{dx^2}\right)$	• Normal strain: $\left(\varepsilon_{xx} = -y\frac{d\phi}{dx}\right)$ • Shear strain: $\left(\gamma_{xy} = \frac{dv}{dx} - \phi\right)$
Stress components	• Normal stress: $\sigma_{xx} = E\varepsilon_{xx}$	• Normal stress: $\sigma_{xx} = E\varepsilon_{xx}$ • Shear stress: $\tau_{xy} = kG\gamma_{xy}$

For deep beams and composite laminated beams, the TB theory, which accounts for the shear-deformation effect, offers results that are more accurate. Some of the major differences between the EB and the TB elements are highlighted in Table 6.1. For the most part, our discussion on this element will be very brief as detailed discussions on the properties of this element can be found, for instance in References 3 and 4.

6.3.1 Displacement and Shape Functions

A careful selection of shape functions is required in the formulation of the TB element to avoid a phenomenon called *shear locking*. Besides, for this element, the rotation of the cross-section ϕ is not the same as the slope of the deflected curve. Consequently, v and ϕ are interpolated independently as shown in Equations 6.24 and 6.25.

$$v = \begin{bmatrix} N_{1v} & N_{ev} & N_{2v} & -N_{ev} \end{bmatrix} \begin{Bmatrix} v_1 \\ \phi_1 \\ v_2 \\ \phi_2 \end{Bmatrix} = [N_v]\{u_e\} \tag{6.24}$$

$$\phi = \begin{bmatrix} 0 & N_{1\phi} & 0 & N_{2\phi} \end{bmatrix} \begin{Bmatrix} v_1 \\ \phi_1 \\ v_2 \\ \phi_2 \end{Bmatrix} = [N_\phi]\{u_e\} \tag{6.25}$$

Following the procedure in Reference 3, we employ the following shape functions:

$$N_{1v} = N_{1\phi} = 1 - \frac{x}{L_e}; \quad N_{2v} = N_{2\phi} = \frac{x}{L_e}; \quad N_{ev} = \frac{x}{2}\left(1 - \frac{x}{L_e}\right) \tag{6.26}$$

6.3.2 Strain–Displacement Relationship and Constitutive Law

As shown in Table 6.1, under the TB theory, two non-zero components of the strain tensor exist: ε_{xx} and γ_{xy}.

By using the vector of shape functions defined in Equations 6.24 and 6.25, one obtains

$$\varepsilon_{xx} = y\frac{d\phi}{dx} = y\underbrace{\begin{bmatrix} 0 & \frac{dN_{1\phi}}{dx} & 0 & \frac{dN_{2\phi}}{dx} \end{bmatrix}}_{[B_b]} \begin{Bmatrix} v_1 \\ \phi_1 \\ v_2 \\ \phi_2 \end{Bmatrix} = y[B_b]\{u_e\}; \tag{6.27}$$

$$\gamma_{xy} = \frac{dv}{dx} - \phi = \underbrace{\begin{bmatrix} \frac{dN_1}{dx} & \frac{dN_e}{dx} - N_{1v} & \frac{dN_2}{dx} & \frac{-dN_e}{dx} - N_{2v} \end{bmatrix}}_{[B_s]} \begin{Bmatrix} v_1 \\ \phi_1 \\ v_2 \\ \phi_2 \end{Bmatrix} = ([B_s])\{u_e\} \tag{6.28}$$

The stress–strain relations for the TB, if we take E and G to be the Young's modulus and shear modulus respectively, are as follows:

$$\sigma_{xx} = E\varepsilon_{xx} = y[B_b]\{u_e\} \tag{6.29}$$

$$\tau_{xy} = kG\gamma_{xy} = kG[B_s]\{u_e\} \tag{6.30}$$

where we have used the parameter k in Equation 6.30 to denote the shear-correction factor.

6.3.3 Minimization of the Total Potential Energy for the Timoshenko Beam Element

Following the steps in Section 6.2.3, we write the strain energy (U) and the work done by external forces (W) on the TB element as

$$U = \frac{1}{2}\int_0^{L_e}\int_A \{\sigma\}^T\{\varepsilon\}\,dx\,dA + \frac{1}{2}\int_0^{L_e}\int_A \{\tau\}^T\{\gamma\}\,dx\,dA \tag{6.31}$$

$$W = \{u_e\}^T \{f_e\} \tag{6.32}$$

Making use of Equations 6.28 and 6.29, the strain-energy expression simplifies to

$$U = \frac{EI}{2}\int_0^{L_e} \left([B_b]\{u_e\}\right)^T \left([B_b]\{u_e\}\right) dx + \frac{kGA}{2}\int_0^{L_e} \left([B_s]\{u_e\}\right)^T \left([B_s]\{u_e\}\right) dx \tag{6.33}$$

As has been the practice, we substitute Equations 6.31 and 6.32 into Equation 6.14 and then minimize P_t with respect to u_e to obtain

$$\frac{\partial}{\partial \{u_e\}^T} P_t = 0 = \left(EI\int_0^{L_e} [B_b]^T [B_b] dx + KGA\int_0^{L_e} [B_s]^T [B_s] dx \right)\{u_e\} - \{f_e\} \tag{6.34}$$

Which leads to the desired relation as

$$\left(EI\int_0^{L_e} [B_b]^T [B_b] dx + KGA\int_0^{L_e} [B_s]^T [B_s] dx \right)\{u_e\} = \{f_e\} = 0 \tag{6.35}$$

In full form, the finite matrix equation for the TB element then takes the following form:

$$\begin{Bmatrix} q_1 \\ m_1 \\ q_2 \\ m_2 \end{Bmatrix} = \begin{bmatrix} \frac{kGA}{L_e} & \frac{kGA}{2} & -\frac{kGA}{L_e} & \frac{kGA}{2} \\ \frac{kGA}{2} & \left(\frac{EI}{L_e} + \frac{kGAL_e}{4}\right) & -\frac{kGA}{2} & \left(\frac{EI}{L_e} + \frac{kGAL_e}{4}\right) \\ -\frac{kGA}{L_e} & -\frac{kGA}{2} & \frac{kGA}{L_e} & -\frac{kGA}{2} \\ \frac{kGA}{2} & -\left(\frac{EI}{L_e} + \frac{kGAL_e}{4}\right) & -\frac{kGA}{2} & \left(\frac{EI}{L_e} + \frac{kGAL_e}{4}\right) \end{bmatrix} \begin{Bmatrix} v_1 \\ \phi \\ v_2 \\ \dot{\varphi}_2 \end{Bmatrix} \tag{6.36}$$

6.4 Equivalent Nodal Forces and Moments for Beams with Distributed Loads

The beam elements derived in the preceding sections are suitable for analyzing beams subjected to concentrated loads. Oftentimes, beams are also designed to support distributed loads in many practical situations. When this happens, a vector of additional load terms (called *equivalent nodal loads*) needs to be added to the finite element matrix equation.

In practice, the distributed load manifests in different forms, sometimes as a uniformly distributed load (UDL) and at other times as a triangular distributed loads (TDL). The work done by a uniform distributed force of magnitude $w(x)$ over the length of an element of length L can generally be written as

$$F_{\text{UDL}} = \int_0^{L_e} w(x) v dx = \int_0^{L_e} w(x)[N]\{u_e\} dx = \{u_e\}^T \begin{Bmatrix} -\dfrac{wL_e}{2} \\ -\dfrac{wL_e^2}{12} \\ -\dfrac{wL_e}{2} \\ \dfrac{1}{12} wL_e^2 \end{Bmatrix} \tag{6.37}$$

Consequently, the equivalent nodal loads to be added to either Equation 6.23 or Equation 6.36 are obtained by minimizing Equation 6.37 with respect to $\{u_e\}$, which gives

$$F_{\text{UDL}}^{eq} = \frac{\partial F_{\text{UDL}}}{\partial \{u_e\}^T} = \begin{Bmatrix} -\dfrac{wL_e}{2} \\ -\dfrac{wL_e^2}{12} \\ -\dfrac{wL_e}{2} \\ \dfrac{1}{12} wL_e^2 \end{Bmatrix} \tag{6.38}$$

For a TDL, if the height of the distributed force is $q(x)$, then the work done over an element of length L is

$$F_{\text{TDL}} = \int_0^{L_e} \frac{w(x)}{L} v dx = \int_0^{L_e} \frac{w(x)}{L}[N]\{u_e\} dx = \{u_e\}^T \begin{Bmatrix} -\dfrac{3wL_e}{20} \\ -\dfrac{1}{30} wL_e^2 \\ -\dfrac{7wL_e}{20} \\ \dfrac{wL_e^2}{20} \end{Bmatrix} \tag{6.39}$$

Hence for the TDL, the equivalent nodal load to be added to either Equation 6.23 or Equation 6.36 is obtained by minimizing Equation 6.39 with respect to $\{u_e\}$ to obtain

$$F_{\text{TDL}}^{eq} = \frac{\partial F_{\text{TDL}}}{\partial \{u_e\}^T} = \begin{Bmatrix} -\dfrac{3wL_e}{20} \\ -\dfrac{1}{30}wL_e^2 \\ -\dfrac{7wL_e}{20} \\ \dfrac{wL_e^2}{20} \end{Bmatrix} \tag{6.40}$$

6.5 Computer Implementations

For a system of connected planar beams, the global matrix equation relates the vector of global nodal loads $\{\boldsymbol{F}\}$, the global stiffness matrix $[\boldsymbol{K}]$, and the vector of global nodal degrees of freedom $\{\boldsymbol{U}\}$ to form the global matrix equation:

$$\{\boldsymbol{F}\} = [\boldsymbol{K}]\{\boldsymbol{U}\} \tag{6.41}$$

Where the matrix equation for each member of the connected beams is formed by using either the EB beam element:

$$\begin{Bmatrix} q_1 \\ m_1 \\ q_2 \\ m_2 \end{Bmatrix} + \begin{Bmatrix} -\dfrac{wL_e}{2} \\ -\dfrac{wL_e^2}{12} \\ -\dfrac{wL_e}{2} \\ \dfrac{1}{12}wL_e^2 \end{Bmatrix} = \frac{EI}{L_e^3}\begin{bmatrix} 12 & 6L_e & -12 & 6L_e \\ 6L_e & 4L_e^2 & -6L_e & 2L_e^2 \\ -12 & -6L_e & 12 & -6L_e \\ 6L_e & 2L_e^2 & -6L_e & 4L_e^2 \end{bmatrix}\begin{Bmatrix} v_1 \\ \theta_1 \\ v_2 \\ \theta_2 \end{Bmatrix} \tag{6.42}$$

Or the TB element:

$$\begin{Bmatrix} q_1 \\ m_1 \\ q_2 \\ m_2 \end{Bmatrix} + \begin{Bmatrix} -\dfrac{wL_e}{2} \\ -\dfrac{wL_e^2}{12} \\ -\dfrac{wL_e}{2} \\ \dfrac{1}{12}wL_e^2 \end{Bmatrix} = \begin{bmatrix} \dfrac{kGA}{L_e} & \dfrac{kGA}{2} & -\dfrac{kGA}{L_e} & \dfrac{kGA}{2} \\ \dfrac{kGA}{2} & \left(\dfrac{EI}{L_e}+\dfrac{kGAL_e}{4}\right) & -\dfrac{kGA}{2} & \left(\dfrac{EI}{L_e}+\dfrac{kGAL_e}{4}\right) \\ -\dfrac{kGA}{L_e} & -\dfrac{kGA}{2} & \dfrac{kGA}{L_e} & -\dfrac{kGA}{2} \\ \dfrac{kGA}{2} & -\left(\dfrac{EI}{L_e}+\dfrac{kGAL_e}{4}\right) & -\dfrac{kGA}{2} & \left(\dfrac{EI}{L_e}+\dfrac{kGAL_e}{4}\right) \end{bmatrix}\begin{Bmatrix} v_1 \\ \phi \\ v_2 \\ \phi_2 \end{Bmatrix} \tag{6.43}$$

6.5.1 Systematic Procedure for Analyses of Beam Systems

6.5.1.1 Precomputation Phase

- Discretize the structure into finite beam elements.
- Label the nodes and elements (this helps one to form the connectivity table).

6.5.1.2 Computation Phase

1. Establish the finite element matrix equation (or just the stiffness matrix) for each beam element. For a beam element with distributed load(s), determine the column matrix of the equivalent nodal loads (by using, e.g., Equations 6.38 and 6.40). This will also have to be expanded accordingly.
2. Determine the global degree of freedom of the connected beams, and carry out the expansion of the matrix equation for each element.
3. Form the global stiffness matrix and the global matrix equation for the system.
4. Identify the essential boundary condition(s) and apply the boundary condition(s) to the global matrix equation to obtain a reduced matrix equation. Programmatically, this step involves two actions:
   a. Obtain the reduced load vector, which will be equal to reduced load vector from concentrated applied loads *plus* the reduced load vector from the expanded vector of equivalent nodal loads.
   b. Use the identified boundary condition to obtain the reduced stiffness matrix.
5. Determine the unknown global nodal parameters (displacements and slopes) by using the reduced matrix equation. For further details, one may also
   a. Obtain the global nodal loads by using the vector of all global nodal displacements (which combines the determined displacements from Step 6 along with the displacement boundary conditions) and the global stiffness matrix from Step 4.
   b. Obtain other parameters such as the bending stress developed in each member (if desired).

6.5.2 Implemented R Functions for the Euler–Bernoulli Beam Elements

```
#*********************************************#
# R functions for the analysis of             #
#   Euler-Bernoulli Beam element              #
#                                             #
#       KB Mustapha                           #
#*********************************************#
EulerBeam_Element_Matrix = function(DOF = 4,YoungMod,
MomentI, Length)
{

  L=Length;
  stiffness=YoungMod*MomentI/Length^3;
  coefficients=c(12,6*L,-12,6*L,6*L,4*L^2,-6*L,2*L^2,-12,
               -6*L,12,-6*L,6*L,2*L^2,-6*L,4*L^2);
  ematrix=stiffness*matrix(coefficients, nrow=DOF, byrow=T);
  return(ematrix)
}


EulerBeam_ExpandedElement_Matrix = function(TDOF,
eMatrix, i, j)
{

  r1=(i-1)+i;r2=(i-1)+(i+1);r3=(j-2)+(j+1);r4=(j-2)+(j+2);
  bigMatrix=matrix(vector(l=TDOF*TDOF),nrow=TDOF, byrow=T);
  bigMatrix[c(r1,r2,r3,r4),c(r1,r2,r3,r4)]=eMatrix;
  return (bigMatrix)
}

EulerBeam_ReducedStiffnessMatrix = function(bigKmatrix,
knownloadnodes)
{
  reducedK = bigKmatrix[c(knownloadnodes),(knownloadnodes)]
  return(reducedK)
}

EulerBeam_ReducedLoadVector = function(loadvector)
{
```

```
  reducedloadvector = matrix(loadvector,ncol = 1)
  return(reducedloadvector)
}

BeamUDL_Matrix= function(DOF=4,LoadMagnitude,Length)
{
  L=Length;
  DistributedLoad=LoadMagnitude*
    matrix(c(-L/2,-(L^2)/12,-L/2,(L^2)/12),nrow=DOF,
    byrow=T)
  return (DistributedLoad)
}

BeamUDL_ExpandedMatirx = function(TDOF, LoadColumnMatrix,
i, j)
{
  r1=(i-1)+i;r2=(i-1)+(i+1);r3=(j-2)+(j+1);r4=(j-2)+(j+2);
  bigColumnMatrix=matrix(vector(l=TDOF),nrow=TDOF,
  byrow=T);
  bigColumnMatrix[c(r1,r2,r3,r4)]=LoadColumnMatrix;
  return (bigColumnMatrix)
}

EulerBeam_NodalDisplacement = function(reducedmatrix,
vec_reducedload)
{
  return(solve(reducedmatrix,vec_reducedload))
}

EulerBeam_Global_ForcesMoments = function(bigKmatrix,vec_
globalnodaldisp)
{
  columndof=matrix(vec_globalnodaldisp, byrow = T)
  global_forces = bigKmatrix %*% vec_globalnodaldisp
  return(round(global_forces))
}

EulerBeam_Local_ForcesMoments = function(YoungMod,
MomentI,Length,vec_globalnodaldisp,vec_distriload)
{
  L=Length;
  DOF=4;
  bendingstiff=YoungMod*MomentI/Length^3;
  stiffcoeff=c(12,6*L,-12,6*L,6*L,4*L^2,-6*L,2*L^2,-12,
            -6*L,12,-6*L,6*L,2*L^2,-6*L,4*L^2);
```

```
    ematrix=bendingstiff*matrix(stiffcoeff,nrow=DOF,
    byrow=T);
    local_loads = (ematrix%*%vec_globalnodaldisp)-
    vec_distributedLoad
    return(local_loads)
  }

  EulerBeamMaxMin_BendingStress = function(YoungMod,
  Length,vec_globalnodaldisp)
  {
  disp_vec=matrix(vec_globalnodaldisp,nrow=length(vec_
  globalnodaldisp),byrow=T, ncol = 1);
    Tmatrix=matrix(c(-m,-n, m, n),nrow=1,byrow=T);
    local_stress=(YoungMod/elem_len)*Tmatrix%*%disp_vec;
    return(local_stress)
  }
```

The explanations of the usage and arguments for the set of functions for the beam elements are provided in Tables 6.2 through 6.4.

TABLE 6.2

Functions for the Beam Element's Matrices and Vectors

Description of Functions
`EulerBeam_Element_Matrix(DOF = 4,YoungMod,MomentI,Length)` This function generates and returns the stiffness matrix for the EB beam element. It accepts three parameters: **`DOF`** (predefined as 4 for the beam element); **`YoungMod`**, which represents a member's Young's modulus; **`MomentI`**, moment of inertia of the cross-section; and **`Length`**.
`EulerBeam_ExpandedElement_Matrix(TDOF,eMatrix,i,j)` This function forms the expanded stiffness matrix for a beam element, and its parameters are of the same form as those defined in the previous chapters.
`EulerBeam_ReducedStiffnessMatrix(bigKmatrix,knownloadnodes)` This function generates a reduced stiffness matrix for the element, and its parameters are of the same form as those defined in previous chapters.
`EulerBeam_ReducedLoadVector(loadvector)` This function generates a reduced load vector needed to find the unknown nodal displacements/rotations. It accepts the parameter **`loadvector`** (a vector of known forces and moments).

TABLE 6.3

Additional Supplementary Functions for Beam Elements with Distributed Load

Description of Functions
`BeamUDL_Matrix(DOF=4,LoadMagnitude,Length)`
This function generates and returns the column matrix for the uniformly distributed load derived in Equation 6.38. The function accepts three parameters: **`DOF`** (predefined as 4 for the beam element); **`LoadMagnitude`**, which represents the magnitude of a rectangular distributed load; and **`Length`**, which denotes a member's geometric length.
`BeamUDL_ExpandedMatirx(TDOF,LoadColumnMatrix,i,j)`
This function forms the expanded column matrix for the distributed load, and its parameters are **`TDOF`** (total degrees of freedom of the system); **`LoadColumnMatrix`** (load column matrix); and **`i, j`**, representing the labels of the *i*th and *j*th nodes of an element.

TABLE 6.4

Functions for the Computation Phase of Analysis of a System of Beam Elements

Description of Functions
`EulerBeam_NodalDisplacement(reducedmatrix,vec_reducedload)`
Given the parameters **`reducedmatrix`** (reduced stiffness matrix) and **`vec_reducedload`** (reduced load vector), this function solves for the unknown nodal displacements.
`EulerBeam_Global_ForcesMoments(bigKmatrix,vec_globalnodaldisp)`
This function calculates the global forces and moments in a system of connected beams. Its parameters are **`bigKmatrix`** (the global stiffness matrix) and **`vec_globalnodaldisp`** (a vector of the global nodal displacement).
`EulerBeam_Local_ForcesMoments(YoungMod, MomentI, Length, vec_globalnodaldisp, vec_distriload)`
This function calculates the local forces and moments in a member of a connected beam system. Its parameters are **`YoungMod`**, which denotes the Young's modulus; **`MomentI`**, which denotes the moment of inertia; **`Length`**, which denotes a member's geometric length; **`vec_globalnodaldisp`** (vector global displacements relevant to that elements); and **`vec_distriload`** (a column matrix of the distributed load, if it exists).

6.6 Examples

6.6.1 Example 6.1

For the three-segment beam depicted in Figure 6.2 with varying cross sections and point loads, determine (1) the transverse nodal displacements at points B and C and (2) the reactions at the end supports.

Take: $E = 200$ GPa; $I_{AB} = 5\times10^{-6}\ \text{m}^4; I_{BC} = 2.5\times10^{-6}\ \text{m}^4; I_{CD} = 6.25\times10^{-7}\ \text{m}^4$.

6.6.1.1 Solution

For this problem, the elements' connectivity is shown in Table 6.5, in which it will be noticed that we have designated the segment AB, BC, and CD as elements 1, 2, and 3, respectively.

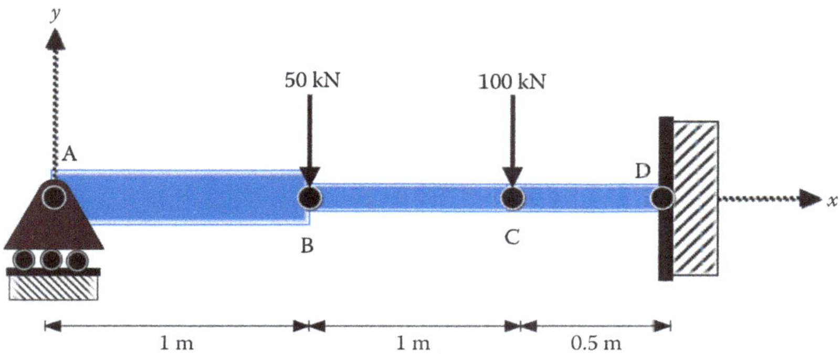

FIGURE 6.2
A three-segment beam systems with point loads.

TABLE 6.5
Connectivity Table from the Discretized System

Elements	Nodes (*i*, *j*)
1	1,2
2	2,3
3	3,4

The rest of the solution steps are as follows:

Step 1: We write the element matrix using the function **`EulerBeam_Element_Matrix()`**.

The properties required by this function are first computed in the following lines of code.

```
YoungMod=200e9;
MomentI1 = 5e-6;MomentI2 = 2.5e-6;MomentI3 = 6.25e-7
L1 = 1;L2 = 1;L3 = 0.5;
```

Based on the above-mentioned data, we obtain the stiffness matrix for each element using the following block of code:

```
k1=EulerBeam_Element_Matrix(4,YoungMod,MomentI1,L1);k1
k2=EulerBeam_Element_Matrix(4,YoungMod,MomentI2,L2);k2
k3=EulerBeam_Element_Matrix(4,YoungMod,MomentI3,L3);k3
```

The output of the above produces the elements stiffness matrices as

```
> k1=EulerBeam_Element_Matrix(4,YoungMod,MomentI1,L1);k1
          [,1]      [,2]      [,3]     [,4]
[1,]  12000000   6000000 -12000000   6000000
[2,]   6000000   4000000  -6000000   2000000
[3,] -12000000  -6000000  12000000  -6000000
[4,]   6000000   2000000  -6000000   4000000
> k2=EulerBeam_Element_Matrix(4,YoungMod,MomentI2,L2);k2
         [,1]     [,2]     [,3]     [,4]
[1,]  6000000  3000000 -6000000  3000000
[2,]  3000000  2000000 -3000000  1000000
[3,] -6000000 -3000000  6000000 -3000000
[4,]  3000000  1000000 -3000000  2000000
> k3=EulerBeam_Element_Matrix(4,YoungMod,MomentI3,L3);k3
          [,1]     [,2]      [,3]     [,4]
[1,]  12000000  3000000 -12000000  3000000
[2,]   3000000  1000000  -3000000   500000
[3,] -12000000 -3000000  12000000 -3000000
[4,]   3000000   500000  -3000000  1000000
```

Step 2: Determine the global degree of freedom of the beam assemblage.

```
n_nodes=4;Total_dof=n_nodes*2;
K1=EulerBeam_ExpandedElement_Matrix(Total_dof,k1,1,2);
K2=EulerBeam_ExpandedElement_Matrix(Total_dof,k2,2,3);
K3=EulerBeam_ExpandedElement_Matrix(Total_dof,k3,3,4);
```

Step 3: Establish the global stiffness matrix by adding the expanded matrices obtained in Step 2.

```
GlobalK=K1+K2+K3;
GlobalK
```

The output of this gives the global stiffness matrix $[K]$ as

```
> GlobalK
          [,1]     [,2]      [,3]     [,4]      [,5]     [,6]      [,7]     [,8]
[1,]  12000000  6000000 -12000000  6000000         0        0         0        0
[2,]   6000000  4000000  -6000000  2000000         0        0         0        0
[3,] -12000000 -6000000  18000000 -3000000  -6000000  3000000         0        0
[4,]   6000000  2000000  -3000000  6000000  -3000000  1000000         0        0
[5,]         0        0  -6000000 -3000000  18000000        0 -12000000  3000000
[6,]         0        0   3000000  1000000         0  3000000  -3000000   500000
[7,]         0        0         0        0 -12000000 -3000000  12000000 -3000000
[8,]         0        0         0        0   3000000   500000  -3000000  1000000
```

Thus,

$$[K]=10^6\begin{bmatrix} 12 & 6 & -12 & 6 & 0 & 0 & 0 & 0 \\ 6 & 4 & -6 & 2 & 0 & 0 & 0 & 0 \\ -12 & -6 & 18 & -3 & -6 & 3 & 0 & 0 \\ 6 & 2 & -3 & 6 & -3 & 1 & 0 & 0 \\ 0 & 0 & -6 & -3 & 18 & 0 & -12 & 3 \\ 0 & 0 & 3 & 1 & 0 & 3 & -3 & 0.5 \\ 0 & 0 & 0 & 0 & -12 & -3 & 12 & -3 \\ 0 & 0 & 0 & 0 & 3 & 0.5 & -3 & 1 \end{bmatrix} \tag{6.44}$$

Step 4: Identify the boundary condition(s) and obtain the reduced stiffness matrix and the reduced force vector. By means of the global stiffness matrix obtained in Step 3, we can write the global matrix equation as

$$\begin{Bmatrix} F_{1Y} \\ M_{1Z} \\ F_{2Y} \\ M_{2Z} \\ F_{3Y} \\ M_{3Z} \\ F_{4Y} \\ M_{4Z} \end{Bmatrix} = 10^6\begin{bmatrix} 12 & 6 & -12 & 6 & 0 & 0 & 0 & 0 \\ 6 & 4 & -6 & 2 & 0 & 0 & 0 & 0 \\ -12 & -6 & 18 & -3 & -6 & 3 & 0 & 0 \\ 6 & 2 & -3 & 6 & -3 & 1 & 0 & 0 \\ 0 & 0 & -6 & -3 & 18 & 0 & -12 & 3 \\ 0 & 0 & 3 & 1 & 0 & 3 & -3 & 0.5 \\ 0 & 0 & 0 & 0 & -12 & -3 & 12 & -3 \\ 0 & 0 & 0 & 0 & 3 & 0.5 & -3 & 1 \end{bmatrix}\begin{Bmatrix} U_{1Y} \\ \theta_{1Z} \\ U_{2Y} \\ \theta_{2Z} \\ U_{3Y} \\ \theta_{3Z} \\ U_{4Y} \\ \theta_{4Z} \end{Bmatrix} \tag{6.45}$$

For the boundary condition, looking at Figure 6.2, we note the following: (1) a roller support at node 1, which eliminates the vertical displacement at node 1 and (2) a fixed support at node 4, which translates into both zero vertical and rotational displacements at node 4. Thus, the following displacement-related boundary conditions hold for this system:

$$U_{1Y}=0;\ U_{4Y}=0;\ \theta_{4Z}=0 \tag{6.46}$$

We know that the system has eight nodal degrees of freedom. From Equation 6.46, we realized that three of these nodal degrees of freedom are known, and the remaining five are unknown. For these five unknown nodal displacements, the nodal loads must be known. In addition, from Figure 6.2, we can identify these known nodal loads as

$$M_{1Z} = 0 \text{ Nm}; \quad F_{2Y} = -50 \text{ kN}; \quad M_{2Z} = 0 \text{ Nm}; \quad F_{3X} = -100 \text{ kN}; \quad M_{3Z} = 0 \text{ Nm} \tag{6.47}$$

The rows of the forces in Equation 6.47 can be identified in the matrix in Equation 6.45 as 2, 3, 4, 5, and 6. By using these rows of the identified known force values, one obtains the reduced stiffness matrix via the function **EulerBeam_ReducedStiffnessMatrix()** as revealed next.

```
vec_rows_of_knownloads=c(2,3,4,5,6);
ReducedK=EulerBeam_ReducedStiffnessMatrix(GlobalK,vec_
rows_of_knownloads);
ReducedK
```

Running the previous block of code yields

```
> ReducedK
         [,1]     [,2]     [,3]     [,4]    [,5]
[1,]  4000000 -6000000  2000000        0       0
[2,] -6000000 18000000 -3000000 -6000000 3000000
[3,]  2000000 -3000000  6000000 -3000000 1000000
[4,]        0 -6000000 -3000000 18000000       0
[5,]        0  3000000  1000000        0 3000000
```

In the next code snippet, we will establish the reduced force vector via the function **EulerBeam_ReducedLoadVector()**.

```
vec_values_of_knownloads=c(0,-50000,0,-100000,0);
ReducedLoadVector=EulerBeam_ReducedLoadVector(vec_values_
of_knownloads);
ReducedLoadVector
```

The output of which shows the column matrix of the load vector as produced in the following:

```
> ReducedLoad=EulerBeam_ReducedLoadVector(vec_values_of_
knownloads);
> ReducedLoad
        [,1]
[1,]       0
[2,]  -50000
[3,]       0
[4,] -100000
[5,]       0
```

Combining the reduced stiffness matrix and the values of the known nodal force leads to the reduced matrix equation in the following form:

$$\begin{Bmatrix} 0 \\ -50,000 \\ 0 \\ -100,000 \\ 0 \end{Bmatrix} = 10^6 \begin{bmatrix} 4 & -6 & 2 & 0 & 0 \\ -6 & 18 & -3 & -6 & 3 \\ 2 & -3 & 6 & -3 & 1 \\ 0 & -6 & -3 & 18 & 0 \\ 0 & 3 & 1 & 0 & 3 \end{bmatrix} \begin{Bmatrix} \theta_{1Z} \\ U_{2Y} \\ \theta_{2Z} \\ U_{3Y} \\ \theta_{3Z} \end{Bmatrix} \tag{6.48}$$

Equation 6.48 is solved for the unknown nodal displacements and rotations in the next step.

Step 5: Determine the unknown nodal displacements and global nodal forces by using the functions:

- **EulerBeam_NodalDisplacement()**
- **EulerBeam_Global_ForcesMoments()**

```
UnknwonNodalDisp=EulerBeam_NodalDisplacement(ReducedK,
ReducedLoad)
UnknwonNodalDisp
```

From the above, we determine the nodal displacements in meters as

```
> UnknwonNodalDisp
           [,1]
[1,] -0.03585526
[2,] -0.03004386
[3,] -0.01842105
[4,] -0.01864035
[5,]  0.03618421
```

The computed result reveals that

$$\begin{Bmatrix} \theta_{1Z} \\ U_{2Y} \\ \theta_{2Z} \\ U_{3Y} \\ \theta_{3Z} \end{Bmatrix} = \begin{Bmatrix} -35.86 \text{ rad} \\ -30.04 \text{ mm} \\ -18.42 \text{ rad} \\ -18.64 \text{ mm} \\ -36.18 \text{ rad} \end{Bmatrix} \tag{6.49}$$

For convenience, we now combine the determined displacements and the boundary conditions (from Equation 6.46).

```
allglobalNodalDisp=matrix(rep(0,Total_dof),byrow=T);
allglobalNodalDisp[vec_rows_of_knownloads]=UnknwonNodalDisp
allglobalNodalDisp
```

By running the above, we will obtain the total vector of nodal displacements for the beam system as

```
> allglobalNodalDisp
           [,1]
[1,]  0.0
[2,] -0.03586
[3,] -0.03004
[4,] -0.01842
[5,] -0.01864
[6,]  0.03618
[7,]  0.0
[8,]  0.0
```

Having determined the nodal displacements, we may now find the global forces and moments. We will use the function **EulerBeam_Global_ForcesMoments()** for this purpose. It is recalled that this function requires the global stiffness matrix (represented by the variable **GlobalK** determined in Step 3) and the vector of global nodal displacements (contained in the variable **allglobalNodalDisp**).

```
GlobalLoads=EulerBeam_Global_ForcesMoments(GlobalK,
allglobalNodalDisp);
GlobalLoads
```

Running the preceding code returns the forces and moments as

```
> GlobalLoads
        [,1]
[1,]   34868
[2,]       0
[3,]  -50000
[4,]       0
[5,] -100000
[6,]       0
[7,]  115132
[8,]  -37829
```

Based on the preceding output, the global nodal forces and moments are as follows:

- $F_{1Y} = 34.868$ kN; $M_{1Z} = 0$ N m; $F_{2Y} = -50$ kN; $M_{2Z} = 0$; $F_{3Y} = -100$ kN
- $M_{3Z} = 0$ N m; $F_{4Y} = 115.132$ kN; $M_{4Z} = -37{,}829$ N m

6.6.2 Example 6.2

A wooden beam is loaded as shown in Figure 6.3. The wood's material has $E = 12$ GPa and a uniform cross section with moment of inertia $I = 10.67 \times 10^{-4}\ \text{m}^4$.

1. Form the global stiffness matrix for the beam.
2. Determine the deflection and slope at point D.
3. Determine the reaction force and bending moment at point A.

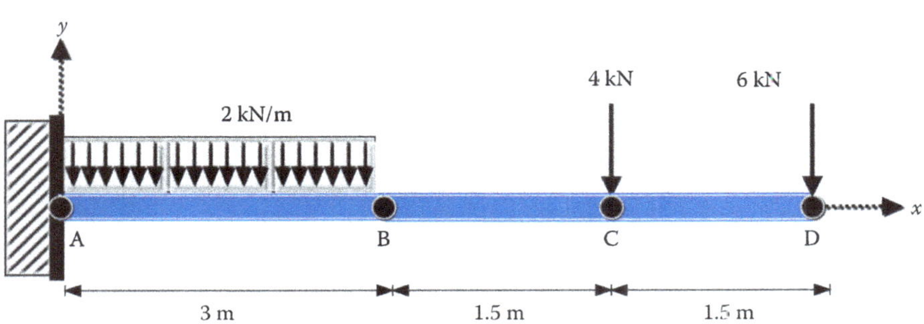

FIGURE 6.3
A three-segment beam with combined distributed and concentrated loads.

6.6.2.1 Solution

To solve the problem, we will designate the segments AB, BC, and CD as elements 1, 2, and 3, respectively. In the same spirit, we denote the points A, B, C, and D as nodes 1, 2, 3, and 4, respectively. With this, we can proceed with the solution as done for the previous example.

Step 1: We write the element matrix.

```
YoungMod=12e9;MomentI=10.67e-4;
n_nodes=4;Total_dof=n_nodes*2;n_ele=2;
L1 = 3.0;
L2 = 1.5;
L3 = 1.5;
k1=EulerBeam_Element_Matrix(4,YoungMod,MomentI,L1);k1
k2=EulerBeam_Element_Matrix(4,YoungMod,MomentI,L2);k2
k3=EulerBeam_Element_Matrix(4,YoungMod,MomentI,L3);k3
```

Running of the above produces the required elements matrices as shown next:

```
> k1=EulerBeam_Element_Matrix(4,YoungMod,MomentI,L1);k1
         [,1]     [,2]     [,3]     [,4]
[1,]  5690667  8536000 -5690667  8536000
[2,]  8536000 17072000 -8536000  8536000
[3,] -5690667 -8536000  5690667 -8536000
[4,]  8536000  8536000 -8536000 17072000
> k2=EulerBeam_Element_Matrix(4,YoungMod,MomentI,L2);k2
          [,1]      [,2]      [,3]      [,4]
[1,]  45525333  34144000 -45525333  34144000
[2,]  34144000  34144000 -34144000  17072000
[3,] -45525333 -34144000  45525333 -34144000
[4,]  34144000  17072000 -34144000  34144000
> k3=EulerBeam_Element_Matrix(4,YoungMod,MomentI,L3);k3
          [,1]      [,2]      [,3]      [,4]
[1,]  45525333  34144000 -45525333  34144000
[2,]  34144000  34144000 -34144000  17072000
[3,] -45525333 -34144000  45525333 -34144000
[4,]  34144000  17072000 -34144000  34144000
```

From the problem statement, it is noticed that the segment AB of the beam (i.e., element 1) is loaded with a UDL. This requires that we add a column matrix of the equivalent nodal loads to its matrix equations.

Accordingly, we will use the function **BeamUDL_Matrix()** to get the equivalent loads as

```
> k1UDL=BeamUDL_Matrix(4,LoadMagnitude = 2000,Length =
L1);k1UDL
```

The output of the above produces the desired column matrix of equivalent load revealed next.

```
> k1UDL=BeamUDL_Matrix(4,LoadMagnitude = 2000,Length =
L1);k1UDL
       [,1]
[1,] -3000
[2,] -1500
[3,] -3000
[4,]  1500
```

From which we write out the matrix equation for element 1 as

$$\begin{Bmatrix} f_{1y}^{1} \\ m_{1z}^{1} \\ f_{2y}^{2} \\ m_{2z}^{2} \end{Bmatrix} + \begin{Bmatrix} -3000 \\ -1500 \\ -3000 \\ 1500 \end{Bmatrix} = 10^{6} \begin{bmatrix} 5.691 & 8.536 & -5.691 & 8.536 \\ 8.536 & 17.072 & -8.536 & 8.536 \\ -5.691 & -8.536 & 5.691 & -8.536 \\ 8.536 & 8.536 & -8.536 & 17.072 \end{bmatrix} \begin{Bmatrix} u_{1y} \\ \theta_{1z} \\ u_{2y} \\ \theta_{2z} \end{Bmatrix} \quad (6.50)$$

Step 2: Determine the global degree of freedom of the assemblage and expand the element matrices accordingly.

```
n_nodes=4;Total_dof=n_nodes*2;
K1=EulerBeam_ExpandedElement_Matrix(Total_dof,k1,1,2);
K2=EulerBeam_ExpandedElement_Matrix(Total_dof,k2,2,3);
K3=EulerBeam_ExpandedElement_Matrix(Total_dof,k3,3,4);
```

For this structure, in addition to the expanded stiffness matrices, it is required to expand the column matrix of the equivalent nodal loads for element 1. This is obtained by using the function **BeamUDL_ExpandedMatirx()**.

```
K1UDL=BeamUDL_ExpandedMatirx(Total_dof,k1UDL,1,2);K1UDL
```

By using the above-mentioned line of code, we obtain the expanded column matrix for the equivalent loads arising from the UDL on element 1 as

```
> K1UDL=BeamUDL_ExpandedMatirx(Total_dof,k1UDL,1,2);K1UDL
      [,1]
[1,] -3000
[2,] -1500
[3,] -3000
[4,]  1500
[5,]     0
[6,]     0
[7,]     0
[8,]     0
```

Step 3: Establish the global stiffness matrix by adding the expanded matrices obtained in Step 2.

```
GlobalK=K1+K2+K3;
GlobalK
```

The output of this gives the global stiffness matrix $[K]$ as

```
         [,1]     [,2]     [,3]     [,4]     [,5]     [,6]     [,7]     [,8]
[1,]  5690667  8536000 -5690667  8536000        0        0        0        0
[2,]  8536000 17072000 -8536000  8536000        0        0        0        0
[3,] -5690667 -8536000 51216000 25608000 -45525333 34144000       0        0
[4,]  8536000  8536000 25608000 51216000 -34144000 17072000       0        0
[5,]        0        0 -45525333 -34144000 91050667       0 -45525333 34144000
[6,]        0        0 34144000 17072000        0 68288000 -34144000 17072000
[7,]        0        0        0        0 -45525333 -34144000 45525333 -34144000
[8,]        0        0        0        0 34144000 17072000 -34144000 34144000
```

For clarity, we rewrite this as

$$[K]=10^6\begin{bmatrix} 5.691 & 8.536 & -5.691 & 8.536 & 0 & 0 & 0 & 0 \\ 8.536 & 17.072 & -8.536 & 8.536 & 0 & 0 & 0 & 0 \\ -5.691 & -8.536 & 51.216 & 25.608 & -45.525 & 34.144 & 0 & 0 \\ 8.536 & 8.536 & 25.608 & 51.216 & -34.144 & 17.072 & 0 & 0 \\ 0 & 0 & -45.525 & -34.144 & 91.051 & 0 & -45.525 & 34.144 \\ 0 & 0 & 34.144 & 17.072 & -45.525 & 68.288 & -34.144 & 17.072 \\ 0 & 0 & 0 & 0 & 68.288 & -34.144 & 45.525 & -34.144 \\ 0 & 0 & 0 & 0 & 34.144 & 17.072 & -34.144 & 34.144 \end{bmatrix} \tag{6.51}$$

Step 4: Identify the boundary condition(s) and obtain the reduced stiffness matrix and the reduced force vector.

Clearly, with the global stiffness matrix obtained in Step 3, we can write the global matrix equation.

$$\begin{Bmatrix} F_{1Y} \\ M_{1Z} \\ F_{2Y} \\ M_{2Z} \\ F_{3Y} \\ M_{3Z} \\ F_{4Y} \\ M_{4Z} \end{Bmatrix} = 10^6 \begin{bmatrix} 5.691 & 8.536 & -5.691 & 8.536 & 0 & 0 & 0 & 0 \\ 8.536 & 17.072 & -8.536 & 8.536 & 0 & 0 & 0 & 0 \\ -5.691 & -8.536 & 51.216 & 25.608 & -45.525 & 34.144 & 0 & 0 \\ 8.536 & 8.536 & 25.608 & 51.216 & -34.144 & 17.072 & 0 & 0 \\ 0 & 0 & -45.525 & -34.144 & 91.051 & 0 & -45.525 & 34.144 \\ 0 & 0 & 34.144 & 17.072 & -45.525 & 68.288 & -34.144 & 17.072 \\ 0 & 0 & 0 & 0 & 68.288 & -34.144 & 45.525 & -34.144 \\ 0 & 0 & 0 & 0 & 34.144 & 17.072 & -34.144 & 34.144 \end{bmatrix} \begin{Bmatrix} U_{1Y} \\ \theta_{1Z} \\ U_{2Y} \\ \theta_{2Z} \\ U_{3Y} \\ \theta_{3Z} \\ U_{4Y} \\ \theta_{4Z} \end{Bmatrix} \tag{6.52}$$

To identify the boundary conditions, we reexamine Figure 6.3, from which it is observed that node 1 is fixed, which enforces zero vertical and rotational displacements. Therefore, the following known displacements or displacement-related boundary conditions are obtained:

$$U_{1Y} = 0; \theta_{1Z} = 0 \tag{6.53}$$

Now, from Figure 6.3, we can identify the known nodal loads as

$$\begin{Bmatrix} F_{1Y} \\ M_{1Z} \\ F_{2Y} \\ M_{2Z} \\ F_{3Y} \\ M_{3Z} \\ F_{4Y} \\ M_{4Z} \end{Bmatrix} = \begin{Bmatrix} F_{1Y} \\ M_{1Z} \\ 0 \\ 0 \\ -4000 \\ 0 \\ -6000 \\ 0 \end{Bmatrix} + \begin{Bmatrix} -3000 \\ -1500 \\ -3000 \\ 1500 \\ 0 \\ 0 \\ 0 \\ 0 \end{Bmatrix} \tag{6.54}$$

The rows of the known loads in Equation 6.54 can be identified as 3, 4, 5, 6, 7, and 8. By using these rows of the identified known loads, one obtains the reduced stiffness matrix via the function **EulerBeam_ReducedStiffnessMatrix()** as done in the following:

```
vec_rows_of_knownloads=c(3,4,5,6,7,8);
ReducedK=EulerBeam_ReducedStiffnessMatrix(GlobalK, vec_
rows_of_knownloads);
ReducedK
```

Running the preceding block of code yields the reduced stiffness matrix as

```
> ReducedK
          [,1]      [,2]      [,3]      [,4]      [,5]      [,6]
[1,]  51216000  25608000 -45525333  34144000         0         0
[2,]  25608000  51216000 -34144000  17072000         0         0
[3,] -45525333 -34144000  91050667         0 -45525333  34144000
[4,]  34144000  17072000         0  68288000 -34144000  17072000
[5,]         0         0 -45525333 -34144000  45525333 -34144000
[6,]         0         0  34144000  17072000 -34144000  34144000
```

In the next code snippet, we will establish the vector of known concentrated loads (along with the equivalent loads from the UDL) by using **EulerBeam_ReducedLoadVector()**.

```
vec_values_of_knownloads=c(0,0,-4000,0,-6000,0)+K1UDL
[c(3,4,5,6,7,8)];
ReducedLoadVector=EulerBeam_ReducedLoadVector(vec_values_
of_knownloads);
ReducedLoadVector
```

Notice that the variable **vec_values_of_knownloads** in the first line of the previous code has two parts, very much like the right-hand side of Equation 6.54. The first part is the vector of the externally applied point loads; and the second part refers to the components 3–8 of the column matrix of the equivalent nodal loads. The output of the block of code shows the column matrix of the nodal loads.

```
> ReducedLoadVector
      [,1]
[1,] -3000
[2,]  1500
[3,] -4000
[4,]     0
[5,] -6000
[6,]     0
```

Combining the reduced stiffness matrix and the values of the known nodal loads leads to the reduced matrix equation, which takes the following form:

$$\begin{Bmatrix} -3000 \\ 1500 \\ -4000 \\ 0 \\ -6000 \\ 0 \end{Bmatrix} = 10^6 \begin{bmatrix} 51.216 & 25.608 & -45.525 & 34.144 & 0 & 0 \\ 25.608 & 51.216 & -34.144 & 17.072 & 0 & 0 \\ -45.525 & -34.144 & 91.051 & 0 & -45.525 & 34.144 \\ 34.144 & 17.072 & 0 & 68.288 & -34.144 & 17.072 \\ 0 & 0 & -45.525 & -34.144 & 45.525 & -34.144 \\ 0 & 0 & 34.144 & 17.072 & -34.144 & 34.144 \end{bmatrix} \begin{Bmatrix} U_{2Y} \\ \theta_{2Z} \\ U_{3Y} \\ \theta_{3Z} \\ U_{4Y} \\ \theta_{4Z} \end{Bmatrix} \tag{6.55}$$

Step 5: Determine the unknown nodal displacements and nodal global loads by using the following functions:

- **EulerBeam_NodalDisplacement()**
- **EulerBeam_Global_ForcesMoments()**.

```
UnknwonNodalDisp=EulerBeam_NodalDisplacement(ReducedK,
ReducedLoad)
round(UnknwonNodalDisp,4)
```

The determined displacements, rounded to the fourth decimal place, are obtained as

```
> round(UnknwonNodalDisp,4)
        [,1]
[1,] -0.0170
[2,] -0.0098
[3,] -0.0335
[4,] -0.0118
[5,] -0.0517
[6,] -0.0123
```

The computed result reveals the values of nodal displacements as

$$\begin{Bmatrix} U_{2Y} \\ \theta_{2Z} \\ U_{3Y} \\ \theta_{3Z} \\ U_{4Y} \\ \theta_{4Z} \end{Bmatrix} = \begin{Bmatrix} -17.0 \text{ mm} \\ -9.8 \text{ rad} \\ -33.5 \text{ mm} \\ -11.8 \text{ rad} \\ -51.7 \text{ mm} \\ -12.3 \text{ rad} \end{Bmatrix} \tag{6.56}$$

From Equation 6.56, we see that point D is vertically displaced by –51.7 mm and has a rotational displacement of –12.3 radians.

In the next line of the code, we combine the determined displacements Equation 6.56 and the boundary conditions (from Equation 6.53).

```
allglobalNodalDisp=matrix(rep(0,Total_dof),byrow=T);
allglobalNodalDisp[vec_rows_of_knownloads]=UnknwonNodalDisp
allglobalNodalDisp
```

Which produces

```
> round(allglobalNodalDisp,4)
        [,1]
[1,]  0.0000
[2,]  0.0000
[3,] -0.0170
[4,] -0.0098
[5,] -0.0335
[6,] -0.0118
[7,] -0.0517
[8,] -0.0123
```

Having determined the nodal displacements, we may now find the global forces and moments (through which the reaction force and moment can be determined). The function **EulerBeam_Global_ForcesMoments()** is suitable for this purpose as done next.

```
PLoadsEffect=EulerBeam_Global_ForcesMoments(GlobalK,
allglobalNodalDisp)
EquivalentLoadEffect=K1UDL
Globalloads=PLoadsEffect-EquivalentLoadEffect
```

In the preceding code snippet, take note that the effect of the equivalent load is subtracted to get the actual point loads. Consequently, running the preceding code returns the forces and moments as

```
> Globalloads
       [,1]
[1,] 16000
[2,] 63000
[3,]     0
[4,]     0
[5,] -4000
[6,]     0
[7,] -6000
[8,]     0
```

Based on the preceding output, the reaction force and moments at point A are $F_{1Y} = 16$ kN and $M_{1Z} = 63{,}000$ Nm.

6.7 Exercises

1. A steel beam is loaded as shown in Figure 6.4. The beam has $E = 200$ GPa and a uniform cross-section with moment of inertia $I = 11.86 \times 10^{-5}\ \text{m}^4$. By using two beam elements:
   a. Form the global stiffness matrix for the beam.
   b. Determine the deflection and slope at points B and C.
   c. Determine the reaction force and moment at point A.

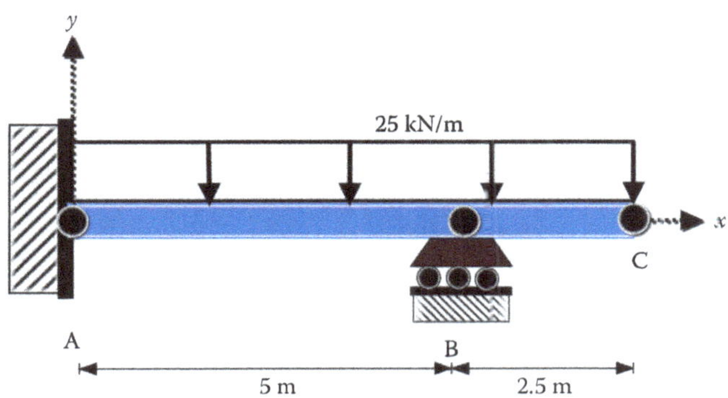

FIGURE 6.4
Exercise 6Q1.

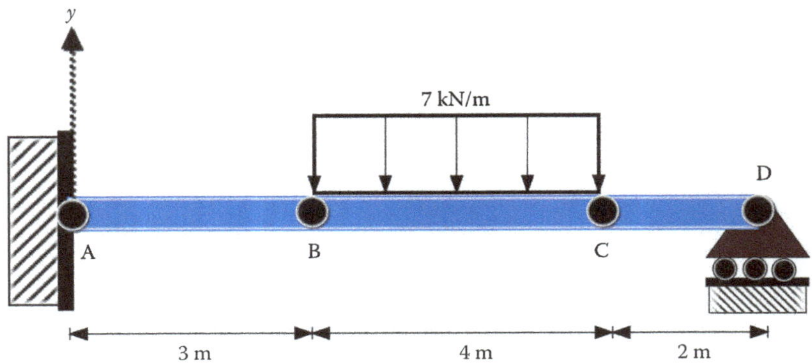

FIGURE 6.5
Exercise 6Q2.

2. For the beam in Figure 6.5, $E = 210$ GPa and inertia $I = 5 \times 10^{-6}\ \mathrm{m}^4$.
   a. Form the global stiffness matrix for the beam.
   b. Determine the deflection and slope at points B and C.
   c. Determine the reactions at points A and D.

References

1. P. P. Benham, R. J. Crawford, C. G. Armstrong, *Mechanics of Engineering Materials*, Longman Group, Harlow, UK, 1996.
2. D. L. Logan, *First Course in the Finite Element Method*, Thomson, Canada, Ontario, 2007.
3. M. A. Bhatti, *Fundamental Finite Element Analysis and Applications: with Mathematica and Matlab Computations*, Wiley, Hoboken, NJ, 2005.
4. A. Öchsner, *Computational Statics and Dynamics: An Introduction Based on the Finite Element Method*, Springer, Singapore, 2016.

7

Planar and Space Frame Elements

7.1 Introduction

The concepts presented in Chapters 5 and 6 will be expanded upon in this chapter to develop elements that combine the degrees of freedom of the beam and truss elements, namely frame elements.

Frame elements are suitable for the static analyses of frame structures consisting of beam members that are connected rigidly or by pins at the joints. Our interest in the analysis of frame structures is analogous to what was discussed for beams. That is, to determine the (a) internal forces (shear forces and bending moments); (b) distribution of the deflection, which includes vertical and rotational displacements at different points along the geometric length of the members of the structures; and (c) internal stresses developed in reaction to applied loads.

As the title indicates, we will consider the computer implementation of both planar and space frame elements (using the shear-rigid theory, in which the effect of shear deformation is neglected) in the sections that follow. In conformity with the previous chapters, we shall start with a brief account of the finite element equation for the planar frame element and then move on to present and illustrate the use of the implemented R functions for both planar and space frame elements.

7.2 Finite Element Equations for Planar Frame Element

7.2.1 Displacement Function

Consider Figure 7.1, which shows an oriented two-node frame element located in the global coordinate system $X-Y$, and having length L_e, area A (in the unloaded state), and a Young's modulus E.

The element is attached with a local coordinate system $x-y$, where x is directed along its length, and θ is the orientation angle between the

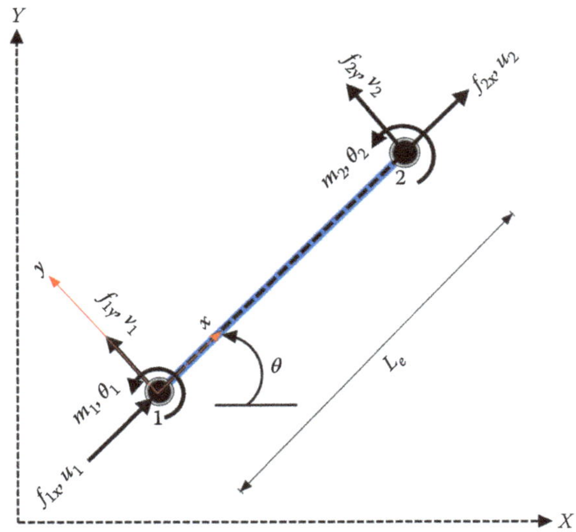

FIGURE 7.1
A plane frame element with the applied nodal loads and degrees of freedom.

coordinate systems Given the nature of the nodal loads, we may prescribe the following functions for the variation of the displacements (translational and rotational) along the length of the element in the form:

$$u = A_1 + A_2 x \tag{7.1}$$

$$v = A_3 + A_4 x + A_5 x^2 + A_6 x^3 \tag{7.2}$$

$$\theta = A_4 + 2A_5 x + 3A_6 x^2 \tag{7.3}$$

where u, v and θ are the axial, transverse, and rotational displacements, respectively. Besides, $A_1 - A_6$ are unknown coefficients, and x is within the range $0 \leq x \leq L_e$. Substituting the boundary conditions (as we did in Chapter 6) into Equations 7.1 through 7.3, one obtains

$$\underbrace{\begin{Bmatrix} u_{1x} \\ u_{1y} \\ \theta_{1z} \\ u_{2x} \\ u_{2y} \\ \theta_{2z} \end{Bmatrix}}_{\{u_e\}} = \underbrace{\begin{bmatrix} 1 & 0 & 0 & 0 & 0 & 0 \\ 0 & 0 & 1 & 0 & 0 & 0 \\ 0 & 0 & 0 & 1 & 0 & 0 \\ 1 & L_e & 0 & 0 & 0 & 0 \\ 0 & 0 & 1 & L_e & L_e^2 & L_e^3 \\ 0 & 0 & 0 & 1 & 2L_e & 3L_e^2 \end{bmatrix}}_{[H]} \underbrace{\begin{Bmatrix} A_1 \\ A_2 \\ A_3 \\ A_4 \\ A_5 \\ A_6 \end{Bmatrix}}_{\{A\}} \quad \text{or} \quad \{u_e\} = [H]\{A\} \tag{7.4}$$

where $\{u_e\}$ is a vector of the nodal displacements, $[H]$ is the coordinate matrix, and $\{A\}$ is the vector of the unknown coefficients. From Equation 7.4, the unknown coefficients are determined as

$$\{A\}=[H]^{-1}\{u_e\} \tag{7.5}$$

We may decompose the nodal degrees of freedom in the local coordinate system into their respective components in the XY coordinate system. To do this, we write

$$u_{1x} = u_{1X}\cos\theta + u_{1Y}\sin\theta \tag{7.6}$$

$$u_{1y} = -u_{1X}\sin\theta + u_{1Y}\cos\theta \tag{7.7}$$

$$\theta_{1z} = \theta_{1Z} \tag{7.8}$$

where u_{1y} and u_{1Y} denote vertical displacements (v) along y and Y respectively. A similar line of analysis (if one considers node 2) leads to

$$u_{2x} = u_{2X}\cos\theta + u_{2Y}\sin\theta \tag{7.9}$$

$$u_{2y} = -u_{2X}\sin\theta + u_{2Y}\cos\theta \tag{7.10}$$

$$\theta_{2z} = \theta_{2Z} \tag{7.11}$$

By using Equations 7.6 through 7.11, we transform the local displacements into global displacements as

$$\{u_e\} = \begin{Bmatrix} u_{1x} \\ u_{1y} \\ \theta_{1z} \\ u_{2x} \\ u_{2y} \\ \theta_{2z} \end{Bmatrix} = \begin{bmatrix} C & S & 0 & C & 0 & 0 \\ -S & C & 0 & 0 & 0 & 0 \\ 0 & 0 & 1 & 0 & 0 & 0 \\ 0 & 0 & 0 & C & S & 0 \\ 0 & 0 & 0 & -S & C & 0 \\ 0 & 0 & 0 & 0 & 0 & 1 \end{bmatrix} \begin{Bmatrix} u_{1X} \\ u_{1Y} \\ \theta_{1Z} \\ u_{2X} \\ u_{2Y} \\ \theta_{2Z} \end{Bmatrix} \tag{7.12}$$

where $C=\cos\theta$ and $S=\sin\theta$. Making use of Equation 7.12 in Equation 7.5 leads to

$$\begin{Bmatrix} A_1 \\ A_2 \\ A_3 \\ A_4 \\ A_5 \\ A_6 \end{Bmatrix} = \begin{bmatrix} 1 & 0 & 0 & 0 & 0 & 0 \\ -\frac{1}{L_e} & 0 & 0 & \frac{1}{L_e} & 0 & 0 \\ 0 & 1 & 0 & 0 & 0 & 0 \\ 0 & 0 & 1 & 0 & 0 & 0 \\ 0 & -\frac{3}{L_e^2} & -\frac{2}{L_e} & 0 & \frac{3}{L_e^2} & -\frac{1}{L_e} \\ 0 & \frac{2}{L_e^3} & \frac{1}{L_e^2} & 0 & -\frac{2}{L_e^2} & \frac{1}{L_e^2} \end{bmatrix} \begin{bmatrix} C & S & 0 & 0 & 0 & 0 \\ -S & C & 0 & 0 & 0 & 0 \\ 0 & 0 & 1 & 0 & 0 & 0 \\ 0 & 0 & 0 & C & S & 0 \\ 0 & 0 & 0 & -S & C & 0 \\ 0 & 0 & 0 & 0 & 0 & 1 \end{bmatrix} \begin{Bmatrix} u_{1X} \\ u_{1Y} \\ \theta_{1Z} \\ u_{2X} \\ u_{2Y} \\ \theta_{2Z} \end{Bmatrix} \tag{7.13}$$

This evaluates to

$$\begin{Bmatrix} A_1 \\ A_2 \\ A_3 \\ A_4 \\ A_5 \\ A_6 \end{Bmatrix} = \begin{bmatrix} C & S & 0 & 0 & 0 & 0 \\ -\frac{C}{L_e} & -\frac{S}{L_e} & 0 & \frac{C}{L_e} & \frac{S}{L_e} & 0 \\ -S & Cc & 0 & 0 & 0 & 0 \\ 0 & 0 & 1 & 0 & 0 & 0 \\ \frac{3S}{L_e^2} & -\frac{3C}{L_e^2} & -\frac{2}{L_e} & -\frac{3S}{L_e^2} & \frac{3C}{L_e^2} & -\frac{1}{L_e} \\ -\frac{2S}{L_e^3} & \frac{2C}{L_e^3} & \frac{1}{L_e^2} & \frac{2S}{L_e^3} & \frac{2C}{L_e^2} & \frac{1}{L_e^2} \end{bmatrix} \begin{Bmatrix} u_{1X} \\ u_{1Y} \\ \theta_{1Z} \\ u_{2X} \\ u_{2Y} \\ \theta_{2Z} \end{Bmatrix} \tag{7.14}$$

We substitute the A_is established in Equation 7.14 into Equations 7.1 and 7.2 to obtain

$$u = \underbrace{\begin{bmatrix} N_{u1} & N_{u2} & N_{u3} & N_{u4} & N_{u5} & N_{u6} \end{bmatrix}}_{N_u} \{u_e\} \tag{7.15}$$

$$v = \underbrace{\begin{bmatrix} N_{v1} & N_{v2} & N_{v3} & N_{v4} & N_{v5} & N_{v6} \end{bmatrix}}_{N_v} \{u_e\} \tag{7.16}$$

where N_u and N_v are the transformed shape functions for the axial and transverse deformations in a frame element. Table 7.1 shows the full form of the shape functions.

TABLE 7.1

The Transformed Shape Functions for the Axial and Transverse Deformations in a Frame Element

Axial Deformation	Transverse Deformation
$N_{u1} = C - \frac{Cx}{L_e}$	$N_{v1} = \left(-S - \frac{2Sx^3}{L_e^3} + \frac{3Sx^2}{L_e^2}\right)$
$N_{u2} = S - \frac{Sx}{L_e}$	$N_{v2} = \left(C + \frac{2Cx^3}{L_e^3} - \frac{3Cx^2}{L_e^2}\right)$
$N_{u3} = 0$	$N_{v3} = \left(x + \frac{x^3}{L_e^2} - \frac{2x^2}{L_e}\right)$
$N_{u4} = \frac{Cx}{L_e}$	$N_{v4} = \left(\frac{2Sx^3}{L_e^3} - \frac{3Sx^2}{L_e^2}\right)$
$N_{u5} = \frac{Sx}{L_e}$	$N_{v5} = \left(-\frac{2Cx^3}{L_e^3} + \frac{3Cx^2}{L_e^2}\right)$
$N_{u6} = 0$	$N_{v6} = \left(\frac{x^3}{L_e^2} - \frac{x^2}{L_e}\right)$

7.2.2 Strain–Displacement Relationship and Constitutive Law

For planar frames, the non-zero component of the strain tensor remains the axial strain (ε_{xx}). By considering the kinematic of the frame's deformation (axial and transverse displacements), ε_{xx} takes the form

$$\varepsilon_{xx} = \frac{\partial u}{\partial x} + y\frac{\partial^2 v}{\partial x^2} \tag{7.17}$$

Bearing in mind Equations 7.15 and 7.16, we rewrite Equation 7.17 as

$$\varepsilon_{xx} = \left(\frac{dN_u}{dx} + y\frac{d^2N_v}{dx^2}\right)\{u_e\} \tag{7.18}$$

Or

$$\varepsilon_{xx} = \left(\left[B_u\right] + y\left[\frac{dB_v}{dx}\right]\right)\{u_e\} \tag{7.19}$$

where B_u and B_v are the first derivatives of N_u and N_v.

The stress–strain relation for the frame then takes the following form:

$$\{\sigma\} = E\left[B_u\right]\{u_e\} + Ey\left[\frac{dB_v}{dx}\right]\{u_e\} \tag{7.20}$$

where the first and second parts of the right-hand side of Equation 7.20 represent the axial and bending stress, respectively.

7.2.3 Minimization of the Total Potential Energy for the Frame Element

Following the steps employed in previous chapters, we write the strain energy (U) of the frame element as

$$U = \frac{1}{2}\int_0^{L_e}\int_A \{\sigma\}^T \{\varepsilon\} dxdA \tag{7.21}$$

With Equation 7.18, the strain-energy expression simplifies to

$$U = \frac{EA}{2}\int_0^{L_e} ([B_u]\{u_e\})^T ([B_u]\{u_e\})dx + \frac{EI}{2}\int_0^{L_e} ([B_b]\{u_e\})^T ([B_b]\{u_e\}) \tag{7.22}$$

where A and I denote cross section area and second moment of inertia of the frame, respectively. Besides, the work done by externally applied concentrated nodal forces (W) is of the following form:

$$W = \{u_e\}^T \begin{Bmatrix} f_{1X} \\ f_{1Y} \\ m_{1Z} \\ f_{2X} \\ f_{2Y} \\ m_{2Z} \end{Bmatrix} \tag{7.23}$$

As has been the practice, we substitute Equations 7.22 and 7.23 into the expression for the potential energy (P_t) and then minimize P_t with respect to u_e to obtain

$$\frac{\partial}{\partial\{u_e\}^T} P_t = 0 = \left(EA\int_0^{L_e} [B_u]^T [B_u] dx + EI\int_0^{L_e} [B_b]^T [B_b] dx \right)\{u_e\} - \{f_e\} \tag{7.24}$$

And consequently, we obtain the desired relation as

$$\left(EA\int_0^{L_e} [B_u]^T [B_u] dx + EI\int_0^{L_e} [B_b]^T [B_b] dx \right)\{u_e\} = \{f_e\} = 0 \tag{7.25}$$

In full form, the matrix equation for the frame element (with no shear-deformation effect) is

$$\begin{Bmatrix} f_{1X} \\ f_{1Y} \\ m_{1Z} \\ f_{2X} \\ f_{2Y} \\ m_{2Z} \end{Bmatrix} = \left[K_{\text{frame}}\right] \begin{Bmatrix} u_{1X} \\ u_{1Y} \\ \theta_{1Z} \\ u_{2X} \\ u_{2Y} \\ \theta_{2Z} \end{Bmatrix} \tag{7.26}$$

where

$$[K_{\text{frame}}] = \frac{E}{L} \begin{bmatrix} AC^2 + \frac{12IS^2}{L_e^2} & CS\left(A - \frac{12I}{L_e^2}\right) & \frac{6IS}{L_e} & -AC^2 - \frac{12IS^2}{L_e^2} & CS\left(-A + \frac{12I}{L_e^2}\right) & \frac{6IS}{L_e} \\ CS\left(A - \frac{12I}{L_e^2}\right) & AS^2 + \frac{12C^2I}{L_e^2} & \frac{6CI}{L_e} & CS\left(-A + \frac{12I}{L_e^2}\right) & -AS^2 - \frac{12C^2I}{L_e^2} & \frac{6CI}{L_e} \\ -\frac{6IS}{L_e} & \frac{6CI}{L_e} & 4I & \frac{6IS}{L_e} & -\frac{6CI}{L_e} & 2I \\ -AC^2 - \frac{12IS^2}{L_e^2} & CS\left(-A + \frac{12I}{L_e^2}\right) & \frac{6IS}{L_e} & AC^2 + \frac{12IS^2}{L_e^2} & CS\left(A - \frac{12I}{L_e^2}\right) & \frac{6IS}{L_e} \\ CS\left(-A + \frac{12I}{L_e^2}\right) & -AS^2 - \frac{12C^2I}{L_e^2} & -\frac{6IS}{L_e} & CS\left(A - \frac{12I}{L_e^2}\right) & AS^2 + \frac{12C^2I}{L_e^2} & -\frac{6CI}{L_e} \\ -\frac{6IS}{L_e} & \frac{6IS}{L_e} & 2I & \frac{6IS}{L_e} & -\frac{6IS}{L_e} & 4I \end{bmatrix} \tag{7.27}$$

Recall also that $C = \cos\theta$ and $S = \sin\theta$.

If we are to deal with a frame element under the influence of uniformly distributed axial load w_1 and uniformly distributed transverse load w_2, the following equivalent nodal loads should be added to Equation 7.26:

$$F_{\text{UDL}}^{eq} = \begin{Bmatrix} \frac{1}{2}L_e\left(-Cw_1 + Sw_2\right) \\ -\frac{1}{2}L_e\left(Sw_1 + Cw_2\right) \\ -\frac{1}{12}L_e^2 w_2 \\ \frac{1}{2}L_e\left(-Cw_1 + Sw_2\right) \\ -\frac{1}{2}L_e\left(Sw_1 + Cw_2\right) \\ \frac{1}{12}L_e^2 w_2 \end{Bmatrix} \tag{7.28}$$

$$F_{\text{TDL}}^{eq} = \begin{Bmatrix} \frac{1}{20} L_e \left(-10Cw_1 + 3Sw_2\right) \\ -\frac{1}{20} L_e \left(10Sw_1 + 3Cw_2\right) \\ -\frac{1}{30} L_e^2 w_2 \\ \frac{1}{20} L_e \left(-10Cw_1 + 7Sw_2\right) \\ -\frac{1}{20} L_e \left(10Sw_1 + 7Cw_2\right) \\ \frac{1}{20} L_e^2 w_2 \end{Bmatrix} \tag{7.29}$$

The final matrix equation for the planar frame element will therefore be of the form

$$\begin{Bmatrix} f_{1X} \\ f_{1Y} \\ m_{1Z} \\ f_{2X} \\ f_{2Y} \\ m_{2Z} \end{Bmatrix} + F_{\text{Distributed}}^{eq} = \left[K_{\text{frame}}\right] \begin{Bmatrix} u_{1X} \\ u_{1Y} \\ \theta_{1Z} \\ u_{2X} \\ u_{2Y} \\ \theta_{2Z} \end{Bmatrix} \tag{7.30}$$

where $F_{\text{Distributed}}^{eq}$ will be equal to (a) zero in the absence of any distributed load; (b) F_{UDL}^{eq} in the presence of uniformly distributed axial and transverse loads w_1 and w_2; and (c) F_{TDL}^{eq} in the presence of a uniformly distributed axial load w_1 and a triangular distributed transverse load w_2.

For space frames, the derivation established for the planar frame can easily be extended to account for three more degrees of freedom at a node. However, we encourage interested readers to consult one of the many excellent specialized texts (such as References 1,2) for this.

7.3 Computer Implementation

7.3.1 Systematic Procedure for Analyses of Plane Frame Structures

7.3.1.1 Precomputation Phase

- Establish a global coordinate system.
- Discretize the frame structure into a number of finite frame elements.
- Label the nodes and elements.

7.3.1.2 Computation Phase

1. Specify and/or determine the geometric lengths and the orientation angles (using the nodal coordinates) if they are not explicitly provided.
2. Establish the stiffness matrix for each element by using Equation 7.27.
3. For an element carrying a distributed load, determine the column matrix of the equivalent nodal loads (by using, e.g., Equations 7.28 and 7.29).
4. Determine the global degree of freedom of the frame structure, and carry out the expansion of
   a. The matrix equation for each element (from Step 1) to conform to the size of the global degree of freedom.
   b. The vector of equivalent nodal loads, for any element with a uniformly distributed load.
5. Form the global stiffness matrix and the global matrix equation.
6. Identify the essential boundary condition(s) and apply the boundary conditions to the global matrix equation to obtain a reduced matrix equation. Programmatically, this step involves two actions:
   a. Obtain the reduced load vector, which will be equal to reduced load vector from concentrated applied loads *plus* the reduced load vector from the expanded vector of equivalent nodal loads.
   b. Use the identified boundary condition to obtain the reduced stiffness matrix.
7. Determine the unknown global nodal degree of freedom (displacements and slopes) by using the reduced matrix equation. After this, one may obtain other parameters such as the bending stress developed in each member (if desired).

7.3.2 Implemented R Functions for the Planar Frame Elements

```
#***********************************************#
# R functionss for the plane frame element      #
# KB Mustapha                                   #
#***********************************************#

PlaneFrame_LengthOrientation = function(vec_nodalcoordinates)
{
  xsquared=(vec_nodalcoordinates[3]-vec_nodal
  coordinates[1])^2
  ysquared=(vec_nodalcoordinates[4]-vec_nodal
  coordinates[2])^2
  length_of_element=sqrt(xsquared+ysquared)
```

```
   ydiff=(vec_nodalcoordinates[4]-vec_nodalcoordinates[2])
   xdiff=(vec_nodalcoordinates[3]-vec_nodalcoordinates[1])
   orientation_of_element=(180/pi)*atan(ydiff/xdiff)
   return(c(length_of_element,orientation_of_element))
}

PlaneFrame_Element_Matrix = function(DOF=6,YoungMod,Area,
momentI,Length,theta)
{
   cx=cos(theta*pi/180);sx=sin(theta*pi/180);
   A=Area;I=momentI;B1=(12*momentI)/Length^2;B2=(6*momentI)/
   Length;
   submatrix1=matrix(c((A*cx^2+B1*sx^2),(A-B1)*cx*sx,
                       (-B2*sx),-(A*cx^2+B1*sx^2),
                       -(A-B1)*cx*sx,(-B2*sx)),nrow=1,
                       byrow=T);
   submatrix2=matrix(c(submatrix1[,2],(A*sx^2+B1*cx^2),
                       (B2*cx),-(A-B1)*cx*sx,-(A*sx^2+B1*
                       cx^2),(B2*cx)),nrow=1,byrow=T);
   submatrix3=matrix(c((-B2*sx),(B2*cx),(4*I),(B2*sx),
                       (-B2*cx),(2*I)),nrow=1,byrow=T);
   submatrix4=matrix(c(submatrix1[,4],submatrix2[,4],
   submatrix3[,4],
   submatrix1[,1],submatrix1[,2],B2*sx),nrow=1,byrow=T);
   submatrix5=matrix(c(submatrix1[,5],submatrix2[,5],
   submatrix3[,5],
   submatrix4[,5],submatrix2[,2],-B2*cx),nrow=1, byrow=T);
   submatrix6=matrix(c(submatrix1[,6],submatrix2[,6],
   submatrix3[,6],
   submatrix4[,6],submatrix5[,6],4*I),nrow=1,byrow=T);
   eMatrix=matrix(vector(l=DOF*DOF),nrow=DOF,ncol=DOF);
   eMatrix[1,]=submatrix1;eMatrix[2,]=submatrix2;
   eMatrix[3,]=submatrix3;eMatrix[4,]=submatrix4;
   eMatrix[5,]=submatrix5;eMatrix[6,]=submatrix6;
   return(round((YoungMod/Length)*eMatrix,3))
}

PlaneFrame_ExpandedElement_Matrix=function(TDOF,eMatrix,
i,j)
{
   r1=2*(i-1)+i;r2=2*(i-1)+(i+1);
   r3=2*(i-1)+(i+2);r4=(j-1)+(j+2)+(j-3);
   r5=(j-1)+(j+3)+(j-3);r6=(j-1)+(j+4)+(j-3);
   bigMatrix=matrix(vector(l=TDOF*TDOF),nrow=TDOF, byrow=T);
   bigMatrix[c(r1,r2,r3,r4,r5,r6),c(r1,r2,r3,r4,r5,r6)]=eMatrix;
   return (bigMatrix)
}
```

```
PlaneFrameUDL_Matrix= function(DOF=6,AxialUDLMagnitude,
TransUDLMagnitude,Length,theta)
{
  L=Length;
  cx=cos(theta*pi/180);sx=sin(theta*pi/180);
  w1=AxialUDLMagnitude;w2=TransUDLMagnitude;
  row1=0.5*L*(-cx*w1+sx*w2);row2=-0.5*L*(sx*w1+cx*w2)
  row3=-(1/12)*(L^2)*w2;row4=0.5*L*(-cx*w1+sx*w2);
  row5=-0.5*L*(sx*w1+cx*w2);row6=(1/12)*(L^2)*w2
equivalentload=matrix(c(row1,row2,row3,row4,row5,row6),
nrow=DOF,byrow=T)
    return (equivalentload)
}

PlaneFrameUDL_ExpandedMatirx = function(TDOF,
LoadColumnMatrix,i,j)
{
  r1=2*(i-1)+i;r2=2*(i-1)+(i+1);
  r3=2*(i-1)+(i+2);r4=(j-1)+(j+2)+(j-3);
  r5=(j-1)+(j+3)+(j-3);r6=(j-1)+(j+4)+(j-3);
  bigColumnMatrix=matrix(vector(l=TDOF),nrow=TDOF,byrow=T);
  bigColumnMatrix[c(r1,r2,r3,r4)]=LoadColumnMatrix;
  return (bigColumnMatrix)
}

PlaneFrame_ReducedStiffnessMatrix = function(bigKmatrix,kn
ownloadnodes)
{
  reducedk = bigKmatrix[c(knownloadnodes),(knownloadnodes)]
  return(reducedk)
}

PlaneFrame_ReducedLoadVector = function(loadvector)
{
  reducedf = matrix(loadvector,ncol = 1)
  return(reducedf)
}

PlaneFrame_NodalDisplacement = function(reducedmatrix,
vec_reducedforce)
{
  return(solve(reducedmatrix,vec_reducedforce))
}

PlaneFrame_GlobalForces = function(bigKmatrix,
vec_globalnodaldisp)
{
  columndof=matrix(vec_globalnodaldisp,byrow = T)
```

```
   globalforces = bigKmatrix %*% vec_globalnodaldisp
   return(round(globalforces))
}

PlaneFrame_LocalForces = function(YoungMod,Area,momentI,
Length,theta,vec_globalnodaldisp)
{
   E1=YoungMod;A=Area;I=momentI;L=Length;
   cx=cos(theta*pi/180);sx=sin(theta*pi/180);
   row1=c(cx,sx,0,0,0,0);row2=c(-sx,cx,0,0,0,0);
   row3=c(0,0,1,0,0,0);row4=c(0,0,0,cx,sx,0);
   row5=c(0,0,0,-sx,cx,0);row6=c(0,0,0,0,0,1);
   tmatrix=matrix(c(row1,row2,row3,row4,row5,row6),nrow=6,
   byrow=T)
   localstiffness=tmatrix%*%PlaneFrame_Element_
   Matrix(6,E1,A,I,L,theta);
   local_forces = localstiffness%*%vec_globalnodaldisp
   return(round(local_forces))
}
```

7.3.3 Implemented R Functions for the Space Frame Elements

```
#**************************************************#
# R functions for the space frame element        #
# KB Mustapha                                     #
#**************************************************#

SpaceFrame_Element_Matrix = function(DOF=12,vec_
Modulus,Area,PolarMoment,vec_MomentI,vec_nodalcoordinates)
{
  xsquared=(vec_nodalcoordinates[4]-vec_nodalcoordinates[1])^2
  ysquared=(vec_nodalcoordinates[5]-vec_nodalcoordinates[2])^2
  zsquared=(vec_nodalcoordinates[6]-vec_nodalcoordinates[3])^2
  L=sqrt(xsquared+ysquared+zsquared)

E=vec_Modulus[1];G=vec_Modulus[2];Imax=vec_MomentI[1];
Imin=vec_MomentI[2];

  coord=vec_nodalcoordinates;
  A1=E*Area/L;G1=G*PolarMoment/L;
  B1=12*(E*Imax)/L^3;B2=6*(E*Imax)/L^2;
  B3=12*(E*Imin)/L^3;B4=6*(E*Imin)/L^2;
  B5=4*(E*Imax)/L;B6=4*(E*Imin)/L;
  row1=c(A1,0,0,0,0,0);row2=c(0,B1,0,0,0,B2);row3=c(0,0,B3,0,-B4,0);
  row4=c(0,0,0,G1,0,0);row5=c(0,0,-B4,0,B6,0);row6=c(0,B2,0,0,0,B5)
  submatrix1=matrix(c(row1,row2,row3,row4,row5,row6),nrow=6, byrow=T);
  row12=c(-A1,0,0,0,0,0);row22=c(0,-B1,0,0,0,B2);row32=c(0,0,
  -B3,0,-B4,0);
  row42=c(0,0,0,-G1,0,0);row52=c(0,0,B4,0,B6/2,0);row62=c(0,-B2,
  0,0,0,B5/2)
```

```
submatrix2=matrix(c(row12,row22,row32,row42,row52,row62),nrow=6,
byrow=T);
row13=c(-A1,0,0,0,0,0);row23=c(0,-B1,0,0,0,-B2);row33=c(0,0,-B3,
0,B4,0);
row43=c(0,0,0,-G1,0,0);row53=c(0,0,-B4,0,B6/2,0);row63=c(0,B2,
0,0,0,B5/2)
submatrix3=matrix(c(row13,row23,row33,row43,row53,row63),nrow=6,
byrow=T);
localstiffness=matrix(rep(0,DOF*DOF),nrow=DOF,ncol=DOF);
row14=c(A1,0,0,0,0,0);row24=c(0,B1,0,0,0,-B2);row34=c(0,0,B3,
0,B4,0);
row44=c(0,0,0,G1,0,0);row54=c(0,0,B4,0,B6,0);row64=c(0,-B2,
0,0,0,B5)
submatrix4=matrix(c(row14,row24,row34,row44,row54,row64),nrow=6,
ncol=6, byrow=T)
localstiffness[1:6,1:6]=submatrix1;
localstiffness[1:6,7:12]=submatrix2;
localstiffness[7:12,1:6]=submatrix3;
localstiffness[7:12,7:12]=submatrix4;
transformationmatrix=matrix(rep(0,DOF*DOF),nrow=DOF,ncol=DOF);
if(coord[1]==coord[4] & coord[2]==coord[5]){

  if(coord[6]>coord[3]) {
        Tsubmatrix=matrix(c(0,0,1,0,1,0,-1,0,0),nrow=3, byrow=T)

     } else
       Tsubmatrix=matrix(c(0,0,-1,0,1,0,1,0,0),nrow=3, byrow=T)
  } else {

     l=(1/L)*(coord[4]-coord[1]);
     m=(1/L)*(coord[5]-coord[2]);
     n=(1/L)*(coord[6]-coord[3]);
     D=sqrt(l^2+m^2);
     Tsubmatrix=matrix(c(l,m,n,-m/D,l/D,0,-l*n/D,-m*n/D,D),nrow=3,
     byrow=T);
  }
transformationmatrix[1:3,1:3]=Tsubmatrix;
transformationmatrix[4:6,4:6]=Tsubmatrix;
transformationmatrix[7:9,7:9]=Tsubmatrix;
transformationmatrix[10:12,10:12]=Tsubmatrix;
globalstiffness=t(transformationmatrix)%*%localstiffness%*%
transformationmatrix
return(globalstiffness)
}

SpaceFrame_ExpandedElement_Matrix = function(TDOF,eMatrix,i,j)
{
      r=vector(length=12);
     for(k in 6:1){
       r[k]=6*i+(k-6);
     }
     for(k in 12:7){
       r[k]=6*j+(k-12);
     }
   bigMatrix=matrix(vector(l=TDOF*TDOF),nrow=TDOF,byrow=T);
   bigMatrix[r[1:12],r[1:12]]=eMatrix;
   return (bigMatrix)
}
```

```
SpaceFrame_ReducedStiffnessMatrix = function(bigKmatrix,knownloadnodes)
{
  reducedk = bigKmatrix[c(knownloadnodes),(knownloadnodes)]
  return(reducedk)
}

SpaceFrame_ReducedLoadVector = function(loadvector)
{
  reducedf = matrix(loadvector,ncol = 1)
  return(reducedf)
}

SpaceFrame_NodalDisplacement = function(reducedmatrix,vec_reducedforce)
{
  return(solve(reducedmatrix,vec_reducedforce))
}

SpaceFrame_GlobalForcesMoments = function(bigKmatrix,
vec_globalnodaldisp)
{
  columndof=matrix(vec_globalnodaldisp,byrow = T)
  globalforces = bigKmatrix %*% vec_globalnodaldisp
  return(round(globalforces))
}
```

The usage of some of these functions is similar (though not the same) to some of those in the previous chapters. For brevity sake, the descriptions of the functions are embedded within the text of the commentaries in the examples that follow.

7.4 Examples

7.4.1 Example 7.1

Figure 7.2 shows a two-member frame structure with a concentrated moment of magnitude 300 kN m. Determine the (1) horizontal, vertical, and rotational displacements of point B; and (2) reactions at points A and C.

Take: $E = 210$ GPa; $I_{AB} = I_{CB} = 300 \times 10^{-6}$ m^4; $A = 2.1$ m^2.

7.4.1.1 Solution

This is a planar frame structure. As such, we shall make use of the functions listed in Section 7.3.2. Besides, we will denote points A, B, and C as nodes 1, 2, and 3, respectively, which means members AB and BC become elements 1 and 2. The elements' connectivity is shown in Table 7.2.

Step 1: Obtain/specify the properties (materials, geometric properties) of each element.

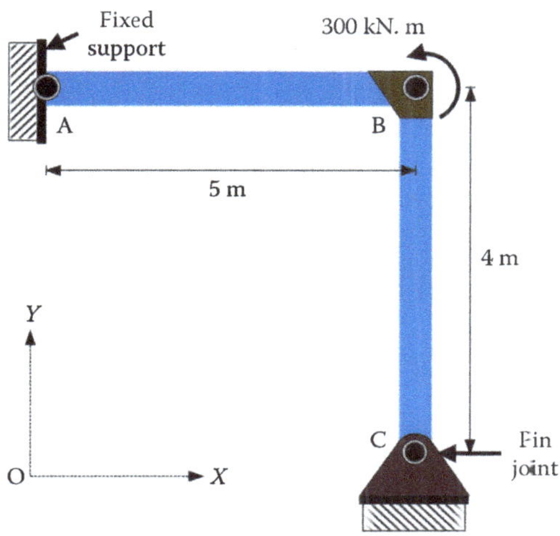

FIGURE 7.2
A two-member frame structure with a concentrated moment.

TABLE 7.2
Connectivity Table from the Discretized System

Elements	Nodes (i,j)	Coordinates (x_i, y_i, x_j, y_j)
1	1,2	(0,4,5,4)
2	2,3	(5,4,5,0)

The material properties, cross-section area, and moment of inertia of the elements are as given in the question. But, for the determination of the length and orientation of each element, we shall be using the function **`PlaneFrame_LengthOrientation()`**. This function returns a vector of two numbers describing the length and orientation of the element as shown in the block of code that follows:

```
E1=E2=210e9;A1=A2=2.1;I1=I2=300e-6

L1T1=PlaneFrame_LengthOrientation(c(0,4,5,4));
L2T2=PlaneFrame_LengthOrientation(c(5,4,5,0));
```

Running the earlier block of code produces

```
> L1T1=PlaneFrame_LengthOrientation(c(0,4,5,4));L1T1
[1] 5 0
> L2T2=PlaneFrame_LengthOrientation(c(5,4,5,0));L2T2
[1]   4 -90
```

From this output, the geometric lengths and orientations of the two members are computed as $L_1 = 5$ m; $\theta_1 = 0°$; $L_2 = 4$ m; $\theta_2 = -90°$.

Step 2: Write the stiffness matrix for the elements using the data from Step 1.

```
k1=PlaneFrame_Element_Matrix(6,E1,A1,I1,L1T1[1],L1T1[2]);k1
k2=PlaneFrame_Element_Matrix(6,E2,A2,I2,L2T2[1],L2T2[2]);k2
```

The stiffness matrices are obtained and displayed as

```
> k1=PlaneFrame_Element_Matrix(6,E1,A1,I1,L1T1[1],L1T1[2]);k1
             [,1]      [,2]      [,3]         [,4]      [,5]      [,6]
[1,]  88200000000         0         0 -88200000000         0         0
[2,]            0   6048000  15120000            0  -6048000  15120000
[3,]            0  15120000  50400000            0 -15120000  25200000
[4,] -88200000000         0         0  88200000000         0         0
[5,]            0  -6048000 -15120000            0   6048000 -15120000
[6,]            0  15120000  25200000            0 -15120000  50400000
> k2=PlaneFrame_Element_Matrix(6,E2,A2,I2,L2T2[1],L2T2[2]);k2
          [,1]          [,2]      [,3]      [,4]          [,5]      [,6]
[1,]  11812500             0  23625000 -11812500             0  23625000
[2,]         0  110250000000         0         0 -110250000000         0
[3,]  23625000             0  63000000 -23625000             0  31500000
[4,] -11812500             0 -23625000  11812500             0 -23625000
[5,]         0 -110250000000         0         0  110250000000         0
[6,]  23625000             0  31500000 -23625000             0  63000000
```

For this frame structure, none of the members carries a distributed load; hence, there is no need to determine the equivalent nodal loads. We therefore move on to the next step.

Step 3: Determine the global degree of freedom of the frame and expand the matrix equation for each element using **PlaneFrame_ExpandedElement_Matrix()**.

```
n_nodes=3;Total_dof=n_nodes*3;
K1=PlaneFrame_ExpandedElement_Matrix(Total_dof,k1,1,2);
K2=PlaneFrame_ExpandedElement_Matrix(Total_dof,k2,2,3);
```

Step 4: Establish the global stiffness matrix by adding the expanded matrices of the elements as shown next.

```
GlobalK=K1+K2;GlobalK
```

The output of this gives the 9×9 global stiffness matrix $[K]$ as

```
> GlobalK=K1+K2;GlobalK
            [,1]     [,2]      [,3]        [,4]         [,5]      [,6]      [,7]         [,8]      [,9]
[1,]  88200000000        0         0 -88200000000            0         0         0            0         0
[2,]            0  6048000  15120000            0     -6048000  15120000         0            0         0
[3,]            0 15120000  50400000            0    -15120000  25200000         0            0         0
[4,] -88200000000        0         0  88211812500            0  23625000 -11812500            0  23625000
[5,]            0 -6048000 -15120000            0 110256048000 -15120000         0 -110250000000         0
[6,]            0 15120000  25200000     23625000    -15120000 113400000 -23625000            0  31500000
[7,]            0        0         0    -11812500            0 -23625000  11812500            0 -23625000
[8,]            0        0         0            0 -110250000000        0         0  110250000000         0
[9,]            0        0         0     23625000            0  31500000 -23625000            0  63000000
```

For reading convenience, we reproduced the following stiffness matrix:

$$[K_{\text{global}}] = 10^6 \begin{bmatrix} 88200. & 0. & 0. & -88200. & 0. & 0. & 0. & 0. & 0. \\ 0. & 6.048 & 15.12. & 0. & -6.048 & 15.12 & 0. & 0. & 0. \\ 0. & 15.12 & 50.4 & 0. & -15.12 & 25.2 & 0. & 0. & 0. \\ -88200. & 0. & 0. & 88211.8125 & 0. & 23.625 & -11.8125 & 0. & 23.625 \\ 0. & -6.048 & -15.12 & 0. & 110256.048 & -15.12 & 0. & -110250 & 0. \\ 0. & 15.12 & 25.2 & 23.625 & -15.12 & 113.4 & -23.625 & 0. & 31.5 \\ 0. & 0. & 0. & -11.8125 & 0. & -23.625 & 11.8125 & 0. & -23.625 \\ 0. & 0. & 0. & 0. & -110250. & 0. & 0. & 110250 & 0. \\ 0. & 0. & 0. & 23.625 & 0. & 31.5 & -23.625 & 0. & 63. \end{bmatrix} \quad (7.31)$$

Step 5: Identify the boundary condition(s) and obtain the reduced stiffness matrix and reduced force vector.

Given the global stiffness matrix obtained in Step 4, we can write the global matrix equation as

$$\begin{Bmatrix} F_{1X} \\ F_{1Y} \\ M_{1Z} \\ F_{2X} \\ F_{2Y} \\ M_{2Z} \\ F_{3X} \\ F_{3Y} \\ M_{3Z} \end{Bmatrix} = \left[K_{\text{global}}\right] \begin{Bmatrix} U_{1X} \\ U_{1Y} \\ \theta_{1Z} \\ U_{2X} \\ U_{2Y} \\ \theta_{2Z} \\ U_{3X} \\ U_{3Y} \\ \theta_{3Z} \end{Bmatrix} \tag{7.32}$$

Identifying the boundary conditions is straightforward if one realizes that the frame is fixed at nodes 1 (eliminating the horizontal, vertical, and rotational displacements) and pinned at node 3 (eliminating only the horizontal and vertical displacements). Therefore, we may write the identified displacement boundary conditions as

$$U_{1X} = 0;\ U_{1Y} = 0;\ \theta_{1Z} = 0;\ U_{3X} = 0;\ U_{3Y} = 0 \tag{7.33}$$

Furthermore, from Figure 7.2, we recognize the known nodal loads as

$$F_{2X} = 0;\ F_{2Y} = 0;\ M_{2Z} = 300\ \text{ kN m};\ M_{3Z} = 0 \tag{7.34}$$

It will be observed that the known loads (forces and moments) stipulated in Equation 7.34 belong to rows 4, 5, 6, and 9.

This set of rows is then used with **Frame_ReducedStiffnessMatrix**() to establish the reduced stiffness matrix as shown next.

```
vec_rows_of_knownloads=c(4:6,9);
ReducedK=PlaneFrame_ReducedStiffnessMatrix(Kglobal,
vec_rows_of_knownloads);
ReducedK
```

Which yields the output:

```
> ReducedK
             [,1]         [,2]      [,3]      [,4]
[1,] 88211812500            0  23625000  23625000
[2,]           0 110256048000 -15120000         0
[3,]    23625000    -15120000 113400000  31500000
[4,]    23625000            0  31500000  63000000
```

In other words, the reduced stiffness matrix is

$$[K_{\text{reduced}}] = 10^6 \begin{bmatrix} 88211.8125 & 0. & 23.625 & 23.625 \\ 0. & 110256.048 & -15.12 & 0. \\ 23.625 & -15.12 & 113.4 & 31.5 \\ 23.625 & 0. & 31.5 & 6.3 \end{bmatrix}$$

In the next block of code, we establish the reduced load vector.

```
vec_values_of_knownloads=c(0,0,300000,0);
Reducedloadvector=PlaneFrame_ReducedLoadVector(vec_
values_of_knownloads);
Reducedloadvector
```

Producing the following:

```
> Reducedloadvector
       [,1]
[1,]      0
[2,]      0
[3,] 300000
[4,]      0
```

Combining the reduced stiffness matrix and the reduced load vector leads to the reduced matrix equation as

$$10^3 \begin{Bmatrix} 0 \\ 0 \\ 300 \\ 0 \end{Bmatrix} = 10^6 \begin{bmatrix} 88211.8125 & 0. & 23.625 & 23.625 \\ 0. & 110256.048 & -15.12 & 0. \\ 23.625 & -15.12 & 113.4 & 31.5 \\ 23.625 & 0. & 31.5 & 6.3 \end{bmatrix} \begin{Bmatrix} U_{2X} \\ U_{2Y} \\ \theta_{2Z} \\ \theta_{3Z} \end{Bmatrix} \tag{7.35}$$

Step 6: Determine the unknown nodal displacements and the global nodal loads.

For the unknown nodal displacements using the function **PlaneFrame_NodalDisplacement()**.

```
UnknwonNodalDisp=PlaneFrame_NodalDisplacement(ReducedK,
Reducedloadvector)
UnknwonNodalDisp
```

From which the determined nodal displacements are obtained as

```
> UnknwonNodalDisp
                    [,1]
[1,] -0.0000004114565
[2,]  0.0000004213225
[3,]  0.0030723116306
[4,] -0.0015360015191
```

The previous box shows that the displacements are approximately

$$\begin{Bmatrix} U_{2X} \\ U_{2Y} \\ \theta_{2Z} \\ \theta_{3Z} \end{Bmatrix} = \begin{Bmatrix} -4.114\times10^{-7}\,\text{m} \\ 4.213\times10^{-7}\,\text{m} \\ 0.00307\,\text{rad} \\ -0.00154\ \ \text{rad} \end{Bmatrix} \tag{7.36}$$

It is preferred to combine the determined displacements and the boundary conditions (in Equation 7.33) in a single variable as done next:

```
allglobalNodalDisp=matrix(rep(0,Total_dof),byrow=T);
allglobalNodalDisp[vec_rows_of_knownloads]=UnknwonNodalDisp
allglobalNodalDisp
```

The output of which reads

```
> allglobalNodalDisp
                    [,1]
 [1,]  0.0
 [2,]  0.0
 [3,]  0.0
 [4,] -0.0000004114565
```

```
[5,]  0.0000004213225
[6,]  0.0030723116306
[7,]  0.0
[8,]  0.0
[9,] -0.0015360015191
```

Next, the global nodal loads are determined via the function **PlaneFrame_GlobalForcesMoments()** as illustrated in the next code snippet.

```
globalLoads=PlaneFrame_GlobalForces(GlobalK,
allglobalNodalDisp)
globalLoads
```

As indicated, this function requires the global stiffness matrix (represented by the variable **GlobalK** determined in Step 4) and the vector of global nodal displacements (contained in the variable **allglobalNodalDisp**). Running the preceding code returns the forces as

```
> globalLoads
       [,1]
 [1,]  36290
 [2,]  46451
 [3,]  77416
 [4,]      0
 [5,]      0
 [6,] 300000
 [7,] -36290
 [8,] -46451
 [9,]      0
```

Based on the preceding output, the required reaction forces and moments at the nodes are listed as follows:

$$\begin{Bmatrix} F_{1X} \\ F_{1Y} \\ M_{1Z} \\ F_{3X} \\ F_{3Y} \end{Bmatrix} = \begin{Bmatrix} 36290\,\text{N} \\ 46451\,\text{N} \\ 77416\,\text{Nm} \\ -36290\,\text{N} \\ -46451\,\text{N} \end{Bmatrix} \tag{7.37}$$

7.4.2 Example 7.2

The frame structure in Figure 7.3 carries a distributed load of magnitude 12 kN/m on member *AB* and a concentrated load at point *C*. Assume joints *A* and *D* are both pinned, determine (1) the displacements of points *A*,*B*,*C*, and *D* and (2) the reactions at points *A* and *D*. Take $E = 200$ GPa; $I_{AB} = I_{BC} = I_{CD} = 60 \times 10^{-6}\,\text{m}^4$; Area $= 20 \times 10^{-3}\,\text{m}^2$.

7.4.2.1 Solution

For the solution, we will treat points *A*,*B*,*C*, and *D* as nodes. This will result in three elements *AB*, *BC*, and *CD*. With this, we formulate the connectivity matrix shown in Table 7.3.

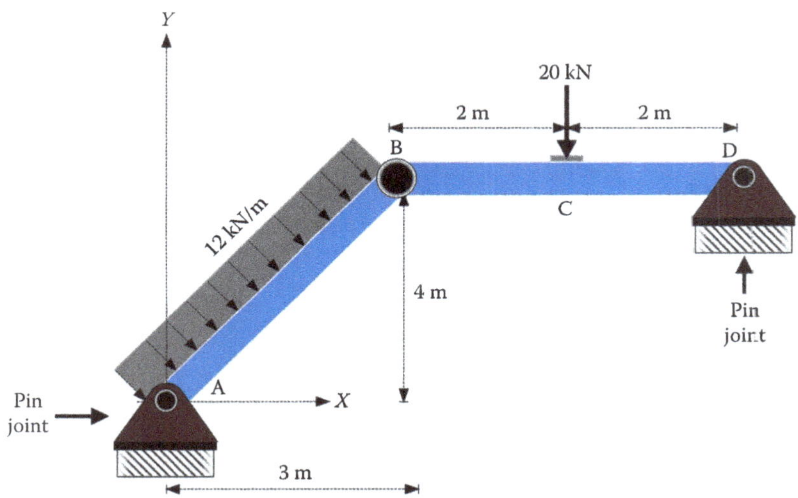

FIGURE 7.3
A frame structure with combined distributed and concentrated loads.

TABLE 7.3
Connectivity Table from the Discretized System

Elements	Nodes (i, j)	Coordinates $(x_i, y_i, x_j \cdot y_j)$
1	1,2	(0,0,3,4)
2	2,3	(3,4,5,4)
3	3,4	(5,4,7,4)

The other steps of the solution are presented next.

Step 1: Obtain the geometric lengths and orientation angles of each elements.

```
E1=E2=E3=200e9;A1=A2=A3=2e-2;I1=I2=I3=350e-6
L1T1=PlaneFrame_LengthOrientation(c(0,0,3,4));L1T1
L2T2=PlaneFrame_LengthOrientation(c(3,4,5,4));L2T2
L3T3=PlaneFrame_LengthOrientation(c(5,4,7,4));L3T3
```

Running the above-mentioned code, we have the output for the geometric lengths and orientations of the members as $L_1 = 5$ m; $\theta_1 = 53.13°$; $L_2 = L_3 = 2$ m; $\theta_2 = \theta_3 = 0°$.

```
> L1T1=PlaneFrame_LengthOrientation(c(0,0,3,4));L1T1
[1]  5.0 53.13
> L2T2=PlaneFrame_LengthOrientation(c(3,4,5,4));L2T2
[1] 2 0
> L3T3=PlaneFrame_LengthOrientation(c(5,4,7,4));L3T3
[1] 2 0
```

Step 2: Write the stiffness matrix for each element using the length and orientation angles.

```
k1=PlaneFrame_Element_Matrix(6,E1,A1,I1,L1T1[1],L1T1[2]);k1
k2=PlaneFrame_Element_Matrix(6,E2,A2,I2,L2T2[1],L2T2[2]);k2
k3=PlaneFrame_Element_Matrix(6,E3,A3,I3,L3T3[1],L3T3[2]);k3
```

The stiffness matrices are obtained and shown in the following block:

```
> k1=PlaneFrame_Element_Matrix(6,E1,A1,I1,L1T1[1],L1T1[2]);k1
           [,1]       [,2]      [,3]       [,4]       [,5]      [,6]
[1,]  292300800  380774400 -13440000 -292300800 -380774400 -13440000
[2,]  380774400  514419200  10080000 -380774400 -514419200  10080000
[3,]  -13440000   10080000  56000000   13440000  -10080000  28000000
[4,] -292300800 -380774400  13440000  292300800  380774400  13440000
[5,] -380774400 -514419200 -10080000  380774400  514419200 -10080000
[6,]  -13440000   10080000  28000000   13440000  -10080000  56000000
> k2=PlaneFrame_Element_Matrix(6,E2,A2,I2,L2T2[1],L2T2[2]);k2
            [,1]       [,2]       [,3]        [,4]       [,5]       [,6]
[1,]  2000000000          0          0 -2000000000          0          0
[2,]           0  105000000  105000000           0 -105000000  105000000
[3,]           0  105000000  140000000           0 -105000000   70000000
[4,] -2000000000          0          0  2000000000          0          0
[5,]           0 -105000000 -105000000           0  105000000 -105000000
[6,]           0  105000000   70000000           0 -105000000  140000000
```

```
> k3=PlaneFrame_Element_Matrix(6,E3,A3,I3,L3T3[1],L3T3[2]);k3
           [,1]      [,2]      [,3]       [,4]       [,5]      [,6]
[1,]  2000000000         0         0 -2000000000          0         0
[2,]           0 105000000 105000000           0 -105000000 105000000
[3,]           0 105000000 140000000           0 -105000000  70000000
[4,] -2000000000         0         0  2000000000          0         0
[5,]           0 -105000000 -105000000         0  105000000 -105000000
[6,]           0 105000000  70000000           0 -105000000 140000000
```

Element 1 (i.e., member AB) carries a distributed load, and we need to determine its equivalent nodal loads. For this, we employ the function **PlaneFrameUDL_Matrix()** as done in the following:

```
k1UDL=PlaneFrameUDL_Matrix(6,0,12000,L1T1[1],L1T1[2]);k1UDL
```

Executing the above produces the output that reads

```
> k1UDL=PlaneFrameUDL _ Matrix(6,0,12000,L1T1[1],L1T1[2]);k1UDL
      [,1]
[1,]  24000
[2,] -18000
[3,] -25000
[4,]  24000
[5,] -18000
[6,]  25000
```

Step 3: Determine the global degree of freedom of the frame and expand the matrix equation for each element.

```
n_nodes=4;Total_dof=n_nodes*3;
K1=PlaneFrame_ExpandedElement_Matrix(Total_dof,k1,1,2);
K2=PlaneFrame_ExpandedElement_Matrix(Total_dof,k2,2,3);
K3=PlaneFrame_ExpandedElement_Matrix(Total_dof,k3,3,4);
K1UDL=PlaneFrameUDL_ExpandedMatirx(Total_dof,k1UDL,1,2);
```

The last line in the preceding block of code is employed for the expansion of the vector of equivalent loads caused by the UDL on element 1.

Step 4: Establish the global stiffness matrix by adding the expanded matrices of the elements as shown next.

```
GlobalK=K1+K2+K3;GlobalK
```

The output of this gives the 12×12 global stiffness matrix $[K]$ as

```
> GlobalK=K1+K2+K3;GlobalK/10^6
            [,1]      [,2]   [,3]       [,4]      [,5]    [,6]  [,7] [,8] [,9] [,10] [,11] [,12]
 [1,]  292.3008  380.7744 -13.44  -292.3008 -380.7744  -13.44     0    0    0     0     0     0
 [2,]  380.7744  514.4192  10.08  -380.7744 -514.4192   10.08     0    0    0     0     0     0
 [3,]  -13.4400   10.0800  56.00    13.4400  -10.0800   28.00     0    0    0     0     0     0
 [4,] -292.3008 -380.7744  13.44  2292.3008  380.7744   13.44 -2000    0    0     0     0     0
 [5,] -380.7744 -514.4192 -10.08   380.7744  619.4192   94.92     0 -105  105     0     0     0
 [6,]  -13.4400   10.0800  28.00    13.4400   94.9200  196.00     0 -105   70     0     0     0
 [7,]    0.0000    0.0000   0.00 -2000.0000    0.0000    0.00  4000    0    0 -2000     0     0
 [8,]    0.0000    0.0000   0.00     0.0000 -105.0000 -105.00     0  210    0     0  -105   105
 [9,]    0.0000    0.0000   0.00     0.0000  105.0000   70.00     0    0  280     0  -105    70
[10,]    0.0000    0.0000   0.00     0.0000    0.0000    0.00 -2000    0    0  2000     0     0
[11,]    0.0000    0.0000   0.00     0.0000    0.0000    0.00     0 -105 -105     0   105  -105
[12,]    0.0000    0.0000   0.00     0.0000    0.0000    0.00     0  105   70     0  -105   140
```

We rewrite the global stiffness matrix for this system in a readable form as

$$
[K_{\text{global}}]=10^{6}\begin{bmatrix}
292.301 & 380.774 & -13.44 & -292.301 & -380.774 & -13.44 & 0. & 0. & 0. & 0. & 0. & 0. \\
380.774 & 514.419 & 10.08 & -380.774 & -514.419 & 10.08 & 0. & 0. & 0. & 0. & 0. & 0. \\
-13.44 & 10.08 & 56. & 13.44 & -10.08 & 28. & 0. & 0. & 0. & 0. & 0. & 0. \\
-292.301 & -380.774 & 13.44 & 2292.301 & 380.774 & 13.44 & -2000. & 0. & 0. & 0. & 0. & 0. \\
-380.774 & -514.419 & -10.08 & 380.774 & 619.419 & 94.92 & 0. & -105. & 105 & 0. & 0. & 0. \\
-13.44 & 10.08 & 28. & 13.44 & 94.92 & 196. & 0. & -105. & 70. & 0. & 0. & 0. \\
0. & 0. & 0. & -2000. & 0. & 0 & 4000. & 0. & 0. & -2000. & 0. & 0. \\
0. & 0. & 0. & 0. & -105. & -105. & 0. & 210. & 0. & 0. & -105. & 105. \\
0. & 0. & 0. & 0. & 105. & 70. & 0. & 0. & 0. & 0. & -105. & 70. \\
0. & 0. & 0. & 0. & 0. & 0. & -2000. & 0. & 0. & 2000. & 0. & 0. \\
0. & 0. & 0. & 0. & 0. & 0. & 0. & -105. & -105. & 0. & 105. & -105 \\
0. & 0. & 0. & 0. & 0. & 0. & 0. & 105. & 70. & 0. & -105. & 140.
\end{bmatrix}
$$

Step 5: Identify the boundary condition(s) and obtain the reduced stiffness matrix and reduced force vector.

For this problem, the global matrix equation will be of the form

$$
\begin{Bmatrix} F_{1X} \\ F_{1Y} \\ M_{1Z} \\ F_{2X} \\ F_{2Y} \\ M_{2Z} \\ F_{3X} \\ F_{3Y} \\ M_{3Z} \\ F_{3X} \\ F_{3Y} \\ M_{3Z} \end{Bmatrix} = [K_{\text{global}}] \begin{Bmatrix} U_{1X} \\ U_{1Y} \\ \theta_{1Z} \\ U_{2X} \\ U_{2Y} \\ \theta_{2Z} \\ U_{3X} \\ U_{3Y} \\ \theta_{3Z} \\ U_{4X} \\ U_{4Y} \\ \theta_{4Z} \end{Bmatrix} \tag{7.38}
$$

The identified displacement boundary conditions for this problem are as follows:

$$
U_{1X} = U_{1Y} = 0; U_{4X} = U_{4Y} = 0 \tag{7.39}
$$

The vector of known nodal loads will be a combination of the applied concentrated loads and the equivalent nodal loads from the applied UDL. That is,

$$\begin{Bmatrix} F_{1X} \\ F_{1Y} \\ M_{1Z} \\ F_{2X} \\ F_{2Y} \\ M_{2Z} \\ F_{3X} \\ F_{3Y} \\ M_{3Z} \\ F_{4X} \\ F_{4Y} \\ M_{4Z} \end{Bmatrix} = \begin{Bmatrix} F_{1X} \\ F_{1Y} \\ 0 \\ 0 \\ 0 \\ 0 \\ 0 \\ -20000 \\ 0 \\ F_{4Y} \\ F_{4Y} \\ 0 \end{Bmatrix}^{con} + \begin{Bmatrix} 24000 \\ -18000 \\ -25000 \\ 24000 \\ -18000 \\ 25000 \\ 0. \\ 0. \\ 0. \\ 0. \\ 0. \\ 0. \end{Bmatrix}^{eq} \quad (7.40)$$

We note that the second vector on the right-hand side of Equation 7.40 is obtained from the variable **K1UDL** in Step 3. The rows of the known loads in Equation 7.40 are thus 3,4,5,6,7,8,9, and 12. By using this set of rows, one obtains the reduced stiffness matrix via the function **PlaneFrame_ ReducedStiffnessMatrix()**, which is shown in the following.

```
vec_rows_of_knownloads=c(3,4,5,6,7,8,9,12);
ReducedK=PlaneFrame_ReducedStiffnessMatrix(GlobalK,
vec_rows_of_knownloads);
ReducedK
```

With the output being,

```
> ReducedK
          [,1]        [,2]       [,3]       [,4]        [,5]       [,6]      [,7]      [,8]
[1,]  56000000    13440000  -10080000   28000000           0          0         0         0
[2,]  13440000  2292300800  380774400   13440000 -2000000000          0         0         0
[3,] -10080000   380774400  619419200   94920000           0 -105000000 105000000         0
[4,]  28000000    13440000   94920000  196000000           0 -105000000  70000000         0
[5,]         0 -2000000000          0          0  4000000000          0         0         0
[6,]         0           0 -105000000 -105000000           0  210000000         0 105000000
[7,]         0           0  105000000   70000000           0          0 280000000  70000000
[8,]         0           0          0          0           0  105000000  70000000 140000000
```

Where the reduced stiffness matrix is rewritten as

$$[K_{\text{reduced}}] = 10^6 \begin{bmatrix} 56. & 13.44 & -10.08 & 28. & 0. & 0. & 0. & 0. \\ 13.44 & 2292.3008 & 380.774 & 13.44 & -2000 & 0. & 0. & 0. \\ -10.08 & 380.7744 & 619.4192 & 94.92 & 0. & -105. & 105. & 0. \\ 28. & 13.44 & 94.92 & 196. & 0. & -105. & 70. & 0. \\ 0. & -2000 & 0. & 0. & 4000. & 0. & 0. & 0. \\ 0. & 0. & -105. & -105. & 0. & 210. & 0. & 105. \\ 0. & 0. & 105. & 70. & 0. & 0. & 280. & 70. \\ 0. & 0. & 0. & 0. & 0. & 105. & 70. & 140. \end{bmatrix}$$

In a like manner, the reduced load vector must also be obtained, and we do this using the function **PlaneFrame_ReducedLoadVector()** in the following code snippet.

```
vec_values_of_knownloads=c(0,0,0,0,0,-20000,0,0)+K1UDL
[c(3,4,5,6,7,8,9,12)];
Reducedloadvector=PlaneFrame_ReducedLoadVector(vec_values_
of_knownloads);
Reducedloadvector
```

The running of which yields

```
> Reducedloadvector
       [,1]
[1,] -25000
[2,]  24000
[3,] -18000
[4,]  25000
[5,]      0
[6,] -20000
[7,]      0
[8,]      0
```

Combining the reduced stiffness matrix and the reduced load vector leads to the reduced matrix equation.

$$\begin{Bmatrix} -25000 \\ 24000 \\ -18000 \\ 25000 \\ 0 \\ -20000 \\ 0 \\ 0 \end{Bmatrix} = 10^6 \begin{bmatrix} 56. & 13.44 & -10.08 & 28. & 0. & 0. & 0. & 0. \\ 13.44 & 2292.3008 & 380.774 & 13.44 & -2000 & 0. & 0. & 0. \\ -10.08 & 380.7744 & 619.4192 & 94.92 & 0. & -105. & 105. & 0. \\ 28. & 13.44 & 94.92 & 196. & 0. & -105. & 70. & 0. \\ 0. & -2000 & 0. & 0. & 4000. & 0. & 0. & 0. \\ 0. & 0. & -105. & -105. & 0. & 210. & 0. & 105. \\ 0. & 0. & 105. & 70. & 0. & 0. & 280. & 70. \\ 0. & 0. & 0. & 0. & 0. & 105. & 70. & 140. \end{bmatrix} \begin{Bmatrix} \theta_{1Z} \\ U_{2X} \\ U_{2Y} \\ \theta_{2Z} \\ U_{3X} \\ U_{3Y} \\ \theta_{3Z} \\ \theta_{4Z} \end{Bmatrix} \tag{7.41}$$

Step 6: Determine the unknown nodal displacements and global nodal loads.

```
UnknwonNodalDisp=PlaneFrame_NodalDisplacement(ReducedK,
Reducedloadvector)
UnknwonNodalDisp
```

The determined displacements are obtained as

```
> UnknwonNodalDisp
              [,1]
[1,] -0.000602883
[2,]  0.000056501
[3,] -0.000116059
[4,]  0.000244007
[5,]  0.000028250
[6,] -0.000063452
[7,] -0.000033573
[8,]  0.000064376
```

The above shows that, to the third decimal place, the displacements are

$$\begin{Bmatrix} \theta_{1Z} \\ U_{2X} \\ U_{2Y} \\ \theta_{2Z} \\ U_{3X} \\ U_{3Y} \\ \theta_{3Z} \\ \theta_{4Z} \end{Bmatrix} = 10^{-4} \begin{Bmatrix} -6.029 \text{ rad} \\ 0.565 \text{ m} \\ -1.161 \text{ m} \\ 2.440 \text{ rad} \\ 0.283 \text{ m} \\ -0.635 \text{ m} \\ -0.336 \text{ rad} \\ 0.644 \text{ rad} \end{Bmatrix}$$

For handiness, we now combine the determined displacements and the boundary condition.

```
allglobalNodalDisp=matrix(rep(0,Total_dof),byrow=T);
allglobalNodalDisp[vec_rows_of_knownloads]=UnknwonNodalDisp
allglobalNodalDisp
```

Finally, we determine the global loads by using the function **PlaneFrame_GlobalForcesMoments()**.

```
concentratedLoads=PlaneFrame_GlobalForcesMoments(GlobalK,
allglobalNodalDisp)
equivalentLoads=K1UDL
globalLoads=concentratedLoads-equivalentLoads
globalLoads
```

Running the preceding code of block returns the forces and moments as

```
> globalLoads
         [,1]
 [1,]    8501
 [2,]   52572
 [3,]      -0
 [4,]       0
 [5,]       0
 [6,]       0
 [7,]       0
 [8,]  -20000
 [9,]       0
[10,]  -56501
[11,]    3428
[12,]       0
```

Based on the foregoing output, the reaction forces and moments at the nodes are as follows:

$$\begin{Bmatrix} F_{1X} \\ F_{1Y} \\ M_{1Z} \\ F_{2X} \\ F_{2Y} \\ M_{2Z} \\ F_{3X} \\ F_{3Y} \\ M_{3Z} \\ F_{4X} \\ F_{4Y} \\ M_{4Z} \end{Bmatrix} = \begin{Bmatrix} 8501 \\ 52572 \\ 0 \\ 0 \\ 0 \\ 0 \\ 0 \\ -20000 \\ 0 \\ -56501 \\ 3428 \\ 0 \end{Bmatrix}$$

7.4.3 Example 7.3

The space frame shown in Figure 7.4 has its three elements occupying a three-dimensional space. The XYZ coordinate of each node of the space frame is placed beside it. If nodes 2,3, and 4 are assumed fixed, determine

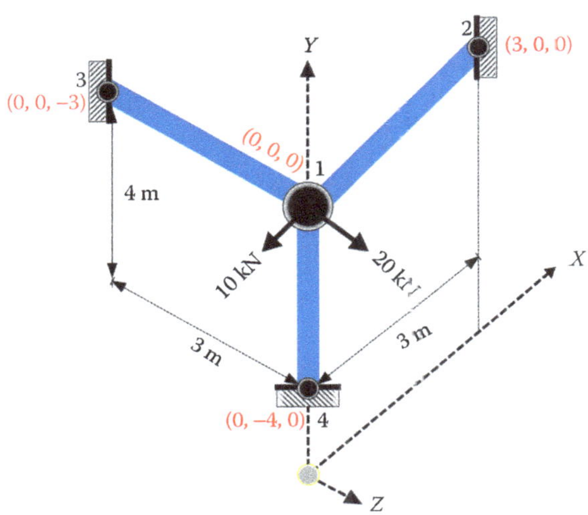

FIGURE 7.4
A three-element space frame.

(1) the unknown displacements of node 1; and (2) the reactions at the fixed nodes. Take, $E = 210\,\text{GPa}$; $G = 84\,\text{GPa}$; $I_z = 200 \times 10^{-6}\,\text{m}^4$; $I_y = 100 \times 10^{-6}\,\text{m}^4$; $I_x = 50 \times 10^{-6}\,\text{m}^4$; $A = 200 \times 10^{-3}\,\text{m}^2$[3].

7.4.3.1 Solution

This is a space frame structure. Thus, we shall make use of the functions listed in Section 7.3.3.

Step 1: Obtain the geometric lengths and orientation angles of each element.

For this example, the geometric lengths and the orientation angles of the elements are generated along with the matrices in the next step.

Step 2: Write the stiffness matrix for each element using the function **SpaceFrame_Element_Matrix()**.

```
A=20e-3;moment=c(200e-6,100e-6);J=50e-6;E=210e9;G=84e9;
e1coord=c(0,0,0,3,0,0);e2coord=c(0,0,0,0,0,-3);e3coord=
c(0,0,0,0,-4,0)

k1=SpaceFrame_Element_Matrix(12,c(E,G),A,J,moment,e1coord);
k1/10^6
k2=SpaceFrame_Element_Matrix(12,c(E,G),A,J,moment,e2coord);
k2/10^6
k3=SpaceFrame_Element_Matrix(12,c(E,G),A,J,moment,e3coord);
k3/10^6
```

The stiffness matrices are obtained and shown in the following block:

```
> k1=SpaceFrame_Element_Matrix(12,c(E,G),A,J,moment,e1coord);k1/10^6
       [,1]    [,2]     [,3] [,4] [,5] [,6]  [,7]    [,8]     [,9] [,10] [,11] [,12]
 [1,]  1400   0.000   0.0000  0.0    0    0 -1400   0.000   0.0000   0.0     0     0
 [2,]     0  18.667   0.0000  0.0    0   28     0 -18.667   0.0000   0.0     0    28
 [3,]     0   0.000   9.3333  0.0  -14    0     0   0.000  -9.3333   0.0   -14     0
 [4,]     0   0.000   0.0000  1.4    0    0     0   0.000   0.0000  -1.4     0     0
 [5,]     0   0.000 -14.0000  0.0   28    0     0   0.000  14.0000   0.0    14     0
 [6,]     0  28.000   0.0000  0.0    0   56     0 -28.000   0.0000   0.0     0    28
 [7,] -1400   0.000   0.0000  0.0    0    0  1400   0.000   0.0000   0.0     0     0
 [8,]     0 -18.667   0.0000  0.0    0  -28     0  18.667   0.0000   0.0     0   -28
 [9,]     0   0.000  -9.3333  0.0   14    0     0   0.000   9.3333   0.0    14     0
[10,]     0   0.000   0.0000 -1.4    0    0     0   0.000   0.0000   1.4     0     0
[11,]     0   0.000 -14.0000  0.0   14    0     0   0.000  14.0000   0.0    28     0
[12,]     0  28.000   0.0000  0.0    0   28     0 -28.000   0.0000   0.0     0    56
> k2=SpaceFrame_Element_Matrix(12,c(E,G),A,J,moment,e2coord);k2/10^6
          [,1]    [,2]  [,3] [,4] [,5] [,6]     [,7]    [,8]  [,9] [,10] [,11] [,12]
 [1,]   9.3333   0.000     0    0  -14  0.0  -9.3333   0.000     0     0   -14   0.0
 [2,]   0.0000  18.667     0   28    0  0.0   0.0000 -18.667     0    28     0   0.0
 [3,]   0.0000   0.000  1400    0    0  0.0   0.0000   0.000 -1400     0     0   0.0
 [4,]   0.0000  28.000     0   56    0  0.0   0.0000 -28.000     0    28     0   0.0
 [5,] -14.0000   0.000     0    0   28  0.0  14.0000   0.000     0     0    14   0.0
 [6,]   0.0000   0.000     0    0    0  1.4   0.0000   0.000     0     0     0  -1.4
 [7,]  -9.3333   0.000     0    0   14  0.0   9.3333   0.000     0     0    14   0.0
 [8,]   0.0000 -18.667     0  -28    0  0.0   0.0000  18.667     0   -28     0   0.0
 [9,]   0.0000   0.000 -1400    0    0  0.0   0.0000   0.000  1400     0     0   0.0
[10,]   0.0000  28.000     0   28    0  0.0   0.0000 -28.000     0    56     0   0.0
```

```
[11,] -14.0000   0.000      0     0    14   0.0 14.0000   0.000      0      0     28   0.0
[12,]   0.0000   0.000      0     0     0  -1.4  0.0000   0.000      0      0      0   1.4
> k3=SpaceFrame_Element_Matrix(12,c(E,G),A,J,moment,e3coord);k3/10^6
        [,1]  [,2]    [,3]    [,4]  [,5]   [,6]    [,7]  [,8]    [,9]  [,10] [,11]  [,12]
 [1,]  7.875     0  0.0000   0.000  0.00  15.75  -7.875     0  0.0000  0.000  0.00  15.75
 [2,]  0.000  1050  0.0000   0.000  0.00   0.00   0.000 -1050  0.0000  0.000  0.00   0.00
 [3,]  0.000     0  3.9375  -7.875  0.00   0.00   0.000     0 -3.9375 -7.875  0.00   0.00
 [4,]  0.000     0 -7.8750  21.000  0.00   0.00   0.000     0  7.8750 10.500  0.00   0.00
 [5,]  0.000     0  0.0000   0.000  1.05   0.00   0.000     0  0.0000  0.000 -1.05   0.00
 [6,] 15.750     0  0.0000   0.000  0.00  42.00 -15.750     0  0.0000  0.000  0.00  21.00
 [7,] -7.875     0  0.0000   0.000  0.00 -15.75   7.875     0  0.0000  0.000  0.00 -15.75
 [8,]  0.000 -1050  0.0000   0.000  0.00   0.00   0.000  1050  0.0000  0.000  0.00   0.00
 [9,]  0.000     0 -3.9375   7.875  0.00   0.00   0.000     0  3.9375  7.875  0.00   0.00
[10,]  0.000     0 -7.8750  10.500  0.00   0.00   0.000     0  7.8750 21.000  0.00   0.00
[11,]  0.000     0  0.0000   0.000 -1.05   0.00   0.000     0  0.0000  0.000  1.05   0.00
[12,] 15.750     0  0.0000   0.000  0.00  21.00 -15.750     0  0.0000  0.000  0.00  42.00
```

Step 3: Determine the global degree of freedom of the space frame and expand the matrix for each element using the function **SpaceFrame_ExpandedElement_Matrix**().

```
n_nodes=4;Total_dof=n_nodes*6;
K1=SpaceFrame_ExpandedElement_Matrix(Total_dof,k1,1,2);
K2=SpaceFrame_ExpandedElement_Matrix(Total_dof,k2,1,3);
K3=SpaceFrame_ExpandedElement_Matrix(Total_dof,k3,1,4);
```

Step 4: Establish the global stiffness matrix by adding the expanded matrices of the elements as shown next.

```
GlobalK=K1+K2+K3;GlobalK
```

The output of this gives the weirdly 24×24 global stiffness matrix $[K_{global}]$ as

$[K_{global}] =$

10^6

1417.208	0.	0.	0.	−14.	15.75	−1400	0.	0.	0.	0.	0.	−9.333	0.	0.	0.	−14.	0.	−7.875	0.	0.	0.	0.	15.75
0.	1087.333	0.	28.	0.	28.	0.	−18.667	0.	0.	0.	28.	0.	−18.667	0.	28.	0.	0.	0.	−10.50	−3.938	−7.875	0.	0.
0.	0.	1413.271	−7.875	−14.	0.	0.	0.	−9.333	0.	−14	0.	0.	0.	−1400	0.	0.	0.	0.	0.	7.875	10.5	0.	0.
0.	28.	−7.875	78.4	0.	0.	0.	0.	0.	−1.4	0.	0.	0.	−28.	0.	28.	0.	0.	0.	0.	0.	0.	0.	0.
−14.	0.	−14.	0.	57.05	0.	0.	0.	14.	0.	14.	0.	14.	0.	0.	0.	14.	0.	0.	0.	0.	0.	−1.05	0.
15.75	28.	0.	0.	0.	99.4	0.	−28.	0.	0.	0.	28.	0.	0.	0.	0.	0.	−1.4	−15.75	0.	0.	0.	0.	21.
−1400.	0.	0.	0.	0.	0	1400.	0.	0.	0.	0.	0.	0.	0.	0.	0.	0.	0.	0.	0.	0.	0.	0.	0.
0.	−18.667	0.	0.	0.	−28.	0.	18.667	0.	0.	0.	−28.	0.	0.	0.	0.	0.	0.	0.	0.	0.	0.	0.	0.
0.	0.	−9.333	0.	14.	0.	0.	0.	9.333	0.	14.	0.	0.	0.	0.	0.	0.	0.	0.	0.	0.	0.	0.	0.
0.	0.	0.	−1.4	0.	0.	0.	0.	0.	1.4	0.	0.	0.	0.	0.	0.	0.	0.	0.	0.	0.	0.	0.	0.
0.	0.	−14.	0.	14.	0.	0.	0.	14.	0.	28.	0.	0.	0.	0.	0.	0.	0.	0.	0.	0.	0.	0.	0.
0.	28.	0.	0.	0	28.	0.	−28.	0.	0.	0.	56.	0.	0.	0.	0.	0.	0.	0.	0.	0.	0.	0.	0.
−9.333	0.	0.	0.	14.	0.	0.	0.	0.	0.	0.	0.	9.333	0.	0.	0.	14.	0.	0.	0.	0.	0.	0.	0.
0.	−18.667	0.	−28.	0.	0.	0.	0.	0.	0.	0.	0.	0.	18.667	0.	−28.	0.	0.	0.	0.	0.	0.	0.	0.
0.	0.	−1400.	0	0.	0.	0.	0.	0.	0.	0.	0.	0.	0.	1400.	0.	0.	0.	0.	0.	0.	0.	0.	0.
0.	28.	0.	28.	0.	0.	0.	0.	0.	0.	0.	0.	0.	−28.	0.	56.	0.	0.	0.	0.	0.	0.	0.	0.
−14.	0.	0.	0.	14.	0.	0.	0.	0.	0.	0.	0.	14.	0.	0.	0.	28.	0.	0.	0.	0.	0.	0.	0.
0.	0.	0.	0.	0.	−1.4	0.	0.	0.	0.	0.	0.	0.	0.	0.	0.	0.	1.4	0.	0.	0.	0.	0.	0.
−7.875	0.	0.	0.	0.	−15.75	0.	0.	0.	0.	0.	0.	0.	0.	0.	0.	0.	0.	7.785	0.	0.	0.	0.	−15.75
0.	−10.50	0.	0.	0.	0.	0.	0.	0.	0.	0.	0.	0.	0.	0.	0.	0.	0.	0.	1050.	0.	0.	0.	0.
0.	0.	−3.938	7.875	0.	0.	0.	0.	0.	0.	0.	0.	0.	0.	0.	0.	0.	0.	0.	0.	3.938	7.875	0.	0.
0.	0.	−7.875	10.5	0.	0.	0.	0.	0.	0.	0.	0.	0.	0.	0.	0.	0.	0.	0.	0.	7.875	21.	0.	0.
0.	0.	0.	0.	−1.05	0.	0.	0.	0.	0.	0.	0.	0.	0.	0.	0.	0.	0.	0.	0.	0.	0.	1.05	0.
15.75	0.	0.	0.	0.	21.	0.	0.	0.	0.	0.	0.	0.	0.	0.	0.	0.	0.	−15.75	0.	0.	0.	0.	42.

Step 5: Identify the boundary condition(s) and obtain the reduced stiffness matrix and reduced force vector. With the global stiffness matrix obtained in Step 3, we can write the global matrix equation as

$$[F]^T = [K_{\text{global}}][U]^T \tag{7.42}$$

where

$[F]^T =$

$[F_{1X}\ F_{1Y}\ F_{1Z}\ M_{1X}\ M_{1Y}\ M_{1Z}\ F_{2X}\ F_{2Y}\ F_{2Z}\ M_{2X}\ M_{2Y}\ M_{2Z}\ F_{3X}\ F_{3Y}\ F_{3Y}\ M_{3X}\ M_{3Y}\ M_{3Z}\ F_{4X}\ F_{4Y}\ F_{4Z}\ M_{4X}\ M_{4Y}\ M_{4Z}]$

$[U]^T =$

$[U_{1X}\ U_{1Y}\ U_{1Z}\ \theta_{1X}\ \theta_{1Y}\ \theta_{1Z}\ U_{2X}\ U_{2Y}\ U_{2Z}\ \theta_{2X}\ \theta_{2Y}\ \theta_{2Z}\ U_{3X}\ U_{3Y}\ U_{3Y}\ \theta_{3X}\ \theta_{3Y}\ \theta_{3Z}\ U_{4X}\ U_{4Y}\ U_{4Z}\ \theta_{4X}\ \theta_{4Y}\ \theta_{4Z}]$

The boundary conditions are derived from the knowledge that the frame is fixed at nodes 2,3, and 4, resulting in zero horizontal, vertical, and rotational displacements at these nodes. Thus, the identified displacement boundary conditions can be written as

$$U_{2X} = U_{2Y} = U_{2Z} = \theta_{2X} = \theta_{2Y} = \theta_{2Z} = 0; \tag{7.43}$$

$$U_{3X} = U_{3Y} = U_{3Z} = \theta_{3X} = \theta_{3Y} = \theta_{3Z} = 0; \tag{7.44}$$

$$U_{4X} = U_{4Y} = U_{4Z} = \theta_{4X} = \theta_{4Y} = \theta_{4Z} = 0; \tag{7.45}$$

Furthermore, from the problem statement we realize the vector of known nodal loads as

$$\begin{Bmatrix} F_{1X} \\ F_{1Y} \\ F_{1Z} \\ M_{1X} \\ M_{1Y} \\ M_{2Z} \end{Bmatrix} = 10^3 \begin{Bmatrix} -10 \\ 0 \\ 20 \\ 0 \\ 0 \\ 0 \end{Bmatrix}, \tag{7.46}$$

The rows of the known loads are thus 1,2,3,4,5,6. By using this set of rows, one obtains the reduced stiffness matrix via the function **`SpaceFrame_ ReducedStiffnessMatrix()`** as done next.

```
vec_rows_of_knownloads=c(1,2,3,4,5,6);
ReducedK=SpaceFrame_ReducedStiffnessMatrix(GlobalK,vec_rows_
of_knownloads);
ReducedK
```

Which then produces

```
> ReducedK
           [,1]       [,2]       [,3]     [,4]      [,5]     [,6]
[1,] 1417208333          0          0        0 -14000000 15750000
[2,]          0 1087333333          0 28000000         0 28000000
[3,]          0          0 1413270833 -7875000 -14000000        0
[4,]          0   28000000   -7875000 78400000         0        0
[5,]  -14000000          0  -14000000        0  57050000        0
[6,]   15750000   28000000          0        0         0 99400000
```

In the next code, we also establish the reduced load vector for the space frame using the function **SpaceFrame_ReducedLoadVector()**.

```
vec_values_of_knownloads=c(-10000,0,20000,0,0,0);
Reducedloadvector=SpaceFrame_ReducedLoadVector(vec_values_
of_knownloads);
Reducedloadvector
```

Producing:

```
> Reducedloadvector
       [,1]
[1,] -10000
[2,]      0
[3,]  20000
[4,]      0
[5,]      0
[6,]      0
```

Step 6: Determine the unknown nodal displacements and nodal global forces by using the **SpaceFrame_NodalDisplacement()** and **SpaceFrame_GlobalForces()**.

```
UnknwonNodalDisp=SpaceFrame_NodalDisplacement(ReducedK,
Reducedloadvector)
UnknwonNodalDisp
```

The determined displacements are obtained as

```
> UnknwonNodalDisp
              [,1]
[1,] -7.0515e-06
[2,] -6.6537e-08
[3,]  1.4177e-05
[4,]  1.4478e-06
[5,]  1.7486e-06
[6,]  1.1361e-06
```

The above shows that, to the third decimal place, the displacements are

$$\begin{Bmatrix} U_{1X} \\ U_{1Y} \\ U_{1Z} \\ \theta_{1X} \\ \theta_{1Y} \\ \theta_{1Z} \end{Bmatrix} = 10^{-4} \begin{Bmatrix} -0.071 \ \text{m} \\ -0.001 \ \text{m} \\ 0.142 \ \text{m} \\ 0.014 \ \text{rad} \\ 0.017 \ \text{rad} \\ 0.011 \ \text{rad} \end{Bmatrix}$$

As has been the practice, the determined displacements and the boundary condition are combined and stored in the variable **allglobalNodalDisp**.

```
allglobalNodalDisp=matrix(rep(0,Total_dof),byrow=T);
allglobalNodalDisp[vec_rows_of_knownloads]=UnknwonNodalDisp
allglobalNodalDisp
```

Finally, the global nodal loads are found via the function **SpaceFrame_GlobalForcesMoments()**, which requires the global stiffness matrix (represented by the variable **GlobalK** determined in Step 4) and the vector of global nodal displacements (contained in the variable **allglobalNodalDisp**).

```
globalLoads=SpaceFrame_GlobalForcesMoments(GlobalK,
allglobalNodalDisp)
globalLoads
```

Running the preceding block of code returns the global forces and moments at the nodes as

```
> globalLoads
        [,1]
 [1,] -10000
 [2,]      0
 [3,]  20000
 [4,]      0
 [5,]      0
 [6,]      0
 [7,]   9872
 [8,]    -31
 [9,]   -108
[10,]     -2
[11,]   -174
[12,]     30
[13,]     90
[14,]    -39
[15,] -19848
[16,]     39
[17,]    123
[18,]     -2
[19,]     38
[20,]     70
[21,]    -44
[22,]    -96
[23,]     -2
[24,]    -87
```

Where the set of forces and moments returned are related to the following vector:

$$[F]^T =$$

$$[F_{1X}\ F_{1Y}\ F_{1Z}\ M_{1X}\ M_{1Y}\ M_{1Z}\ F_{2X}\ F_{2Y}\ F_{2Z}\ M_{2X}\ M_{2Y}\ M_{2Z}\ F_{3X}\ F_{3Y}\ F_{3Y}\ M_{3X}\ M_{3Y}\ M_{3Z}\ F_{4X}\ F_{4Y}\ F_{4Z}\ M_{4X}\ M_{4Y}\ M_{4Z}]$$

7.5 Exercises

1. Figure 7.5 shows a two-member frame with a distributed load of magnitude 3 kN/m. Determine (i) the horizontal, vertical, and rotational displacements of point *B* and (ii) the reactions at points *A* and *C*. Take $E = 200$ GPa; $I_1 = I_2 = 60 \times 10^{-6}\,\text{m}^4$; $A = 6 \times 10^{-4}\,\text{m}^2$.
2. Figure 7.6 shows a frame with a combined load at point *C*. Take $E = 200\,\text{GPa}$, $I = 60 \times 10^{-6}\,\text{m}^4, A = 4 \times 10^{-4}\,\text{m}^2$.

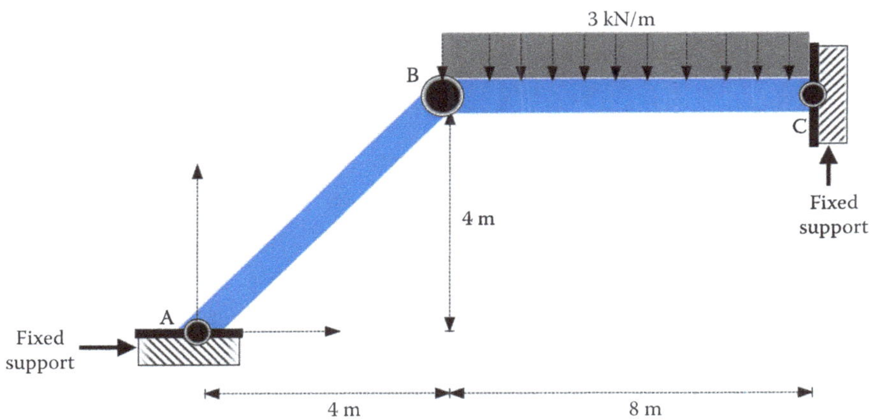

FIGURE 7.5
Exercise 7Q1.

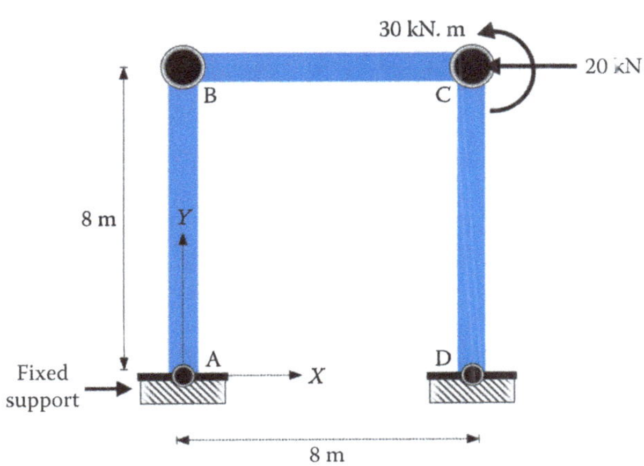

FIGURE 7.6
Exercise 7Q2.

3. Form the global stiffness matrix for this system.
4. Determine the horizontal, vertical, and rotational displacements of points *B* and *C*.
5. Determine the reaction forces at points *A* and *D*.

References

1. D. L. Logan, *First Course in the Finite Element Method*, Thomson, Toronto, ON, 2007.
2. M. A. Bhatti, *Fundamental Finite Element Analysis and Applications: with Mathematica and Matlab Computations*, Wiley, Hoboken, NJ, 2005.
3. P. I. Kattan, *MATLAB Guide to Finite Elements: An Interactive Approach*, Springer, Berlin, Germany, 2010.

8

Two-Dimensional Continuum Elements for Plane Elasticity Problems

8.1 Introduction

A throng of elements exist to aid analysts in the finite element solutions of many practical problems. In Chapters 3 through 7, we explored a limited number of elements, the behaviours of which are analogous to members that have been well scrutinized under the subject of *strength of materials*: bars, beams, truss and frame elements. It would be noticed that each of these aforesaid elements has a single-variable shape functions (mostly a function of the coordinate x), and each represents a one-dimensional approximation of a three-dimensional structure. Consequently, those elements are commonly referred to as one-dimensional structural elements. The one-dimensional structural elements are connected only at nodal points, which tends to occur at locations of concentrated loads, change of cross-section, joints/supports, and so on. In principle, when a structure comprises a series of one-dimensional structural members, the aforesaid elements can be employed to determine the stress and deformation at different locations.

Nonetheless, as attractive as the one-dimensional structural elements are, they do not quite demonstrate the versatility of the finite element method in modeling of structures with irregular shapes and boundaries. For one, the nodal points mentioned in the preceding paragraph does not occur naturally for higher order continuum structures such as flat plates and shells [1]. In such cases, one must artificially discretize the continuum structures into several planar elements before proceeding with an analysis. In this chapter, our goal is to examine two-dimensional (2D) continuum elements that provide convenience for approximate planar analysis of 3D solids subjected to external planar loads. Mainly, we will look at the constant strain triangular (CST) element in this chapter and examine the axisymmetric element in the next chapter. The CST, while not exactly the best 2D planar element, has applications for tackling plane stress analysis (e.g., analysis of thin plates with holes, hooks, brackets, fillets) and plane strain problems (e.g., shafts, pipes, and others).

The basic formulations for continuum elements can be found in numerous existing textbooks [2,3]. Still for enriched presentation, Section 8.2 provides a short account of the element's equations, whereas Section 8.3 contains the implemented R functions for analyses of plane stress or strain problems.

8.2 Finite Element Equations for the Constant Strain Triangular Element

8.2.1 Displacement and Shape Functions

Figure 8.1a is a schematic of a thin rectangular structure that can be analysed with the element developed in this chapter. In Figure 8.1b, we show a structure in the x–y plane representing a 2D state of stress at a point within the

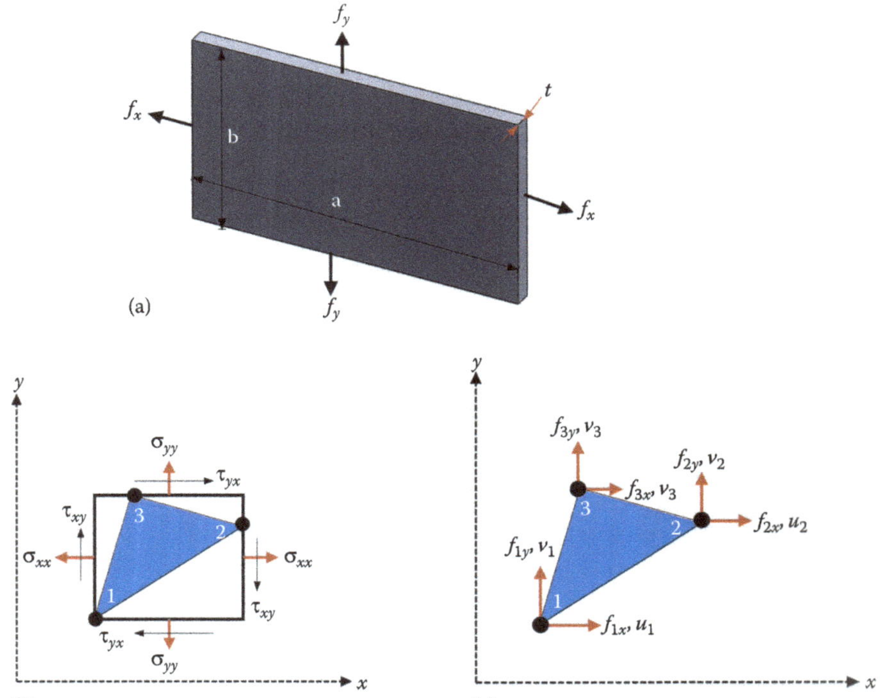

FIGURE 8.1
(a) A thin plate with generic planar forces, (b) an element in a state of plane stress, and (c) an isolated triangular element showing nodal forces and degrees of freedom.

rectangular plate. Furthermore, let the 2D domain be discretized, as demonstrated in Figure 8.1b, into a few small triangular areas, each having three straight edges.

In Figure 8.1c, we isolate the central triangle, which shares an edge with the other triangles, for further analysis as a triangular element. At each node of the isolated element, we indicate

- A rectangular coordinate system
- The nodal displacements $\left(u_i, v_i, u_j, v_j, u_k, v_k\right)$
- The nodal forces $\left(f_{ix}, f_{iy}, f_{jx}, f_{jy}, f_{kx}, f_{ky}\right)$

where the subscripts i, j, and k represent the nodal numbers, for example, 1, 2 and 3. Before we proceed, it is highlighted that the derivation presented here is constrained by certain assumptions. First, the structure to be analysed is planar in the undeformed state and of constant thickness t, where t is much less than the plane dimensions a and b (similar to Figure 8.1). Second, the external forces act only at the boundary parallel to the plane or are distributed uniformly over the thickness. Furthermore, assuming the triangular element experiences only planar deformation when under planar loads, we may represent variation of the displacements at any of its interior point in the following form:

$$u = A_1 + A_2x + A_3y \tag{8.1}$$

$$v = A_4 + A_5x + A_6y \tag{8.2}$$

where $A_1 - A_6$ are unknown coefficients. As explained in the previous chapters, we have chosen the total number of coefficients A_i to be equal to the total number of degrees of freedom associated with the element.

Next, we solve for the unknown coefficients by imposing boundary conditions on Equations 8.1 and 8.2 as follows:

- At node 1, $x = x_1$ & $y = y_1$: $u = u_1$
- At node 1, $x = x_1$ & $y = y_1$: $v = v_1$
- At node 2, $x = x_2$ & $y = y_2$: $u = u_2$
- At node 2, $x = x_2$ & $y = y_2$: $v = v_2$
- At node 3, $x = x_3$ & $y = y_3$: $u = u_3$
- At node 3, $x = x_3$ & $y = y_3$: $v = v_3$

Substituting these boundary conditions into Equations 8.1 and 8.2, one obtains

$$\underbrace{\begin{Bmatrix} u_1 \\ v_1 \\ u_2 \\ v_2 \\ u_3 \\ v_3 \end{Bmatrix}}_{\{u_e\}} = \underbrace{\begin{bmatrix} 1 & x_1 & y_1 & 0 & 0 & 0 \\ 0 & 0 & 0 & 1 & x_1 & y_1 \\ 1 & x_2 & y_2 & 0 & 0 & 0 \\ 0 & 0 & 0 & 1 & x_2 & y_2 \\ 1 & x_3 & y_3 & 0 & 0 & 0 \\ 0 & 0 & 0 & 1 & x_3 & y_3 \end{bmatrix}}_{[H]} \underbrace{\begin{Bmatrix} A_1 \\ A_2 \\ A_3 \\ A_4 \\ A_5 \\ A_6 \end{Bmatrix}}_{\{A\}} \tag{8.3}$$

where $\{u_e\}$ is a vector of the nodal displacements, $[H]$ is the coordinate matrix, and $\{A\}$ is the vector of the unknown coefficients. From Equation 8.3, the unknown coefficients are determined as

$$\{A\} = [H]^{-1}\{u_e\} \tag{8.4}$$

The inversion of matrix $[H]$, carried out within a symbolic computational engine, reads

$$[H]^{-1} = \begin{bmatrix} \frac{-x_3y_2+x_2y_3}{2A} & 0 & \frac{x_3y_1-x_1y_3}{2A} & 0 & \frac{-x_2y_1+x_1y_2}{2A} & 0 \\ \frac{y_2-y_3}{2A} & 0 & \frac{-y_1+y_3}{2A} & 0 & \frac{y_1-y_2}{2A} & 0 \\ \frac{-x_2+x_3}{2A} & 0 & \frac{x_1-x_3}{2A} & 0 & \frac{-x_1+x_2}{2A} & 0 \\ 0 & \frac{-x_3y_2+x_2y_3}{2A} & 0 & \frac{x_3y_1-x_1y_3}{2A} & 0 & \frac{-x_2y_1+x_1y_2}{2A} \\ 0 & \frac{y_2-y_3}{2A} & 0 & \frac{-y_1+y_3}{2A} & 0 & \frac{y_1-y_2}{2A} \\ 0 & \frac{-x_2+x_3}{2A} & 0 & \frac{x_1-x_3}{2A} & 0 & \frac{-x_1+x_2}{2A} \end{bmatrix} \tag{8.5}$$

where $2A = x_3(y_1-y_2)+x_1(y_2-y_3)+x_2(-y_1+y_3)$.

Having found the unknowns constants A, we now rewrite Equations 8.1 and 8.2 as

$$u = \underline{\begin{bmatrix} 1 & x & y & 0 & 0 & 0 \end{bmatrix}[H]^{-1}}\{u_e\} \tag{8.6}$$

$$v = \underline{\begin{bmatrix} 0 & 0 & 0 & 1 & x & y \end{bmatrix}[H]^{-1}}\{u_e\} \tag{8.7}$$

By carrying out the multiplication of the underlined matrix terms in Equations 8.6 and 8.7, one obtains

$$u = \underbrace{\begin{bmatrix} N_{u1} & 0 & N_{u2} & 0 & N_{u3} & 0 \end{bmatrix}}_{[N_u]} \{u_e\} \tag{8.8}$$

$$v = \underbrace{\begin{bmatrix} 0 & N_{v1} & 0 & N_{v2} & 0 & N_{v3} \end{bmatrix}}_{[N_v]} \{u_e\} \tag{8.9}$$

where N_u and N_v are the shape functions for axial and transverse deformations at the interior points of the element. We replaced the nodal numbers 1, 2, and 3 by a generic sequence i, j, and k and show the full form of the shape functions in Table 8.1.

8.2.2 Strain–Displacement Relationship and Constitutive Law

For plane elasticity problems, the non-zero components of the strain measure are related to the displacement derivatives in the form

$$\begin{Bmatrix} \varepsilon_{xx} \\ \varepsilon_{yy} \\ \varepsilon_{xy} \end{Bmatrix} = \begin{Bmatrix} \dfrac{\partial u}{\partial x} \\ \dfrac{\partial v}{\partial y} \\ \dfrac{\partial u}{\partial y} + \dfrac{\partial v}{\partial x} \end{Bmatrix} \tag{8.10}$$

TABLE 8.1

Shape Functions for the Axial and Transverse Deformations in a Constant Strain Triangular (CST) Element

Shape Function	Expression
$N_{ui} = N_{vi}$	$\frac{1}{2A}\left(x_k(y - y_j) + x(y_j - y_k) + x_j(-y + y_k)\right)$
$N_{uj} = N_{vj}$	$\frac{1}{2A}\left(x_k(-y + y_i) + x_i(y - y_k) + x(-y_i + y_k)\right)$
$N_{uk} = N_{vk}$	$\frac{1}{2A}\left(x_j(y - y_i) + x(y_i - y_j) + x_i(-y + y_j)\right)$

Bearing in mind Equations 8.8 and 8.9, we rewrite Equation 8.10 as

$$\begin{Bmatrix} \varepsilon_{xx} \\ \varepsilon_{yy} \\ \varepsilon_{xy} \end{Bmatrix} = \begin{Bmatrix} \dfrac{\partial N_u}{\partial x} \\ \dfrac{\partial N_v}{\partial y} \\ \dfrac{\partial N_u}{\partial y} + \dfrac{\partial N_v}{\partial x} \end{Bmatrix} \tag{8.11}$$

Evaluating the derivatives on the left side of Equation 8.11 leads to

$$\begin{Bmatrix} \varepsilon_{xx} \\ \varepsilon_{yy} \\ \varepsilon_{xy} \end{Bmatrix} = \begin{bmatrix} \dfrac{y_2 - y_3}{2A} & 0 & \dfrac{-y_1 + y_3}{2A} & 0 & \dfrac{y_1 - y_2}{2A} & 0 \\ 0 & \dfrac{-x_2 + x_3}{2A} & 0 & \dfrac{x_1 - x_3}{2A} & 0 & \dfrac{-x_1 + x_2}{2A} \\ \dfrac{-x_2 + x_3}{2A} & \dfrac{y_2 - y_3}{2A} & \dfrac{x_1 - x_3}{2A} & \dfrac{-y_1 + y_3}{2A} & \dfrac{-x_1 + x_2}{2A} & \dfrac{y_1 - y_2}{2A} \end{bmatrix} \begin{Bmatrix} u_1 \\ v_1 \\ u_2 \\ v_2 \\ u_3 \\ v_3 \end{Bmatrix} \tag{8.12}$$

which, in compact form, becomes

$$\{\varepsilon\} = [B]\{u_e\} \tag{8.13}$$

Notice that Equation 8.12 predicts a set of constant strain, because all the terms within the matrix $[B]$ have no dependence on x and y. This explains the naming of this element as a *CST* element [2].

To link the strain to stress, we call on the constitutive law, which for the general plane elasticity problems, is of the form

$$\begin{Bmatrix} \sigma_{xx} \\ \sigma_{yy} \\ \tau_{xy} \end{Bmatrix} = [D] \begin{Bmatrix} \varepsilon_{xx} \\ \varepsilon_{yy} \\ \varepsilon_{xy} \end{Bmatrix} \tag{8.14}$$

Or:

$$\{\sigma\} = [D][B]\{u_e\} \tag{8.15}$$

where $[D]$ expresses the material properties of the element and it depends on the nature of the problem. For plane stress problems,

$$[D] = \frac{E}{1-v} \begin{bmatrix} 1 & v & 0 \\ v & 1 & 0 \\ 0 & 0 & \frac{1}{2}(1-v) \end{bmatrix} \tag{8.16}$$

And for the plane strain problems, we have

$$[D]=\frac{E(1-v)}{(1+v)(1-2v)}\begin{bmatrix} 1 & \frac{v}{1-v} & 0 \\ \frac{v}{1-v} & 1 & 0 \\ 0 & 0 & \frac{1-2v}{2(1-v)} \end{bmatrix} \tag{8.17}$$

8.2.3 Minimization of the Total Potential Energy for the Constant Strain Triangular Element

Following the steps employed in the previous chapters, we write the strain energy (U) of the CST element as

$$U=\frac{1}{2}\int_{\text{vol}}\{\sigma\}^T\{\varepsilon\}dV \tag{8.18}$$

With Equations 8.13 and 8.15, the strain-energy expression simplifies to

$$U=\frac{1}{2}\int_{vol}([D][B]\{u_e\})^T([B]\{u_e\})dV \tag{8.19}$$

The work done by externally applied concentrated nodal forces (W) is of the form

$$W=\{u_e\}^T\begin{Bmatrix} f_{1x} \\ f_{1y} \\ f_{2x} \\ f_{2y} \\ f_{3x} \\ f_{3y} \end{Bmatrix} \tag{8.20}$$

To obtain the stiffness matrix, we substitute Equations 8.19 and 8.20 into the expression for the potential energy P_t, and then minimize P_t with respect to u_e to obtain

$$\frac{\partial}{\partial\{u_e\}^T}P_t=\left(\int_0^t\int_A [B]^T[D][B]dtdA\right)\{u_e\}-\{f_e\}=0 \tag{8.21}$$

Recalling that the matrix $[D]$ and $[B]$ have no dependence on x,y and z, the integration simplifies to

$$\left([B]^T[D][B]\int_0^t\int_A dtdA\right)\{u_e\}-\{f_e\}=0 \tag{8.22}$$

And the desired finite element matrix equation for the CST is obtained as

$$\{f_e\}=\left(At[B]^T[D][B]\right)\{u_e\} \tag{8.23}$$

In full form, Equation 8.23 becomes:

$$\begin{Bmatrix} f_{1x} \\ f_{1y} \\ f_{2x} \\ f_{2y} \\ f_{3x} \\ f_{3y} \end{Bmatrix} = [K_{CST}] \begin{Bmatrix} u_1 \\ v_1 \\ u_2 \\ v_2 \\ u_3 \\ v_3 \end{Bmatrix} \tag{8.24}$$

where

$$[K_{CST}]=[B]^T[D][B] \tag{8.25}$$

Given the nodal coordinates of a triangular element, the full form of the stiffness matrix in Equation 8.25 is straightforwardly obtained numerically.

8.3 Equivalent Nodal Forces for the Constant Strain Triangular with Distributed Loads

The matrix model of Equation 8.24 is suitable for analysing triangular elements under the influence of concentrated nodal loads. In the presence of a surface traction p (N/m^2), the equivalent nodal loads must be evaluated, as done for the beam element in Chapter 6.

The equivalent nodal loads derived from a uniform, tensile surface traction force $p_x\left(N/m^2\right)$ acting normal to the z–y plane and a uniform, shear traction force $p_y\left(N/m^2\right)$ acting tangential to the z–y plane is

$$f_s^{eq} = \begin{Bmatrix} f_{s1x} \\ f_{s1y} \\ f_{s2x} \\ f_{s2y} \\ f_{s3x} \\ f_{s3y} \end{Bmatrix} = \begin{Bmatrix} 0 \\ 0 \\ \dfrac{p_x Lt}{2} \\ \dfrac{p_y Lt}{2} \\ \dfrac{p_x Lt}{2} \\ \dfrac{p_y Lt}{2} \end{Bmatrix} \tag{8.26}$$

where surface traction forces have been assumed to act on edge 2−3 and covering the area of thickness t and edge length L.

8.4 Computer Implementation

8.4.1 Systematic Procedure for the Analyses of Structures with Constant Strain Triangular Elements

8.4.1.1 Precomputation Phase

- Establish a global coordinate system.
- Discretize the continuum structure into finite triangular elements.
- Label the nodes and elements (this helps one to form the connectivity table).

8.4.1.2 Computation Phase

1. Specify nodal coordinate (based on the connectivity table) and material properties.
2. Establish the finite element matrix equation for each triangular element by using Equation 8.24 or simply form the stiffness matrix for each element by using Equation 8.25.
3. Determine the global degree of freedom of the discretized continuum, and carry out the expansion of the matrix equation for each element.
4. For a triangular element with a surface traction, determine the vector of the equivalent nodal loads by using Equation 8.26.
5. Form the global stiffness matrix and the global matrix equation.

6. Identify the essential boundary condition(s) and apply the boundary conditions to the global matrix equation to obtain a reduced matrix equation. Programmatically, this step involves two actions:
   a. Obtain the reduced load vector, which will be equal to reduced load vector from concentrated applied loads *plus* the reduced load vector from the expanded vector of equivalent nodal loads.
   b. Use the identified boundary condition to obtain the reduced stiffness matrix.
7. Determine the unknown nodal displacements by using the reduced matrix equation. With the aid of the nodal displacements, one may then obtain other variables such the nodal loads and stress.

8.4.2 Implemented R Functions for the Constant Strain Triangular Element

```
#********************************************************************#
# R functions for the CSTriangular element                           #
# KB Mustapha                                                        #
#********************************************************************#

CSTriangular_Element_Matrix = function(DOF=6,YoungMod,Nu,thickness,
                                  vec_nodalcoordinates,case)
{
  x1=vec_nodalcoordinates[1];x2=vec_nodalcoordinates[3];
  x3=vec_nodalcoordinates[5];
  y1=vec_nodalcoordinates[2];y2=vec_nodalcoordinates[4];
  y3=vec_nodalcoordinates[6];
  A=(x3*(y1-y2)+x2*(y3-y1)+x1*(y2-y3))/2;

  B1=y2-y3;B2=y3-y1;B3=y1-y2;
  G1=x3-x2;G2=x1-x3;G3=x2-x1;
  row1=c(B1,0,B2,0,B3,0);
  row2=c(0,G1,0,G2,0,G3)
  row3=c(G1,B1,G2,B2,G3,B3)
  p1=c(1,Nu,0,Nu,1,0,0,0,(1-Nu)/2)
  p2=c(1-Nu,Nu,0,Nu,1-Nu,0,0,0,(1-2*Nu)/2)

  BMatrix=(1/(2*A))*matrix(c(row1,row2,row3),nrow=3,byrow=T);
  DPlaneStress=(YoungMod/(1-Nu^2))*matrix(p1,nrow=3,byrow=T);
  DPlaneStrain=(YoungMod/((1+Nu)*(1-2*Nu)))*matrix(p2,nrow=3,byrow=T)

  eMatrix=matrix(vector(l=DOF*DOF),nrow=DOF,ncol=DOF);
  if(case==1){
    eMatrix=(thickness*A)*t(BMatrix)%*%DPlaneStress%*%BMatrix
  }
  if(case==2){
    eMatrix=(thickness*A)*t(BMatrix)%*%DPlaneStress%*%BMatrix
  }else{eMatrix=(thickness*A)*t(BMatrix)%*%DPlaneStress%*%BMatrix}

  return(eMatrix)
}
```

```
CSTriangular_ExpandedElement_Matrix = function(TDOF,eMatrix,i,j,k)
{
  r1=2*i-1; r2=2*i
  r3=2*j-1; r4=2*j
  r5=2*k-1; r6=2*k
  bigMatrix=matrix(vector(l=TDOF*TDOF),nrow=TDOF,byrow=T);
  bigMatrix[c(r1,r2,r3,r4,r5,r6),c(r1,r2,r3,r4,r5,r6)]=eMatrix;
  return (bigMatrix)
}
CSTriangular_SF = function(DOF=6,SFtensile,SFshear,Length,thickness)
{
  px=SFtensile;py=SFshear;
  L=Length;b=thickness;
equivalentload=matrix(c(0,0,px*L*b/2,py*L*b/2,px*L*b/2,py*L*b/2),nrow=
DOF,byrow=T)  return (equivalentload)
}

CSTriangular_ExpandedSF = function(TDOF,LoadColumnMatrix,i,j,k)
{
  r1=2*i-1; r2=2*i
  r3=2*j-1; r4=2*j
  r5=2*k-1; r6=2*k
  bigColumnMatrix=matrix(vector(l=TDOF),nrow=TDOF,byrow=T);
  bigColumnMatrix[c(r1,r2,r3,r4,r5,r6)]=LoadColumnMatrix;
  return (bigColumnMatrix)
}

CSTriangular_ReducedStiffnessMatrix = function(bigKmatrix,
knownloadnodes)
{
  reducedk = bigKmatrix[c(knownloadnodes),(knownloadnodes)]
  return(reducedk)
}

CSTriangular_ReducedLoadVector = function(loadvector)
{
  reducedf = matrix(loadvector,ncol = 1)
  return(reducedf)
}

CSTriangular_NodalDisplacement = function(reducedmatrix,vec_reducedforce)
{
  return(solve(reducedmatrix,vec_reducedforce))
}

CSTriangular_GlobalForces = function(bigKmatrix,vec_globalnodaldisp)
{
  columndof=matrix(vec_globalnodaldisp,byrow = T)
  globalforces = bigKmatrix %*% vec_globalnodaldisp
  return(round(globalforces))
}

CSTriangular_LocalForces = function(ematrix,vec_globalnodaldisp,i,j,k)
{
  r1=2*i-1; r2=2*i
  r3=2*j-1; r4=2*j
  r5=2*k-1; r6=2*k
```

```
    localforces = ematrix%*%vec_globalnodaldisp[c(r1,r1,r3,r4,r5,r6)]
    return(round(localforces))
}

CSTriangular_Stresses = function(YoungMod,Nu,thickness,vec_nodal
coord,case,vec_globalnodaldisp,i,j,k)
{
    x1=vec_nodalcoord[1];x2=vec_nodalcoord[3];
    x3=vec_nodalcoord[5];
    y1=vec_nodalcoord[2];y2=vec_nodalcoord[4];
    y3=vec_nodalcoord[6];
    A=(x3*(y1-y2)+x2*(y3-y1)+x1*(y2-y3))/2;
    B1=y2-y3;B2=y3-y1;B3=y1-y2;
    G1=x3-x2;G2=x1-x3;G3=x2-x1;
    row1=c(B1,0,B2,0,B3,0);
    row2=c(0,G1,0,G2,0,G3)
    row3=c(G1,B1,G2,B2,G3,B3)
    p1=c(1,Nu,0,Nu,1,0,0,0,(1-Nu)/2)
    p2=c(1-Nu,Nu,0,Nu,1-Nu,0,0,0,(1-2*Nu)/2)
    BMatrix=(1/(2*A))*matrix(c(row1,row2,row3),nrow=3,byrow=T);
    DPlaneStress=(YoungMod/(1-Nu^2))*matrix(p1,nrow=3,byrow=T);
    DPlaneStrain=(YoungMod/((1+Nu)*(1-2*Nu)))*matrix(p2,nrow=3,byrow=T)
    r1=2*i-1; r2=2*i;  r3=2*j-1; r4=2*j
    r5=2*k-1; r6=2*k
    if(case==1){
      localstresses=DPlaneStress%*%
        BMatrix%*%vec_globalnodaldisp[c(r1,r2,r3,r4,r5,r6)]
    }
    if(case==2){
      localstresses=DPlaneStrain%*%
        BMatrix%*%vec_globalnodaldisp[c(r1,r2,r3,r4,r5,r6)]
    }else{localstresses=DPlaneStress%*%
            BMatrix%*%vec_globalnodaldisp[c(r1,r2,r3,r4,r5,r6)]
      }
    return(localstresses)
}
```

The explanations of the usage and arguments for the set of functions for the CST elements are provided in Tables 8.2 through 8.4.

We will put the listed functions to use in the next section. But it must be stressed that the functions presented are applicable only to elements, the edges of which are parallel to the global axes for the whole structure.

TABLE 8.2

Functions for CST Element's Matrices and Vectors

Description of Functions
`CSTriangular_Element_Matrix(DOF=6,YoungMod,Nu,thickness, vec_nodalcoordinates,case)` This function generates and returns the stiffness matrix for the CST element. It accepts six parameters: **`DOF`** (predefined as 6); **`YoungMod`**, which represents Young's modulus; **`Nu`**, Poisson's ratio; **`vec_nodalcoordinates`**, a vector of nodal coordinate in the form $(x_i, y_i, x_j, y_j, x_k, y_k)$ and **`case`**, which should either be 1 for a plane stress problem or 2 for a plane strain problem.
`CSTriangular_ExpandedElement_Matrix(TDOF, eMatrix, i,j,k)` This function forms the expanded stiffness matrix for the CST element, and its parameters are of the same form as those defined in previous chapters. That is, the value of the total degrees of freedom (**`TDOF`**), element matrix (**`eMatrix`**) and the nodal labels i, j and k.
`CSTriangular_ReducedStiffnessMatrix(bigKmatrix, knownloadnodes)` This function generates a reduced stiffness matrix for the element, and its parameters are of the same form as those defined in previous chapters.
`CSTriangular_ReducedLoadVector(loadvector)` This function generates a reduced load vector needed to find the unknown nodal displacements. It accepts the parameter **`loadvector`** (a vector of known nodal forces) and transforms this vector to a column matrix.

TABLE 8.3

Additional Functions for CST Elements with Distributed Surface Tractions

Description of Functions
`CSTriangular_SF(DOF=6,SFtensile, SFshear, Length, thickness)` This function generates and returns the column matrix for the uniformly distributed load derived in Equation 8.26. It accepts five parameters: **`DOF`** (predefined as 6 for CST element); **`SFtensile`**, which represents the magnitude of a positive tensile surface traction on edge $j-k$; **`SFshear`**, which represents the magnitude of a positive (upward-facing) shear surface traction on edge $j-k$, **`Length`**, which denotes a member's geometric length and **`thickness`**, which denotes the element's thickness.
`CSTriangular_ExpandedSF(TDOF, LoadColumnMatrix, i,j,k)` This function forms the expanded column matrix for the surface traction, and its parameters are **`TDOF`** (total degrees of freedom of the system); **`LoadColumnMatrix`** (load column matrix, obtained in row 1 of this table); and **`i,j,k`** representing the ith, jth and kth nodes of the element with the surface traction.

TABLE 8.4

Functions for the Solution Phase in the Finite Element Analyses of Triangulated Structures

Description of Functions
`CSTriangular_NodalDisplacement(reducedmatrix, vec_reducedforce)` Given the parameters **`reducedmatrix`** (reduced stiffness matrix) and **`reducedforcevec`** (reduced force vector), this function solves for the unknown nodal displacements.
`CSTriangular_GlobalForces(bigKmatrix, vec_globalnodaldisp)` This function calculates the global forces (and global reaction forces). Its parameters are **`bigKmatrix`** (the global stiffness matrix) and **`vec_globalnodaldisp`** (a vector of the global nodal displacement).
`CSTriangular_Stresses(YoungMod, Nu, thickness, vec_nodalcoord, case, vec_globalnodaldisp, i,j,k)` This function returns the stress developed in a triangular element. Its parameters are **`YoungMod,`** which represents the member's Young's modulus; **`Nu`**, Poisson's ratio; **`vec_nodalcoord`**, a vector of nodal coordinate in the form $(x_i, y_i, x_j, y_j, x_k, y_k)$; **`case`**, which should either be 1 for a plane stress problem and 2 for a plane strain problem; **`vec_globalnodaldisp`** (a vector of the global nodal displacement); and **`i,j,k`** representing the ith, jth and kth nodes of the element.

8.5 Examples

8.5.1 Example 8.1

Figure 8.2 shows a thin plate pinned at three of its vertices and a planar inclined load applied at the unrestrained vertex. Given that $E = 210$ GPa, $v = 0.3$, and $t = 0.5$ mm [2]:

- Establish the global stiffness matrix for the structure.
- Determine the nodal displacements at the point of the applied load.
- Determine the reactions at the pin supports.
- Determine the planar stresses engendered by the applied load.

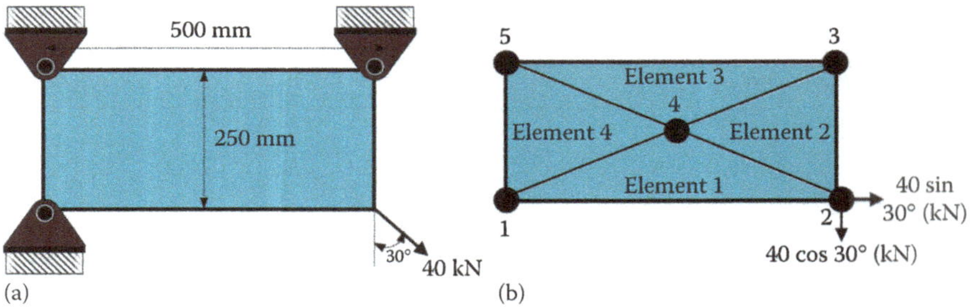

FIGURE 8.2
(a) A thin plate with a corner load and (b) triangulation of the rectangular domain.

TABLE 8.5

Connectivity Table from the Discretized System in Figure 8.2b

Elements	Nodes (i, j, k)	Coordinates $(x_i, y_i, x_j, y_j, x_k, y_k)$
1	1,2,4	(0,0,0.5,0,0.25,0.125)
2	2,3,4	(0.5,0,0.5,0.25,0.25,0.125)
3	4,3,5	(0.25,0.125,0.5,0.25,0,0.25)
4	1,4,5	(0,0,0.25,0.125,0,0.25)

8.5.1.1 Solution

Following the triangulation of the thin plate shown in Figure 8.2b, we form the element's connectivity of Table 8.5.

With the above-shown connectivity table, the rest of the solution steps are provided in the following.

Step 1: We specify the nodal coordinate values (based on the connectivity table) and the properties of each element.

All parameters supplied are in SI units.

```
e1coord=c(0,0,0.5,0,0.25,0.125)
e2coord=c(0.5,0,0.5,0.25,0.25,0.125)
e3coord=c(0.25,0.125,0.5,0.25,0,0.25)
e4coord=c(0,0,0.25,0.125,0,0.25)
E1=E2=E3=E4=210e9;t1=t2=t3=t4=0.5e-3;Nu=0.3;
```

Step 2: Write the stiffness matrix for each element using the coordinates and properties specified in Step 1 as arguments to the function **CSTriangular_Element_Matrix()**.

```
k1=CSTriangular_Element_Matrix(6,E1,Nu,t1,e1coord,1);k1
k2=CSTriangular_Element_Matrix(6,E2,Nu,t2,e2coord,1);k2
k3=CSTriangular_Element_Matrix(6,E3,Nu,t3,e3coord,1);k3
k4=CSTriangular_Element_Matrix(6,E4,Nu,t4,e4coord,1);k4
```

It is recalled that the last parameter of the function **CSTriangular_Element_Matrix()** refers to case, which should either be 1 (for plane stress problems) or 2 (for plane strain problems). The stiffness matrices are obtained and shown as

```
> k1=CSTriangular_Element_Matrix(6,E1,Nu,t1,e1coord,1);k1
          [,1]      [,2]      [,3]      [,4]      [,5]       [,6]
[1,]  34615385  18750000   5769231  -1442308 -40384615  -17307692
[2,]  18750000  62740385   1442308  52644231 -20192308 -115384615
[3,]   5769231   1442308  34615385 -18750000 -40384615   17307692
```

```
[4,]  -1442308   52644231 -18750000   62740385  20192308 -115384615
[5,] -40384615  -20192308 -40384615   20192308  80769231          0
[6,] -17307692 -115384615  17307692 -115384615         0  230769231
> k2=CSTriangular_Element_Matrix(6,E2,Nu,t2,e2coord,1);k2
          [,1]      [,2]      [,3]      [,4]      [,5]      [,6]
[1,]  34615385 -18750000  -5769231  -1442308 -28846154  20192308
[2,] -18750000  62740385   1442308 -52644231  17307692 -10096154
[3,]  -5769231   1442308  34615385  18750000 -28846154 -20192308
[4,]  -1442308 -52644231  18750000  62740385 -17307692 -10096154
[5,] -28846154  17307692 -28846154 -17307692  57692308         0
[6,]  20192308 -10096154 -20192308 -10096154         0  20192308
> k3=CSTriangular_Element_Matrix(6,E3,Nu,t3,e3coord,1);k3
          [,1]       [,2]      [,3]       [,4]      [,5]       [,6]
[1,]  80769231          0 -40384615  -20192308 -40384615   20192308
[2,]         0  230769231 -17307692 -115384615  17307692 -115384615
[3,] -40384615  -17307692  34615385   18750000   5769231   -1442308
[4,] -20192308 -115384615  18750000   62740385   1442308   52644231
[5,] -40384615   17307692   5769231    1442308  34615385  -18750000
[6,]  20192308 -115384615  -1442308   52644231 -18750000   62740385
> k4=CSTriangular_Element_Matrix(6,E4,Nu,t4,e4coord,1);k4
          [,1]      [,2]      [,3]      [,4]      [,5]      [,6]
[1,]  34615385  18750000 -28846154 -20192308  -5769231   1442308
[2,]  18750000  62740385 -17307692 -10096154  -1442308 -52644231
[3,] -28846154 -17307692  57692308         0 -28846154  17307692
[4,] -20192308 -10096154         0  20192308  20192308 -10096154
[5,]  -5769231  -1442308 -28846154  20192308  34615385 -18750000
[6,]   1442308 -52644231  17307692 -10096154 -18750000  62740385
```

None of the triangular elements carries a distributed load, so we moved on to form the global stiffness matrix.

Step 3: Determine the global degree of freedom of the continuum and expand the matrix equation for each triangular element.

The expanded matrices for each element are obtained via **CSTriangular_ExpandedElement_Matrix()**, which requires **TDOF** (the total degrees of freedom), **ematrix** (the element stiffness matrix) and **i,j,k** (respective nodal numbers of the element).

```
n_nodes=5;Total_dof=n_nodes*2;
K1=CSTriangular_ExpandedElement_Matrix(Total_dof,k1,1,2,4);
K2=CSTriangular_ExpandedElement_Matrix(Total_dof,k2,2,3,4);
K3=CSTriangular_ExpandedElement_Matrix(Total_dof,k3,4,3,5);
K4=CSTriangular_ExpandedElement_Matrix(Total_dof,k4,1,4,5);
```

Step 4: Establish the global stiffness matrix by adding the expanded matrices of the elements as shown next.

```
GlobalK=K1+K2+K3+K4;GlobalK
```

The output of this gives the 10×10 global stiffness matrix $[K]$ as

```
> GlobalK=K1+K2+K3+K4;GlobalK
          [,1]       [,2]      [,3]       [,4]      [,5]       [,6]      [,7]       [,8]      [,9]      [,10]
 [1,]  69230769   37500000   5769231   -1442308         0          0 -69230769  -37500000  -5769231    1442308
 [2,]  37500000  125480769   1442308   52644231         0          0 -37500000 -125480769  -1442308  -52644231
 [3,]   5769231    1442308  69230769  -37500000  -5769231   -1442308 -69230769   37500000         0          0
 [4,]  -1442308   52644231 -37500000  125480769   1442308  -52644231  37500000 -125480769         0          0
 [5,]         0          0  -5769231    1442308  69230769   37500000 -69230769  -37500000   5769231   -1442308
 [6,]         0          0  -1442308  -52644231  37500000  125480769 -37500000 -125480769   1442308   52644231
 [7,] -69230769  -37500000 -69230769   37500000 -69230769  -37500000 276923077          0 -69230769   37500000
 [8,] -37500000 -125480769  37500000 -125480769 -37500000 -125480769         0  501923077  37500000 -125480769
 [9,]  -5769231   -1442308         0          0   5769231    1442308 -69230769   37500000  69230769  -37500000
[10,]   1442308  -52644231         0          0  -1442308   52644231  37500000 -125480769 -37500000  125480769
```

A readable form of the global stiffness matrix is thus

$$[K_{global}] = 10^6 \begin{bmatrix} 69.2308 & 37.5 & 5.7692 & -1.4423 & 0. & 0. & -69.2308 & -37.5 & -5.7692 & 1.4423 \\ 37.5 & 125.4808 & 1.4423 & 52.6442 & 0. & 0. & -37.5 & -125.4808 & -1.4423 & -52.6442 \\ 5.7692 & 1.4423 & 69.2308 & -37.5 & -5.7692 & -1.4423 & -69.2308 & 37.5 & 0. & 0. \\ -1.4423 & 52.6442 & -37.5 & 125.4808 & 1.4423 & -52.6442 & 37.5 & -125.4808 & 0. & 0. \\ 0. & 0. & -5.7692 & 1.4423 & 69.2308 & 37.5 & -69.2308 & -37.5 & 5.7692 & -1.4423 \\ 0. & 0. & -1.4423 & -52.6442 & 37.5 & 125.4808 & -37.5 & -125.4808 & 1.4423 & 52.6442 \\ -69.2308 & -37.5 & -69.2308 & 37.5 & -69.2308 & -37.5 & 276.9231 & 0. & -69.2308 & 37.5 \\ -37.5 & -125.4808 & 37.5 & -125.4808 & -37.5 & -125.4808 & 0. & 501.9231 & 37.5 & -125.4808 \\ -5.7692 & -1.4423 & 0. & 0. & 5.7692 & 1.4423 & -69.2308 & 37.5 & 69.2308 & -37.2 \\ 1.4423 & -52.6442 & 0. & 0. & -1.4423 & 52.6442 & 37.5 & -125.4808 & -37.5 & 125.4808 \end{bmatrix}$$

Step 5: Identify the boundary condition(s) and obtain the reduced stiffness matrix and reduced force vector.

By using the global stiffness matrix obtained in Step 4, we can write the global matrix equation as

$$\begin{Bmatrix} F_{1X} \\ F_{1Y} \\ F_{2X} \\ F_{2Y} \\ F_{3X} \\ F_{3Y} \\ F_{4X} \\ F_{4Y} \\ F_{5X} \\ F_{5Y} \end{Bmatrix} = \left[K_{\text{global}}\right] \begin{Bmatrix} U_{1X} \\ U_{1Y} \\ U_{2X} \\ U_{2Y} \\ U_{3X} \\ U_{3Y} \\ U_{4X} \\ U_{4Y} \\ U_{5X} \\ U_{5Y} \end{Bmatrix} \tag{8.27}$$

From Figure 8.2, we notice that the plate is pinned at nodes 1, 3, and 5 (translating to zero horizontal and vertical displacements at these nodes). Therefore, the identified displacement boundary conditions will be as follows:

$$U_{1X} = U_{1Y} = U_{3X} = U_{3Y} = U_{5X} = U_{5Y} = 0 \tag{8.28}$$

From Equation 8.28, it is observed that the system has six known nodal degrees of freedom (out of 10), leaving us with four unknown nodal degrees of freedom. Realistically, for the four unknown nodal displacements, the nodal load values must be known. The set of known nodal loads, as observed from the problem statement, is

$$F_{2X} = 40,000 \sin 30°; \;\; F_{2Y} = 40,000 \cos 30°; \;\; F_{4X} = F_{4Y} = 0 \tag{8.29}$$

The rows of the forces in Equation 8.29 can be identified as 3, 4, 7, and 8. We will use these numbers along with the function **CSTriangular_ReducedStiffnessMatrix()** to obtain the reduced stiffness matrix as done in the following:

```
vec_rows_of_knownloads=c(3,4,7,8);
ReducedK=CSTriangular_ReducedStiffnessMatrix(GlobalK,
vec_rows_of_knownloads);
ReducedK
```

Which then produces

```
> ReducedK
          [,1]       [,2]       [,3]       [,4]
[1,]  69230769  -37500000 -69230769   37500000
[2,] -37500000  125480769  37500000 -125480769
[3,] -69230769   37500000 276923077          0
[4,]  37500000 -125480769         0  501923077
```

In the next block of code, we establish the reduced force vector, which is needed to supplement the reduced stiffness matrix.

```
vec_values_of_knownloads=c(40000*sin(30*pi/180),-
40000*cos(30*pi/180),0,0)
Reducedloadvector=CSTriangular_ReducedLoadVector(vec_
values_of_knownloads);
Reducedloadvector
```

With the output displayed in the following:

```
> Reducedloadvector
          [,1]
[1,]  20000.00
[2,] -34641.02
[3,]      0.00
[4,]      0.00
```

Combining the reduced stiffness matrix and the reduced vector of applied loads, one obtains the reduced matrix equation for the problem.

$$10^3\begin{Bmatrix} 20 \\ -34.641 \\ 0 \\ 0 \end{Bmatrix} = 10^6\begin{bmatrix} 69.231 & -37.5 & -69.231 & 37.5 \\ -37.5 & 125.481 & 37.5 & -125.481 \\ -69.231 & 37.5 & 276.923 & 0. \\ 37.5 & -125.481 & 0. & 501.923 \end{bmatrix}\begin{Bmatrix} U_{2X} \\ U_{2Y} \\ U_{4X} \\ U_{4Y} \end{Bmatrix} \quad (8.30)$$

Step 6: Determine the unknown nodal displacements and nodal global forces by using **CSTriangular_NodalDisplacement()** and **CSTriangular_GlobalForces()**.

```
UnknwonNodalDisp=CSTrangular_NodalDisplacement(ReducedK,
Reducedloadvector)
UnknwonNodalDisp
```

The determined displacements are obtained as

```
> UnknwonNodalDisp
               [,1]
[1,]  0.0002812474
[2,] -0.0003298531
[3,]  0.0001149795
[4,] -0.0001034760
```

The above shows that the unknown nodal displacements are as follows:

$$\begin{Bmatrix} U_{2X} \\ U_{2Y} \\ U_{4X} \\ U_{4Y} \end{Bmatrix} = \begin{Bmatrix} 0.2812\text{ mm} \\ -0.3298\text{ mm} \\ 0.11497\text{ mm} \\ -0.10347\text{ mm} \end{Bmatrix} \quad (8.31)$$

We are now able to combine the determined displacements and the boundary conditions (from Equation 8.28).

```
allglobalNodalDisp=matrix(rep(0,Total_dof),byrow=T);
allglobalNodalDisp[vec_rows_of_knownloads]=UnknwonNodalDisp
allglobalNodalDisp
```

The output of which takes the form

```
> round(allglobalNodalDisp,8)
                 [,1]
 [1,]   0.0
 [2,]   0.0
 [3,]   0.00028125
 [4,]  -0.00032985
 [5,]   0.0
 [6,]   0.0
 [7,]   0.00011498
 [8,]  -0.00010348
 [9,]   0.0
[10,]   0.0
```

The global loads, which will include the reaction forces at nodes 1, 3 and 5, are obtained by using the function **CSTriangular_GlobalForces()**. This function requires the global stiffness matrix (represented by the variable **GlobalK** determined in Step 4) and the vector of global nodal displacements (contained in the variable **allglobalNodalDisp**).

```
globalLoads=CSTriangular_GlobalForces(GlobalK,allglobal
NodalDisp)
globalLoads
```

Running the preceding block of code returns the global nodal forces as

```
> globalLoads
          [,1]
 [1,]   -1981
 [2,]   -8287
 [3,]   20000
 [4,]  -34641
 [5,]   -6178
 [6,]   25632
 [7,]       0
 [8,]       0
 [9,]  -11840
[10,]   17296
```

Hence, the reactions at the pin supports are

$$F_{1X} = -1981\,\text{N};\ F_{1Y} = -8287\,\text{N};\ F_{3X} = -6178\,\text{N};$$
$$F_{3Y} = 25632\,\text{N};\ F_{5X} = -11840\,\text{N};\ F_{5Y} = 17296\,\text{N} \tag{8.32}$$

Furthermore, we obtain the stresses at the centroid of each element using the function **CSTriangular_Stresses()**.

```
Element1_Stress=CSTriangular_Stresses(E1,Nu,t1,e1coord,1,
allglobalNodalDisp,1,2,4)
Element1_Stress/10^6
Element2_Stress=CSTriangular_Stresses(E2,Nu,t2,e2coord,1,
allglobalNodalDisp,2,3,4)
Element2_Stress/10^6
Element3_Stress=CSTriangular_Stresses(E3,Nu,t3,e3coord,1,
allglobalNodalDisp,4,3,5)
Element3_Stress/10^6
Element4_Stress=CSTriangular_Stresses(E4,Nu,t4,e4coord,1,
allglobalNodalDisp,1,4,5)
Element4_Stress/10^6
```

The output returns the stresses σ_{xx},σ_{yy}, and τ_{xy} (in MPa) for each element as demonstrated.

```
> Element1_Stress/10^6
          [,1]
[1,] 163.84065
[2,] 152.38908
[3,] -69.85409
> Element2_Stress=CSTriangular_Stresses(E2,Nu,t2,e2coord,
1,allglobalNodalDisp,2,3,4)
> Element2_Stress/10^6
          [,1]
[1,]  115.0155
[2,]  311.5812
[3,] -110.7178
> Element3_Stress=CSTriangular_Stresses(E3,Nu,t3,e3coord,1,
allglobalNodalDisp,4,3,5)
> Element3_Stress/10^6
          [,1]
[1,]  57.30979
[2,] 191.03262
[3,] -74.29442
> Element4_Stress=CSTriangular_Stresses(E4,Nu,t4,e4coord,1,
allglobalNodalDisp,1,4,5)
> Element4_Stress/10^6
          [,1]
[1,] 106.13489
[2,]  31.84047
[3,] -33.43071
```

The solutions obtained for the nodal displacements and stresses agree with the values provided in Reference [2].

8.5.2 Example 8.2

The thin L-shaped plate shown in Figure 8.3 has $E = 210\text{GPa}$ and $v = 0.3$. The thickness of the plate is 25 mm[4]. By using the 12-element-triangulated domain shown in Figure 8.3b:

- Determine the nodal displacements,
- Determine the nodal reaction forces.
- Determine the stresses at the centroid of elements 1, 2, 3 and 10.

8.5.2.1 Solution

The connectivity for the elements depicted in Figure 8.3b is presented in Table 8.6. The complete solution to the problem is then obtained by using the implemented functions.

With the connectivity table established, we move on to the rest of the computations.

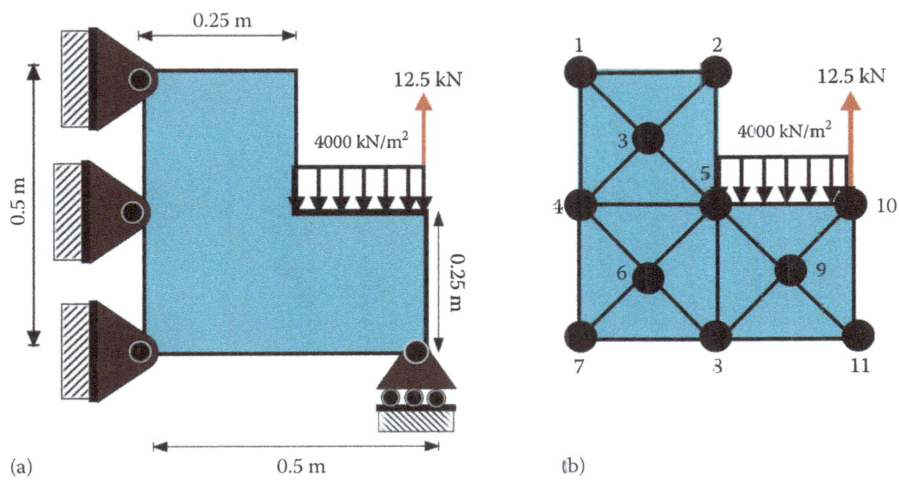

FIGURE 8.3
(a) An L-shaped thin plate with a surface traction and (b) triangulation of the domain.

TABLE 8.6

Connectivity Table from the Discretized System in Figure 8.3b

Elements	Nodes (i, j, k)	Coordinates $(x_i, y_i, x_j, y_j, x_k, y_k)$
1	1,3,2	(0,0.5,0.125,0.375,0.25,0.5,1)
2	1,4,3	(0,0.5,0,0.25,0.125,0.375,1)
3	3,5,2	(0.125,0.375,0.25,0.25,0.25,0.5,1)
4	3,4,5	(0.125,0.375,0,0.25,0.25,0.25,1)
5	4,6,5	(0,0.25,0.125,0.125,0.25,0.25,1)
6	4,7,6	(0,0.25,0,0,0.125,0.125,1)
7	5,6,8	(0.25,0.25,0.125,0.125,0.25,0,1)
8	6,7,8	(0.125,0.125,0,0,0.25,0,1)
9	5,8,9	(0.25,0.25,0.25,0,0.375,0.125,1)
10	5,9,10	(0.25,0.25,0.375,0.125,0.5,0.25,1)
11	8,11,9	(0.25,0,0.5,0,0.375,0.125,1)
12	9,11,10	(0.375,0.125,0.5,0,0.5,0.25,1)

Step 1: We specify the nodal coordinate values (based on the connectivity table) and the properties of each element.

```
e1coord=c(0,0.5,0.125,0.375,0.25,0.5,1)
e2coord=c(0,0.5,0,0.25,0.125,0.375,1)
e3coord=c(0.125,0.375,0.25,0.25,0.25,0.5,1)
e4coord=c(0.125,0.375,0,0.25,0.25,0.25,1)
e5coord=c(0,0.25,0.125,0.125,0.25,0.25,1)
e6coord=c(0,0.25,0,0,0.125,0.125,1)
e7coord=c(0.25,0.25,0.125,0.125,0.25,0,1)
e8coord=c(0.125,0.125,0,0,0.25,0,1)
e9coord=c(0.25,0.25,0.25,0,0.375,0.125,1)
e10coord=c(0.25,0.25,0.375,0.125,0.5,0.25,1)
e11coord=c(0.25,0,0.5,0,0.375,0.125,1)
e12coord=c(0.375,0.125,0.5,0,0.5,0.25,1)
YoungMod=E=210e9;
thickness=0.025;Nu=0.3;
```

All parameters used are in SI units.

Step 2: Write the stiffness matrix for each element using the coordinates and properties specified in Step 1 as arguments here.

```
k1=CSTriangular_Element_Matrix(6,E,Nu,thickness,
e1coord,1);k1
k2=CSTriangular_Element_Matrix(6,E,Nu,thickness,
e2coord,1);k2
k3=CSTriangular_Element_Matrix(6,E,Nu,thickness,
e3coord,1);k3
k4=CSTriangular_Element_Matrix(6,E,Nu,thickness,
e4coord,1);k4
k5=CSTriangular_Element_Matrix(6,E,Nu,thickness,
e5coord,1);k5
k6=CSTriangular_Element_Matrix(6,E,Nu,thickness,
e6coord,1);k6
k7=CSTriangular_Element_Matrix(6,E,Nu,thickness,
e7coord,1);k7
k8=CSTriangular_Element_Matrix(6,E,Nu,thickness,
e8coord,1);k8
k9=CSTriangular_Element_Matrix(6,E,Nu,thickness,
e9coord,1);k9
k10=CSTriangular_Element_Matrix(6,E,Nu,thickness,
e10coord,1);k10
k11=CSTriangular_Element_Matrix(6,E,Nu,thickness,
e11coord,1);k11
k12=CSTriangular_Element_Matrix(6,E,Nu,thickness,
e12coord,1);k12
```

The computed stiffness matrices are as follows:

```
> k1=CSTriangular_Element_Matrix(6,E,Nu,thickness,e1coord,1);k1
            [,1]        [,2]        [,3]        [,4]        [,5]        [,6]
[1,]  1947115385  -937500000 -1009615385   865384615  -937500000    72115385
[2,]  -937500000  1947115385  1009615385 -2884615385   -72115385   937500000
[3,] -1009615385  1009615385  2019230769           0 -1009615385 -1009615385
[4,]   865384615 -2884615385           0  5769230769  -865384615 -2884615385
[5,]  -937500000   -72115385 -1009615385  -865384615  1947115385   937500000
[6,]    72115385   937500000 -1009615385 -2884615385   937500000  1947115385
> k2=CSTriangular_Element_Matrix(6,E,Nu,thickness,e2coord,1);k2
            [,1]        [,2]        [,3]        [,4]        [,5]        [,6]
[1,]  1947115385  -937500000   937500000   -72115385 -2884615385  1009615385
[2,]  -937500000  1947115385    72115385  -937500000   865384615 -1009615385
[3,]   937500000    72115385  1947115385   937500000 -2884615385 -1009615385
[4,]   -72115385  -937500000   937500000  1947115385  -865384615 -1009615385
[5,] -2884615385   865384615 -2884615385  -865384615  5769230769           0
[6,]  1009615385 -1009615385 -1009615385 -1009615385           0  2019230769
> k3=CSTriangular_Element_Matrix(6,E,Nu,thickness,e3coord,1);k3
            [,1]        [,2]        [,3]        [,4]        [,5]        [,6]
[1,]  5769230769           0 -2884615385   865384615 -2884615385  -865384615
[2,]           0  2019230769  1009615385 -1009615385 -1009615385 -1009615385
[3,] -2884615385  1009615385  1947115385  -937500000   937500000   -72115385
[4,]   865384615 -1009615385  -937500000  1947115385    72115385  -937500000
[5,] -2884615385 -1009615385   937500000    72115385  1947115385   937500000
[6,]  -865384615 -1009615385   -72115385  -937500000   937500000  1947115385
> k4=CSTriangular_Element_Matrix(6,E,Nu,thickness,e4coord,1);k4
            [,1]        [,2]        [,3]        [,4]        [,5]        [,6]
[1,]  2019230769           0 -1009615385 -1009615385 -1009615385  1009615385
[2,]           0  5769230769  -865384615 -2884615385   865384615 -2884615385
[3,] -1009615385  -865384615  1947115385   937500000  -937500000   -72115385
```

```
[4,] -1009615385 -2884615385   937500000  1947115385    72115385   937500000
[5,] -1009615385   865384615  -937500000    72115385  1947115385  -937500000
[6,]  1009615385 -2884615385   -72115385   937500000  -937500000  1947115385
> k5=CSTriangular_Element_Matrix(6,E,Nu,thickness,e5coord,1);k5
            [,1]        [,2]        [,3]        [,4]        [,5]        [,6]
[1,]  1947115385  -937500000 -1009615385   865384615  -937500000    72115385
[2,]  -937500000  1947115385  1009615385 -2884615385   -72115385   937500000
[3,] -1009615385  1009615385  2019230769           0 -1009615385 -1009615385
[4,]   865384615 -2884615385           0  5769230769  -865384615 -2884615385
[5,]  -937500000   -72115385 -1009615385  -865384615  1947115385   937500000
[6,]    72115385   937500000 -1009615385 -2884615385   937500000  1947115385
> k6=CSTriangular_Element_Matrix(6,E,Nu,thickness,e6coord,1);k6
            [,1]        [,2]        [,3]        [,4]        [,5]        [,6]
[1,]  1947115385  -937500000   937500000   -72115385 -2884615385  1009615385
[2,]  -937500000  1947115385    72115385  -937500000   865384615 -1009615385
[3,]   937500000    72115385  1947115385   937500000 -2884615385 -1009615385
[4,]   -72115385  -937500000   937500000  1947115385  -865384615 -1009615385
[5,] -2884615385   865384615 -2884615385  -865384615  5769230769           0
[6,]  1009615385 -1009615385 -1009615385 -1009615385           0  2019230769
> k7=CSTriangular_Element_Matrix(6,E,Nu,thickness,e7coord,1);k7
            [,1]        [,2]        [,3]        [,4]        [,5]        [,6]
[1,]  1947115385   937500000 -2884615385 -1009615385   937500000    72115385
[2,]   937500000  1947115385  -865384615 -1009615385   -72115385  -937500000
[3,] -2884615385  -865384615  5769230769           0 -2884615385   865384615
[4,] -1009615385 -1009615385           0  2019230769  1009615385 -1009615385
[5,]   937500000   -72115385 -2884615385  1009615385  1947115385  -937500000
[6,]    72115385  -937500000   865384615 -1009615385  -937500000  1947115385
```

```
> k8=CSTriangular_Element_Matrix(6,E,Nu,thickness,e8coord,1);k8
            [,1]        [,2]        [,3]        [,4]        [,5]        [,6]
[1,]  2019230769           0 -1009615385 -1009615385 -1009615385  1009615385
[2,]           0  5769230769  -865384615 -2884615385   865384615 -2884615385
[3,] -1009615385  -865384615  1947115385   937500000  -937500000   -72115385
[4,] -1009615385 -2884615385   937500000  1947115385    72115385   937500000
[5,] -1009615385   865384615  -937500000    72115385  1947115385  -937500000
[6,]  1009615385 -2884615385   -72115385   937500000  -937500000  1947115385
> k9=CSTriangular_Element_Matrix(6,E,Nu,thickness,e9coord,1);k9
            [,1]        [,2]        [,3]        [,4]        [,5]        [,6]
[1,]  1947115385  -937500000   937500000   -72115385 -2884615385  1009615385
[2,]  -937500000  1947115385    72115385  -937500000   865384615 -1009615385
[3,]   937500000    72115385  1947115385   937500000 -2884615385 -1009615385
[4,]   -72115385  -937500000   937500000  1947115385  -865384615 -1009615385
[5,] -2884615385   865384615 -2884615385  -865384615  5769230769           0
[6,]  1009615385 -1009615385 -1009615385 -1009615385           0  2019230769
> k10=CSTriangular_Element_Matrix(6,E,Nu,thickness,e10coord,1);k10
            [,1]        [,2]        [,3]        [,4]        [,5]        [,6]
[1,]  1947115385  -937500000 -1009615385   865384615  -937500000    72115385
[2,]  -937500000  1947115385  1009615385 -2884615385   -72115385   937500000
[3,] -1009615385  1009615385  2019230769           0 -1009615385 -1009615385
[4,]   865384615 -2884615385           0  5769230769  -865384615 -2884615385
[5,]  -937500000   -72115385 -1009615385  -865384615  1947115385   937500000
[6,]    72115385   937500000 -1009615385 -2884615385   937500000  1947115385
```

```
> k11=CSTriangular_Element_Matrix(6,E,Nu,thickness,e11coord,1);k11
            [,1]        [,2]        [,3]        [,4]        [,5]        [,6]
[1,]  1947115385   937500000  -937500000   -72115385 -1009615385  -865384615
[2,]   937500000  1947115385    72115385   937500000 -1009615385 -2884615385
[3,]  -937500000    72115385  1947115385  -937500000 -1009615385   865384615
[4,]   -72115385   937500000  -937500000  1947115385  1009615385 -2884615385
[5,] -1009615385 -1009615385 -1009615385  1009615385  2019230769           0
[6,]  -865384615 -2884615385   865384615 -2884615385           0  5769230769
> k12=CSTriangular_Element_Matrix(6,E,Nu,thickness,e12coord,1);k12
            [,1]        [,2]        [,3]        [,4]        [,5]        [,6]
[1,]  5769230769           0 -2884615385   865384615 -2884615385  -865384615
[2,]           0  2019230769  1009615385 -1009615385 -1009615385 -1009615385
[3,] -2884615385  1009615385  1947115385  -937500000   937500000   -72115385
[4,]   865384615 -1009615385  -937500000  1947115385    72115385  -937500000
[5,] -2884615385 -1009615385   937500000    72115385  1947115385   937500000
[6,]  -865384615 -1009615385   -72115385  -937500000   937500000  1947115385
```

Element 10 carries a traction force of magnitude 4000 kN/m^2. We determine the equivalent nodal loads for this element by using the function **CSTriangular_SF()**. It is good to highlight here that this function accepts six parameters:

- The first is the degree of freedom of the element (predefined as 6).
- The second and third parameters represent the magnitude of surface tensile and surface shearing force, respectively. For this problem, **SFtensile** is replaced by zero as no surface tensile is present and **SFshear** is -4000×10^3 due to the presence of the uniform transverse force on this element.
  - The fourth and fifth parameters are length of the edge on which the traction load acts and thickness of the element, respectively.
  - The sixth parameter, **case,** is used to indicate the edge on which the load is applied. For instance, if the traction load is on the edge $i-j$ (case = 1), or on edge $j-k$ (case = 2) or on edge $i-k$ (case = 3).

By using these information, we now compute the equivalent nodal loads from the surface traction on element 10 as

```
k10SurfaceForce=CSTriangular_SF(6,0,-4000e3,0.25,
thickness,3);
k10SurfaceForce
```

The output of this reads

```
> k10SurfaceForce
       [,1]
[1,]      0
[2,] -12500
[3,]      0
[4,]      0
[5,]      0
[6,] -12500
```

Step 3: Determine the global degree of freedom of the continuum and expand the matrix equation for each triangular element.

The expanded matrices for each element is obtained via **CSTriangular_ExpandedElement_Matrix()**, which requires TDOF (the total degrees of freedom), **ematrix** (the element stiffness matrix) and **i,j,k** (the respective nodal numbers of the element). We also need to expand the equivalent load vector via **CSTriangular_ExpandedSF()**.

```
n_nodes=11;Total_dof=n_nodes*2;
K1=CSTriangular_ExpandedElement_Matrix(Total_dof,k1,1,3,2);
K2=CSTriangular_ExpandedElement_Matrix(Total_dof,k2,1,4,3);
K3=CSTriangular_ExpandedElement_Matrix(Total_dof,k3,3,5,2);
K4=CSTriangular_ExpandedElement_Matrix(Total_dof,k4,3,4,5);
K5=CSTriangular_ExpandedElement_Matrix(Total_dof,k5,4,6,5);
K6=CSTriangular_ExpandedElement_Matrix(Total_dof,k6,4,7,6);
K7=CSTriangular_ExpandedElement_Matrix(Total_dof,k7,5,6,8);
K8=CSTriangular_ExpandedElement_Matrix(Total_dof,k8,6,7,8);
K9=CSTriangular_ExpandedElement_Matrix(Total_dof,k9,
5,8,9);
K10=CSTriangular_ExpandedElement_Matrix(Total_dof,k10,
5,9,10);
K11=CSTriangular_ExpandedElement_Matrix(Total_dof,k11,
8,11,9);
K12=CSTriangular_ExpandedElement_Matrix(Total_dof,k12,
9,11,10);
K10SF=CSTriangular_ExpandedSF(Total_dof,k10SurfaceForce,
5,9,10);K10SF
```

Step 4: The global stiffness matrix is determined by adding the expanded matrices of the elements as shown next.

```
GlobalK=K1+K2+K3+K4+K5+K6+K7+K8+K9+K10+K11+K12;GlobalK
```

The output of this gives the 22×22 global stiffness matrix $[K]$, which is too big to be displayed here.

Step 5: Identify the boundary condition(s) and obtain the reduced stiffness matrix and the reduced force vector.

For this system, the global matrix equation will be of the form

$$[F_{1X},F_{1Y},\ldots,F_{11X},F_{11Y}]^T=[K_{\text{global}}][U_{1X},U_{1Y},\ldots,U_{11X},U_{11Y}]^T \tag{8.33}$$

To establish the boundary condition, we notice that the plate is pinned at nodes 1, 4, and 7 (translating to zero horizontal and vertical displacements at these nodes). It has a horizontal roller support at node 11 (which restrains only the vertical movement). Based on this information, the identified displacement boundary conditions will be as follows:

$$U_{1X}=U_{1Y}=U_{4X}=U_{4Y}=U_{7X}=U_{7Y}=U_{11Y}=0; \tag{8.34}$$

Equation 8.34 reveals only 7 known nodal degrees of freedom, leaving us with 15 unknown nodal degrees of freedom. We have the following set of known nodal loads:

$$F_{2X} = F_{2Y} = F_{3X} = F_{3Y} = F_{6X} = F_{6Y} = F_{8X} = F_{8Y} = F_{9X} = F_{9Y} = F_{11X} = 0 \quad (8.35)$$

To estimate the known nodal loads at nodes 5 and 10, we must add the concentrated load at these two nodes to the equivalent nodal loads from the surface traction. Hence,

$$F_{5X} = 0;\ F_{5Y} = -12,500;\ F_{10X} = 0;\ F_{10Y} = -12,500+12,500=0 \quad (8.36)$$

where it is shown in Equation 8.36 that the load at node 10 equals zero (because of the equal magnitude but different directions of the equivalent nodal load and the concentrated load).

By using the function **CSTriangular_ReducedStiffnessMatrix()** we will extract the reduced stiffness matrix.

```
vec_rows_of_knownloads=c(3,4,5,6,9,10,11,12,15,16,17,18,
19,20,21);
ReducedK=CSTriangular_ReducedStiffnessMatrix(GlobalK,
vec_rows_of_knownloads);
ReducedK
```

In the previous code snippet, the **vec_rows_of_knownloads** contains the rows of the known nodal loads. The reduced stiffness matrix is displayed as follows:

```
> ReducedK/10^6
          [,1]      [,2]    [,3]    [,4]      [,5]      [,6]    [,7]    [,8]    [,9]     [,10]   [,11]   [,12]     [,13]     [,14]     [,15]
 [1,]  3894.231  1875.000 -3894.2 -1875.0   937.500    72.115     0.0     0.0     0.0     0.000     0.0     0.0     0.000     0.000     0.000
 [2,]  1875.000  3894.231 -1875.0 -3894.2   -72.115  -937.500     0.0     0.0     0.0     0.000     0.0     0.0     0.000     0.000     0.000
 [3,] -3894.231 -1875.000 15576.9     0.0 -3894.231  1875.000     0.0     0.0     0.0     0.000     0.0     0.0     0.000     0.000     0.000
 [4,] -1875.000 -3894.231     0.0 15576.9  1875.000 -3894.231     0.0     0.0     0.0     0.000     0.0     0.0     0.000     0.000     0.000
 [5,]   937.500   -72.115 -3894.2  1875.0 11682.692 -1875.000 -3894.2 -1875.0  1875.0     0.000 -3894.2  1875.0  -937.500    72.115     0.000
 [6,]    72.115  -937.500  1875.0 -3894.2 -1875.000 11682.692 -1875.0 -3894.2     0.0 -1875.000  1875.0 -3894.2   -72.115   937.500     0.000
 [7,]     0.000     0.000     0.0     0.0 -3894.231 -1875.000 15576.9     0.0 -3894.2  1875.000     0.0     0.0     0.000     0.000     0.000
 [8,]     0.000     0.000     0.0     0.0 -1875.000 -3894.231     0.0 15576.9  1875.0 -3894.231     0.0     0.0     0.000     0.000     0.000
 [9,]     0.000     0.000     0.0     0.0  1875.000     0.000 -3894.2  1875.0  7788.5     0.000 -3894.2 -1875.0     0.000     0.000  -937.500
[10,]     0.000     0.000     0.0     0.0     0.000 -1875.000  1875.0 -3894.2     0.0  7788.462 -1875.0 -3894.2     0.000     0.000    72.115
[11,]     0.000     0.000     0.0     0.0 -3894.231  1875.000     0.0     0.0 -3894.2 -1875.000 15576.9     0.0 -3894.231 -1875.000 -3894.231
[12,]     0.000     0.000     0.0     0.0  1875.000 -3894.231     0.0     0.0 -1875.0 -3894.231     0.0 15576.9 -1875.000 -3894.231  1875.000
[13,]     0.000     0.000     0.0     0.0  -937.500   -72.115     0.0     0.0     0.0     0.000 -3894.2 -1875.0  3894.231  1875.000   937.500
[14,]     0.000     0.000     0.0     0.0    72.115   937.500     0.0     0.0     0.0     0.000 -1875.0 -3894.2  1875.000  3894.231   -72.115
[15,]     0.000     0.000     0.0     0.0     0.000     0.000     0.0     0.0  -937.5    72.115 -3894.2  1875.0   937.500   -72.115  3894.231
```

The 15×15reduced stiffness matrix (K_{red}) is reproduced in the following:

$$10^6 \begin{bmatrix} 3894.231 & 1875. & -3894.2 & -1875. & 937.5 & 72.115 & 0. & 0. & 0. & 0. & 0. & 0. & 0. & 0. & 0. \\ 1875. & 3894.231 & -1875. & -3894.2 & -72.115 & -937.5 & 0. & 0. & 0. & 0. & 0. & 0. & 0. & 0. & 0. \\ -3894.231 & -1875. & 15576.9 & 0. & -3894.231 & 1875. & 0. & 0. & 0. & 0. & 0. & 0. & 0. & 0. & 0. \\ -1875. & -3894.231 & 0. & 15576.9 & 1875. & -3894.231 & 0. & 0. & 0. & 0. & 0. & 0. & 0. & 0. & 0. \\ 937.5 & -72.115 & -3894.2 & 1875. & 11682.692 & -1875. & -3894.2 & -1875. & 1875. & 0. & -3894.2 & 1875. & -937.5 & 72.115 & 0. \\ 72.115 & -937.5 & 1875. & -3894.2 & -1875. & 11682.692 & -1875. & -3894.2 & 0. & -1875. & 1875. & -3894.2 & -72.115. & 937.5 & 0. \\ 0. & 0. & 0. & 0. & -3894.231 & -1875. & 15576.9 & 0. & -3894.2 & 1875. & 0. & 0. & 0. & 0. & 0. \\ 0. & 0. & 0. & 0. & -1875. & -3894.231 & 0. & 15576.9 & 1875. & -3894.231 & 0. & 0. & 0. & 0. & 0. \\ 0. & 0. & 0. & 0. & 1875. & 0. & -3894.2 & 1875. & 7788.5 & 0. & -3894.2 & -1875. & 0. & 0. & -937.5 \\ 0. & 0. & 0. & 0. & 0. & -1875. & 1875. & -3894.2 & 0. & 7788.462 & -1875. & -3894.2 & 0. & 0. & 72.115 \\ 0. & 0. & 0. & 0. & -3894.231 & 1875. & 0. & 0. & -3894.2 & -1875. & 15576.9 & 0. & -3894.231 & -1875. & -3894.231 \\ 0. & 0. & 0. & 0. & 1875. & -3894.231 & 0. & 0. & -1875. & -3894.231 & 0. & 15576.9 & -1875. & -3894.231 & 187.5 \\ 0. & 0. & 0. & 0. & -937.5 & -72.115 & 0. & 0. & 0. & 0. & -3894.2 & -1875. & 3894.231 & 1875. & 937.5 \\ 0. & 0. & 0. & 0. & 72.115 & 937.5 & 0. & 0. & 0. & 0. & -1875. & -3894.2 & 1875. & 3894.231 & -72.115 \\ 0. & 0. & 0. & 0. & 0. & 0. & 0. & 0. & -937.5 & 72.115 & -3894.2 & 1875. & 937.5 & -72.115 & 3894.231 \end{bmatrix}$$

In the next code, the set of known loads identified in Equations 8.36 is transformed into a column matrix via the function **CSTriangular_ReducedLoadVector()**.

```
vec_values_of_knownloads=c(rep(0,13),12.5e3,0)+K10SF
[vec_rows_of_knownloads];
Reducedloadvector=CSTriangular_ReducedLoadVector
(vec_values_of_knownloads);
Reducedloadvector
```

The first line of the preceding code snippet has two parts. The first part is the vector of the known concentrated loads, whereas the second part represents the equivalent nodal loads. Besides, we used the in-built function **rep()** (explained in Chapter 2) to repeat zero 13 times for the first part and we extracted the required element of the expanded equivalent load matrix in the second part (i.e. **K10SF[vec_rows_of_knownloads]**). The output of the preceding code reads:

```
> Reducedloadvector
        [,1]
 [1,]      0
 [2,]      0
 [3,]      0
 [4,]      0
 [5,]      0
 [6,] -12500
 [7,]      0
 [8,]      0
 [9,]      0
[10,]      0
[11,]      0
[12,]      0
[13,]      0
[14,]      0
[15,]      0
```

Combining the reduced stiffness matrix and the reduced vector of applied loads, one obtains the reduced matrix equation.

$$10^3\begin{Bmatrix} 0 \\ 0 \\ 0 \\ 0 \\ 0 \\ -12.5 \\ 0 \\ 0 \\ 0 \\ 0 \\ 0 \\ 0 \\ 0 \\ 0 \\ 0 \end{Bmatrix} = \left[K_{red}\right]\begin{Bmatrix} U_{2X} \\ U_{2Y} \\ U_{3X} \\ U_{3Y} \\ U_{5X} \\ U_{5Y} \\ U_{6X} \\ U_{6Y} \\ U_{8X} \\ U_{8Y} \\ U_{9X} \\ U_{9Y} \\ U_{10X} \\ U_{10Y} \\ U_{11X} \end{Bmatrix} \tag{8.37}$$

Step 6: Determine the unknown nodal displacements and nodal global forces by using the **CSTriangular_NodalDisplacement()**.

```
UnknwonNodalDisp=CSTriangular_NodalDisplacement(ReducedK,
Reducedloadvector)
UnknwonNodalDisp
```

From which one obtains the unknown nodal displacements as

```
> UnknwonNodalDisp
                  [,1]
 [1,]  0.000000919847
 [2,] -0.000002304396
 [3,]  0.000000211765
 [4,] -0.000001195602
 [5,] -0.000000481172
 [6,] -0.000003152579
 [7,] -0.000000214159
```

```
 [8,] -0.000001415990
 [9,]  0.000000058328
[10,] -0.000002251624
[11,] -0.000000050671
[12,] -0.000001628664
[13,] -0.000001099235
[14,] -0.000000336534
[15,]  0.000001047638
```

The above shows that, to the sixth decimal place, the nodal displacements are

$$\begin{Bmatrix} U_{2X} \\ U_{2Y} \\ U_{3X} \\ U_{3Y} \\ U_{5X} \\ U_{5Y} \\ U_{6X} \\ U_{6Y} \\ U_{8X} \\ U_{8Y} \\ U_{9X} \\ U_{9Y} \\ U_{10X} \\ U_{10Y} \\ U_{11X} \end{Bmatrix} = \begin{Bmatrix} 0.000920 \\ -0.002304 \\ 0.000212 \\ -0.001196 \\ -0.000481 \\ -0.003153 \\ -0.000214 \\ -0.001416 \\ 0.000058 \\ -0.002252 \\ -0.000051 \\ -0.001629 \\ -0.001099 \\ -0.000337 \\ 0.001048 \end{Bmatrix} \text{mm} \tag{8.38}$$

For further usage, we combine the determined displacements and the boundary condition in the next lines of code.

```
allglobalNodalDisp=matrix(rep(0,Total_dof),byrow=T);
allglobalNodalDisp[vec_rows_of_knownloads]= UnknwonNodal
Disp
allglobalNodalDisp
```

which produces

```
> allglobalNodalDisp
                  [,1]
 [1,]  0.0
 [2,]  0.0
 [3,]  0.000000919847
 [4,] -0.000002304396
 [5,]  0.000000211765
 [6,] -0.000001195602
 [7,]  0.0
 [8,]  0.0
 [9,] -0.000000481172
[10,] -0.000003152579
[11,] -0.000000214159
[12,] -0.000001415990
[13,]  0.0
[14,]  0.0
[15,]  0.000000058328
[16,] -0.000002251624
[17,] -0.000000050671
[18,] -0.000001628664
[19,] -0.000001099235
[20,] -0.000000336534
[21,]  0.000001047638
[22,]  0.0
```

The global forces, which will include the reaction forces at nodes 1, 4, 7, and 11, are obtained by using the function **CSTriangular_GlobalForces()**, as done next.

```
ConcglobalLoads=CSTriangular_GlobalForces(GlobalK,all
globalNodalDisp)
EquivalentSF=K10SF
globalLoads=ConglobalLoads-EquivalentSF;
round(globalLoads,6)
```

Running the preceding code returns the global reaction forces (after subtracting the effect of the equivalent load from the surface traction) as

```
> round(globalLoads,6)
          [,1]
 [1,]  -4094.96
 [2,]   2826.31
 [3,]      0.00
 [4,]      0.00
 [5,]      0.00
 [6,]      0.00
 [7,]    498.29
 [8,]   3460.45
 [9,]      0.00
[10,] -12500.00
[11,]      0.00
[12,]      0.00
[13,]   3596.66
[14,]   3809.05
[15,]      0.00
[16,]      0.00
[17,]      0.00
[18,]      0.00
[19,]      0.00
[20,]      0.00
[21,]      0.00
[22,]   2404.19
```

The stresses at the centroid of elements 1, 2, 3, and 10 are obtained by using the function **CSTriangular_Stresses()** as demonstrated in the next block of code.

```
Element1_Stress=CSTriangular_Stresses(E,Nu,thickness,e1co
ord,1,allglobalNodalDisp,1,3,3)
Element1_Stress/10^6
Element2_Stress=CSTriangular_Stresses(E,Nu,thickness,e2co
ord,1,allglobalNodalDisp,1,4,3)
Element2_Stress/10^6
Element3_Stress=CSTriangular_Stresses(E,Nu,thickness,e3co
ord,1,allglobalNodalDisp,3,5,2)
Element3_Stress/10^6
Element10_Stress=CSTriangular_Stresses(E,Nu,thickness,e4
coord,1,allglobalNodalDisp,5,9,10)
Element10_Stress/10^6
```

The output returns:

```
> Element1_Stress/10^6
          [,1]
[1,]  0.52657
[2,]  1.16228
[3,] -0.45469
> Element2_Stress=CSTriangular_Stresses(E,Nu,thickness,e2
coord,1,allglobalNodalDisp,1,4,3)
> Element2_Stress/10^6
          [,1]
[1,]  0.39095
[2,]  0.11729
[3,] -0.77254
> Element3_Stress=CSTriangular_Stresses(E,Nu,thickness,e3
coord,1,allglobalNodalDisp,3,5,2)
> Element3_Stress/10^6
          [,1]
[1,]  0.24886
[2,]  0.78713
[3,] -0.53784
> Element10_Stress=CSTriangular_Stresses(E,Nu,thickness,
e4coord,1,allglobalNodalDisp,5,9,10)
> Element10_Stress/10^6
          [,1]
[1,] -2.16974
[2,] -4.29649
[3,]  0.47805
```

Where the computed stresses (σ_{xx}, σ_{yy} and τ_{xy}) obtained earlier (in MPa) are those experienced at the centroid of elements 1, 2, 3, and 10.

8.6 Exercises

1. A thin triangular plate with thickness 10 mm is constrained along the lower edge and loaded as shown in Figure 8.4. Determine the deflections at nodes 3 and 4. Take $E = 210$ GPa, $v = 0.3$.
2. For the thin plate shown in Figure 8.5, given that $E = 210$ GPa, $v = 0.3$, and $t = 0.25$ mm: (a) Establish the global stiffness matrix for the structure; (b) determine the unknown nodal displacements; and (c) compute the stress developed at the centroid of each triangular elements.

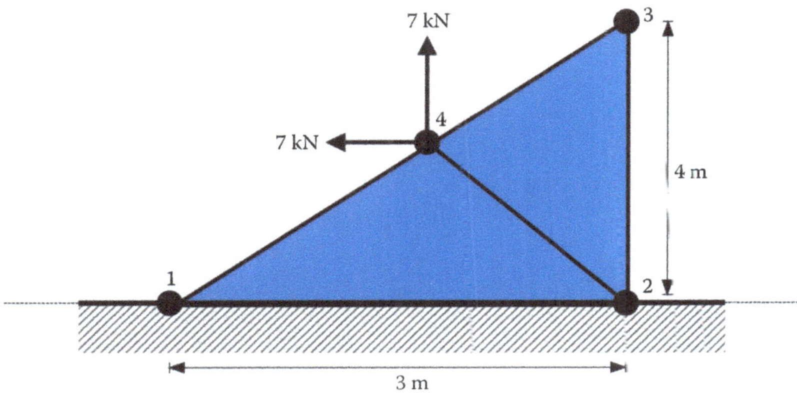

FIGURE 8.4
Exercise 8Q1.

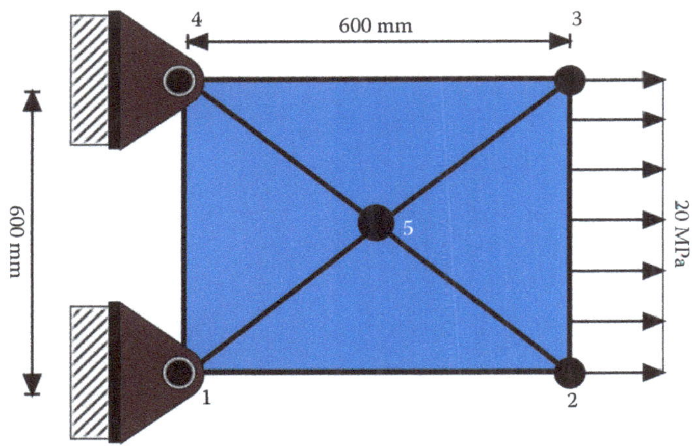

FIGURE 8.5
Exercise 8Q2.

References

1. T. H. G. Megson, *Aircraft Structures for Engineering Students*, Edward Arnold, London, UK, 1999.
2. D. L. Logan, *First Course in the Finite Element Method*, Thomson, Toronto, ON, 2007.
3. T. R. Chandrupatla, A. D. Belegundu, *Introduction to Finite Elements in Engineering*, Prentice Hall, Upper Saddle River, NJ, 1997.
4. P. I. Kattan, *MATLAB Guide to Finite Elements: An Interactive Approach*, Springer, Berlin, Germany, 2010.

9

Two-Dimensional Continuum Elements for Axisymmetric Body Problems

9.1 Introduction

Planar approximations of axisymmetric elastic bodies having geometry, constraint, and material properties with rotational symmetries about an axis of revolution appear in the solutions of many problems of practical importance. Hence, given their applications, few axisymmetric finite element models have been developed in the literature. However, our focus in this chapter is on the axisymmetric triangular element with three nodes.

The general applications of axisymmetric elements can be found in the analyses of hyperboloid nuclear reactor, circular tunnels, flywheels, pressure vessels, engine valve stem, water tank, piston head, turbine disk, rocket motor casing, hyperbolic cooling towers, synovial joints and so on [1–5]. Similar to the other chapters, in the succeeding sections, a brief account of the equations describing the axisymmetric triangular element will be presented along with the R functions developed for using this element in the analyses of axisymmetric elastic bodies in Sections 9.2 and 9.3. Example problems are given in Section 9.4, whereas exercises are given in Section 9.5. The chapter ends with a reference section.

9.2 Finite Element Equations for the Triangular Axisymmetric Element

9.2.1 Displacement and Shape Functions

Figure 9.1a depicts the three-dimensional model of a toroid, whereas the schematic in Figure 9.1b shows the outline of the same toroid attached with a reference cylindrical coordinates r,θ,z, and a triangular element that can be

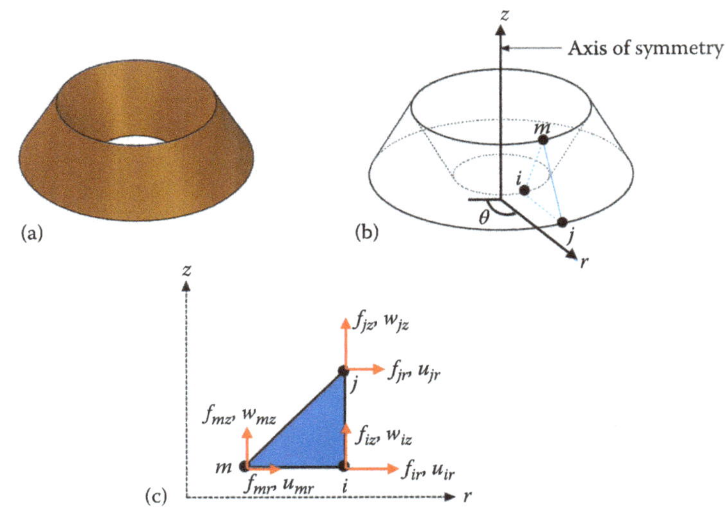

FIGURE 9.1
(a) A 3D model of a toroid, (b) schematics of the toroid revealing the triangular cross-section, and (c) a triangular element showing the nodal degrees of freedom.

revolved to form the 3D solid.[1] In Figure 9.1c, the isolated triangular element is shown, and at each of its node, we indicate

- A 2D coordinate system, which forms part of the cylindrical coordinate $r\theta z$
- The nodal displacements $\left(u_{ir}, w_{iz}, u_{jr}, w_{jz}, u_{mr}, w_{mz}\right)$
- The nodal forces $\left(f_{ir}, f_{iz}, f_{jr}, f_{jz}, f_{mr}, f_{mz}\right)$

where the subscripts i, j and m represent the nodal labels of the element.

Following the same line of thought presented in Chapter 8, we assume the triangular element experiences only planar deformations, and we thus represent variation of the displacements at its interior points with the functions:

$$u = A_1 + A_2 r + A_3 z \tag{9.1}$$

$$w = A_4 + A_5 r + A_6 z \tag{9.2}$$

where $A_1 - A_6$ are unknown coefficients, whereas u and w represent the radial and axial displacements, respectively. Next, we solve for the unknown coefficients by imposing boundary conditions on Equations 9.1 and 9.2 as follows:

---

[1] Although we have used a solid formed from a triangular revolving plane here, any revolving 2D domain can be discretized via triangulation into several smaller areas of triangular elements, each having three straight edges.

- At node i, $r = r_i$ & $z = z_i$: $u = u_i$
- At node i, $r = r_i$ & $z = z_i$: $w = w_i$
- At node j, $r = r_j$ & $z = z_j$: $u = u_j$
- At node j, $r = r_j$ & $z = z_j$: $w = w_j$
- At node m, $r = r_m$ & $z = z_m$: $u = u_m$
- At node m, $r = r_m$ & $z = z_m$: $w = w_m$

Substituting these boundary conditions into Equations 9.1 and 9.2, one obtains

$$\underbrace{\begin{Bmatrix} u_i \\ w_i \\ u_j \\ w_j \\ u_m \\ w_m \end{Bmatrix}}_{\{u_e\}} = \underbrace{\begin{bmatrix} 1 & r_i & z_i & 0 & 0 & 0 \\ 0 & 0 & 0 & 1 & r_i & z_i \\ 1 & r_j & z_j & 0 & 0 & 0 \\ 0 & 0 & 0 & 1 & r_j & z_j \\ 1 & r_m & z_m & 0 & 0 & 0 \\ 0 & 0 & 0 & 1 & r_m & z_m \end{bmatrix}}_{[H]} \underbrace{\begin{Bmatrix} A_1 \\ A_2 \\ A_3 \\ A_4 \\ A_5 \\ A_6 \end{Bmatrix}}_{\{A\}} \tag{9.3}$$

From Equation 9.3, the unknown coefficients are determined in terms of the coordinate matrix and the vector of nodal displacements as

$$\{A\} = [H]^{-1}\{u_e\} \tag{9.4}$$

The inversion of matrix $[H]$, carried out within a symbolic computational engine, reads

$$[H]^{-1} = \begin{bmatrix} \frac{-r_m z_j + r_j z_m}{2A} & 0 & \frac{r_m z_i - r_i z_m}{2A} & 0 & \frac{-r_j z_i + r_i z_j}{2A} & 0 \\ \frac{z_j - z_m}{2A} & 0 & \frac{-z_i + z_m}{2A} & 0 & \frac{z_i - z_j}{2A} & 0 \\ \frac{-r_j + r_m}{2A} & 0 & \frac{r_i - r_m}{2A} & 0 & \frac{-r_i + r_j}{2A} & 0 \\ 0 & \frac{-r_m z_j + r_j z_m}{2A} & 0 & \frac{r_m z_i - r_i z_m}{2A} & 0 & \frac{-r_j z_i + r_i z_j}{2A} \\ 0 & \frac{z_j - z_m}{2A} & 0 & \frac{-z_i + z_m}{2A} & 0 & \frac{z_i - z_j}{2A} \\ 0 & \frac{-r_j + r_m}{2A} & 0 & \frac{r_i - r_m}{2A} & 0 & \frac{-r_i + r_j}{2A} \end{bmatrix} \tag{9.5}$$

where $2A = r_m(z_i - z_j) + r_i(z_j - z_m) + r_j(-z_i + z_m)$. We may now rewrite Equations 9.1 and 9.2 as

$$u = \begin{bmatrix} 1 & r & z & 0 & 0 & 0 \end{bmatrix}[H]^{-1}\{u_e\} \tag{9.6}$$

$$w = \begin{bmatrix} 0 & 0 & 0 & 1 & r & z \end{bmatrix}[H]^{-1}\{u_e\} \tag{9.7}$$

The multiplication of the first two matrix terms in Equations 9.6 and 9.7 produces

$$u = \underbrace{\begin{bmatrix} N_{u1} & 0 & N_{u2} & 0 & N_{u3} & 0 \end{bmatrix}}_{[N_u]}\{u_e\} \tag{9.8}$$

$$v = \underbrace{\begin{bmatrix} 0 & N_{w1} & 0 & N_{w2} & 0 & N_{w3} \end{bmatrix}}_{[N_w]}\{u_e\} \tag{9.9}$$

The underlined term is the vector of shape function for the axisymmetric element. For a generic element with arbitrary nodal labels i, j and m, the full form of the shape functions will take the form itemized in Table 9.1.

9.2.2 Strain–Displacement Relationship and Constitutive Law

For the axisymmetric problem, the non-zero components of the strain measure are related to the displacement derivatives in the form

$$\begin{Bmatrix} \varepsilon_{rr} \\ \varepsilon_{zz} \\ \varepsilon_{\theta\theta} \\ \gamma_{rz} \end{Bmatrix} = \begin{Bmatrix} \dfrac{\partial u}{\partial r} \\ \dfrac{\partial w}{\partial z} \\ \dfrac{u}{r} \\ \dfrac{\partial u}{\partial z} + \dfrac{\partial w}{\partial r} \end{Bmatrix} = \begin{Bmatrix} \dfrac{\partial N_u}{\partial r} \\ \dfrac{\partial N_w}{\partial z} \\ \dfrac{N_u}{r} \\ \dfrac{\partial N_u}{\partial z} + \dfrac{\partial N_w}{\partial r} \end{Bmatrix} \tag{9.10}$$

TABLE 9.1

Shape Functions for the Radial and Axial Deformations in an Axisymmetric Triangular Element

Shape Functions	Expression
$N_{ui} = N_{wi}$	$\frac{1}{2A}\left(r_m(z - z_j) + r(z_j - z_m) + r_j(-z + z_m)\right)$
$N_{uj} = N_{wj}$	$\frac{1}{2A}\left(r_m(-z + z_i) + r_i(z - z_m) + r(-z_i + z_m)\right)$
$N_{um} = N_{wm}$	$\frac{1}{2A}\left(r_j(z - z_i) + r(z_i - z_j) + r_i(-z + z_j)\right)$

Evaluating the derivatives on the right side of Equation 9.10 leads to

$$\begin{Bmatrix} \varepsilon_{rr} \\ \varepsilon_{zz} \\ \varepsilon_{\theta\theta} \\ \gamma_{rz} \end{Bmatrix} = \begin{bmatrix} \frac{z_j - z_m}{2A} & 0 & \frac{-z_i + z_m}{2A} & 0 & \frac{z_i - z_j}{2A} & 0 \\ 0 & \frac{-r_j + r_m}{2A} & 0 & \frac{r_i - r_m}{2A} & 0 & \frac{-r_i + r_j}{2A} \\ \frac{\begin{pmatrix} r_m(z - z_j) + \\ r(z_j - z_m) + \\ r_j(-z + z_m) \end{pmatrix}}{2Ar} & 0 & \frac{\begin{pmatrix} r_m(-z + z_j) + \\ r_i(z - z_m) + \\ r(-z_i + z_m) \end{pmatrix}}{2Ar} & 0 & \frac{\begin{pmatrix} r_j(z - z_i) + \\ r(z_i - z_j) + \\ r_i(-z + z_j) \end{pmatrix}}{2Ar} & 0 \\ \frac{-r_j + r_m}{2A} & \frac{z_i - z_m}{2A} & \frac{r_i - r_m}{2A} & \frac{-z_i + z_m}{2A} & \frac{-r_i + r_j}{2A} & \frac{z_i - z_j}{2A} \end{bmatrix} \begin{Bmatrix} u_i \\ w_i \\ u_j \\ w_j \\ u_m \\ w_m \end{Bmatrix} \tag{9.11}$$

which, in compact form, becomes

$$\{\varepsilon\} = [B]\{u_e\} \tag{9.12}$$

Besides, the constitutive rule that helps one to establish the stress–strain relation for the axisymmetric problem is

$$\begin{Bmatrix} \sigma_{rr} \\ \sigma_{zz} \\ \sigma_{\theta\theta} \\ \tau_{rz} \end{Bmatrix} = [D] \begin{Bmatrix} \varepsilon_{rr} \\ \varepsilon_{zz} \\ \varepsilon_{\theta\theta} \\ \gamma_{rz} \end{Bmatrix} \tag{9.13}$$

Or:

$$\{\sigma\} = [D]\,[B]\{u_e\} \tag{9.14}$$

where $[D]$ takes the following form for the axisymmetric problem [1]:

$$[D] = \frac{E}{(1+v)(1-2v)} \begin{bmatrix} 1-v & v & v & 0 \\ v & 1-v & v & 0 \\ v & v & 1-v & 0 \\ 0 & 0 & 0 & \frac{1-2v}{2} \end{bmatrix} \tag{9.15}$$

9.2.3 Minimization of the Total Potential Energy for the Triangular Axisymmetric Element

Recalling the definition of the strain energy (U):

$$U = \frac{1}{2}\iiint_{\text{vol}} \{\sigma\}^T \{\varepsilon\} dV \tag{9.16}$$

With Equations 9.12 and 9.14, the strain-energy expression for the axisymmetric problem simplifies to

$$U = \frac{1}{2}\iiint_V ([D][B]\{u_e\})^T ([B]\{u_e\}) dV \tag{9.17}$$

The work done by externally applied concentrated nodal forces (W) is of the form

$$W = \{u_e\}^T \begin{Bmatrix} f_{ir} \\ f_{iz} \\ f_{jr} \\ f_{jz} \\ f_{mr} \\ f_{mz} \end{Bmatrix} \tag{9.18}$$

We substitute Equations 9.17 and 9.18 into the expression for the potential energy P_t, and then minimize P_t with respect to u_e to obtain

$$\frac{\partial}{\partial \{u_e\}^T} P_t = \left(\iiint_{\text{vol}} [B]^T [D][B] dV\right)\{u_e\} - \{f_e\} = 0 \tag{9.19}$$

Simplifying Equation 9.19 further, we obtain

$$\left(\iiint_{\text{vol}} [B]^T [D][B] dV\right)\{u_e\} = \{f_e\} \tag{9.20}$$

The matrix $[B]$ (as observed from Equation 9.15) is a function of the coordinates r and z; hence, the integrand of Equation 9.20 cannot be taken out of the integral sign as done in Chapter 8. A way around this is to evaluate the matrix $[B]$ at the centroid of the triangular element [3]. That is,

$$[B] = [B]|_{r=\bar{r},z=\bar{z}} = [\underline{B}] \tag{9.21}$$

where

$$\bar{r} = \frac{1}{3}\left(r_i + r_j + r_k\right);\ \bar{z} = \frac{1}{3}\left(z_i + z_j + z_k\right) \tag{9.22}$$

With this approximate evaluation of the matrix $[B]$, the stiffness matrix for the axisymmetric element is obtained as

$$\left[K_{\text{AXISY}}\right] = \left[\underline{B}\right]^T \left[D\right]\left[\underline{B}\right] \iiint_{\text{vol}} dV = 2\pi \bar{r} A \left[\underline{B}\right]^T \left[D\right]\left[\underline{B}\right] \tag{9.23}$$

And the full equation for the three-node triangular axisymmetric element takes the form

$$\begin{Bmatrix} f_{ir} \\ f_{iz} \\ f_{jr} \\ f_{jz} \\ f_{mr} \\ f_{mz} \end{Bmatrix} = \left[K_{\text{AXISY}}\right] \begin{Bmatrix} u_i \\ w_i \\ u_j \\ w_j \\ u_n \\ w_n \end{Bmatrix} \tag{9.24}$$

9.2.4 Equivalent Body and Surface Forces

In certain axisymmetric problems, contributions to the elastic deformation arise from the presence of body forces (e.g., centrifugal forces from rotation and gravitational sag) and surface forces (forces per unit area) [2]. The effects of these forces can easily be considered by using the shape functions expressed in Table 9.1. For the surface load, we consider the schematics in Figure 9.2.

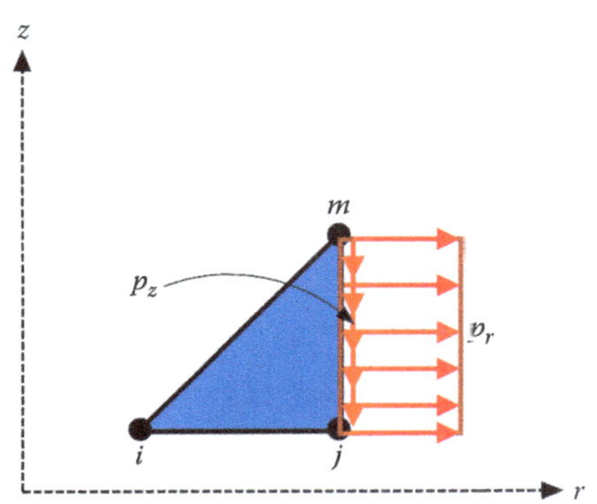

FIGURE 9.2
An axisymmetric element with axial and radial surface loads

The equivalent nodal forces, induced by the presence of surface loads p_r and p_z acting along the vertical edge $j-m$ of the element in Figure 9.2, then turn out to be

$$f_{\text{surf}}^{eq} = \iint_{\text{surf}} \begin{bmatrix} N_{ui} & 0 \\ 0 & N_{wi} \\ N_{uj} & 0 \\ 0 & N_{wj} \\ N_{um} & 0 \\ 0 & N_{wm} \end{bmatrix} \begin{Bmatrix} p_r \\ p_z \end{Bmatrix} dS = \int_{S_{jm}} \begin{bmatrix} N_{ui} & 0 \\ 0 & N_{wi} \\ N_{uj} & 0 \\ 0 & N_{wj} \\ N_{um} & 0 \\ 0 & N_{wm} \end{bmatrix} \begin{Bmatrix} p_r \\ p_z \end{Bmatrix} 2\pi r_j dz = \pi r_j S_{jm} \begin{Bmatrix} 0 \\ 0 \\ p_r \\ p_z \\ p_r \\ p_z \end{Bmatrix} \tag{9.25}$$

where the shape functions in the integrand have been evaluated at $r = r_j = r_m$ and the surface $s_{jm} = (z_m - z_j)$.

Following [2], the equivalent nodal forces at the nodes i, j and m of an element under the effect of rotating and gravity-induced body forces can also be obtained in a similar manner. An approximate expression for this is given as

$$f_{\text{body}}^{eq} = \frac{2\pi\bar{r}A}{3} \begin{Bmatrix} \bar{R}_b \\ Z_b \\ \bar{R}_b \\ Z_b \\ \bar{R}_b \\ Z_b \end{Bmatrix} \tag{9.26}$$

where, in the case of a structure with mass density per unit volume (ρ) rotating with a constant angular velocity (ω) about the axis of symmetry, we have

$$\bar{R}_b = \omega^2 \rho \bar{r} \tag{9.27}$$

and Z_b is the body force per unit volume (e.g., due to gravity).

9.3 Computer Implementation

9.3.1 Systematic Procedure for Analyses of Axisymmetric Elastic Structures

The procedure for analyses of axisymmetric elastic structures is analogous to that provided in Chapter 8. That said, the addition of the body and surface forces to the nodal load vector during computations must be carried out with care.

9.3.2 Implemented R Functions for the Axisymmetric Triangular Element

```
#**************************************************#
# R functions for the axisymmetric element         #
#                                                  #
# KB Mustapha                                      #
#**************************************************#

AxisymmetricT_Element_Matrix = function(DOF=6,YoungMod,Nu,
vec_nodalcoordinates)
{
  r1=vec_nodalcoordinates[1];r2=vec_nodalcoordinates[3];
  r3=vec_nodalcoordinates[5];z1=vec_nodalcoordinates[2];
  z2=vec_nodalcoordinates[4];z3=vec_nodalcoordinates[6];

  Area=abs(r3*(z1-z2)+r2*(z3-z1)+r1*(z2-z3))*0.5;
  rbar=(1/3)*(r1+r2+r3);zbar=(1/3)*(z1+z2+z3);

  a1=r2*z3-z2*r3;
  a2=r3*z1-z3*r1;
  a3=r1*z2-z1*r2;
  B1=z2-z3;B2=z3-z1;B3=z1-z2;
  G1=r3-r2;G2=r1-r3;G3=r2-r1;
  row1=c(B1,0,B2,0,B3,0);
  row2=c(0,G1,0,G2,0,G3);
  row3=c(((a1/rbar)+B1+(G1*zbar/rbar)),0,((a2/rbar)+B2+(G
  2*zbar/rbar)),0,((a3/rbar)+B3+(G3*zbar/rbar)),0);
  row4=c(G1,B1,G2,B2,G3,B3);
  constitutiveprop=c(1-Nu,Nu,Nu,0,Nu,1-Nu,Nu,0,Nu,Nu,
  1-Nu,0,0,0,0,(1-2*Nu)/2);

  BMatrix=(1/(2*Area))*matrix(c(row1,row2,row3,row4),nrow
  =4,byrow=T);
  DMatrix=YoungMod/((1+Nu)*(1-2*Nu))*matrix(constitutive
  prop,nrow=4,byrow=T);
  eMatrix=(2*pi*rbar*Area)*t(BMatrix)%*%DMatrix%*%BMatrix;
  return(eMatrix)
}

AxisymmetricT_ExpandedElement_Matrix = function(TDOF,
eMatrix,i,j,k)
{
  r1=2*i-1; r2=2*i;
  r3=2*j-1; r4=2*j;
  r5=2*k-1; r6=2*k;
  bigMatrix=matrix(vector(l=TDOF*TDOF),nrow=TDOF,byrow=T);
```

```
    bigMatrix[c(r1,r2,r3,r4,r5,r6),c(r1,r2,r3,r4,r5,r6)]=
    eMatrix;
    return (bigMatrix)
}

AxisymmetricT_ReducedStiffnessMatrix = function(bigKmatrix,
knownloadnodes)
{
    reducedk = bigKmatrix[c(knownloadnodes),(knownloadnodes)]
    return(reducedk)
}

AxisymmetricT_ReducedLoadVector = function(loadvector)
{
    reducedf = matrix(loadvector,ncol = 1)
    return(reducedf)
}

AxisymmetricT_SF = function(DOF=6,pr,pz,case,
vec_nodalcoordinates)
{
    #1=J-M;#2=I-M;#3=I-J
    r1=vec_nodalcoordinates[1];r2=vec_nodalcoordinates[3];
    r3=vec_nodalcoordinates[5];z1=vec_nodalcoordinates[2];
    z2=vec_nodalcoordinates[4];z3=vec_nodalcoordinates[6];
    if(case==1){
        equivalentsfload=pi*r2*abs((z3-z2))*matrix(c(0,0,pr,
        pz,pr,pz),nrow=DOF,byrow=T)
    }
    if(case==2){
        equivalentsfload=pi*r1*abs((z3-z1))*matrix(c(pr,pz,0,
        0,pr,pz),nrow=DOF,byrow=T)
    }
    if(case==3){
        equivalentsfload=pi*r1*abs((z2-z1))*matrix(c(pr,pz,pr,
        pz,0,0),nrow=DOF,byrow=T)
    }
    return (equivalentsfload)
}

AxisymmetricT_BF = function(DOF=6,Rb,Zb,vec_nodalcoordinates)
{
    r1=vec_nodalcoordinates[1];r2=vec_nodalcoordinates[3];
    r3=vec_nodalcoordinates[5];z1=vec_nodalcoordinates[2];
    z2=vec_nodalcoordinates[4];z3=vec_nodalcoordinates[6];
    rbar=(1/3)*(r1+r2+r3);
    Area=abs(r3*(z1-z2)+r2*(z3-z1)+r1*(z2-z3))*0.5;
```

```
  equivalentbodyload =(2/3)*pi*rbar*Area*matrix(c(Rb,Zb,Rb,
  Zb,Rb,Zb),nrow=DOF,byrow=T)
  return (equivalentbodyload)
}

AxisymmetricT_ExpandedSFBF = function(TDOF,
LoadColumnMatrix,i,j,k)
{
  r1=2*i-1; r2=2*i
  r3=2*j-1; r4=2*j
  r5=2*k-1; r6=2*k
  bigColumnMatrix=matrix(vector(l=TDOF),nrow=TDOF,
  byrow=T);
  bigColumnMatrix[c(r1,r2,r3,r4,r5,r6)]=LoadColumnMatrix;
  return (bigColumnMatrix)
}

AxisymmetricT_NodalDisplacement =
function(reducedmatrix,vec_reducedforce)
{
  return(solve(reducedmatrix,vec_reducedforce))
}

AxisymmetricT_GlobalForces = function(bigKmatrix,
vec_globalnodaldisp)
{
  columndof=matrix(vec_globalnodaldisp,byrow = T)
  globalforces = bigKmatrix %*% vec_globalnodaldisp
  return(globalforces)
}

AxisymmetricT_Centroid = function(vec_nodalcoordinates)
{
  r1=vec_nodalcoordinates[1];r2=vec_nodalcoordinates[3];
  r3=vec_nodalcoordinates[5];z1=vec_nodalcoordinates[2];
  z2=vec_nodalcoordinates[4];z3=vec_nodalcoordinates[6];
  rbar=(1/3)*(r1+r2+r3);
  zbar=(1/3)*(z1+z2+z3);
  return(c(rbar,zbar))
}

AxisymmetricT_LocalForces = function(ematrix,
vec_globalnodaldisp,i,j,k)
{
  r1=2*i-1; r2=2*i
  r3=2*j-1; r4=2*j
  r5=2*k-1; r6=2*k
```

```
    localforces = ematrix%*%vec_globalnodaldisp[c(r1,r1,r3,
    r4,r5,r6)]
    return(round(localforces))
}

AxisymmetricT_Stresses = function(YoungMod,Nu,
vec_nodalcoord,vec_globalnodaldisp,i,j,k)
{
    r1=vec_nodalcoord[1];r2=vec_nodalcoord[3];
    r3=vec_nodalcoord[5];z1=vec_nodalcoord[2];
    z2=vec_nodalcoord[4];z3=vec_nodalcoord[6];

    Area=abs(r3*(z1-z2)+r2*(z3-z1)+r1*(z2-z3))*0.5;
    rbar=(1/3)*(r1+r2+r3);
    zbar=(1/3)*(z1+z2+z3);

    a1=r2*z3-z2*r3;
    a2=r3*z1-z3*r1;
    a3=r1*z2-z1*r2;
    B1=z2-z3;B2=z3-z1;B3=z1-z2;
    G1=r3-r2;G2=r1-r3;G3=r2-r1;
    row1=c(B1,0,B2,0,B3,0);
    row2=c(0,G1,0,G2,0,G3);
    row3=c(((a1/rbar)+B1+(G1*zbar/rbar)),0,((a2/rbar)+B2+
    (G2*zbar/rbar)),0,((a3/rbar)+B3+(G3*zbar/rbar)),0);
    row4=c(G1,B1,G2,B2,G3,B3);
    constitutiveprop=c(1-Nu,Nu,Nu,0,Nu,1-Nu,Nu,0,Nu,Nu,1-Nu,
    0,0,0,0,(1-2*Nu)/2)

    BMatrix=(1/(2*Area))*matrix(c(row1,row2,row3,row4),
    nrow=4,byrow=T);
    DMatrix=YoungMod/((1+Nu)*(1-2*Nu))*matrix(constitutive
    prop,nrow=4,byrow=T);
    r1=2*i-1; r2=2*i
    r3=2*j-1; r4=2*j
    r5=2*k-1; r6=2*k
    localstresses=DMatrix%*%BMatrix%*%vec_globalnodaldisp
    [c(r1,r2,r3,r4,r5,r6)]
    return(localstresses)
}
```

The descriptions for the listed functions are provided in Tables 9.2 through 9.4.

TABLE 9.2

Functions for the Element's Matrices and Vectors

AxisymmetricT_Element_Matrix(DOF=6,YoungMod,Nu,vec_nodalcoordinates) This function generates and returns the stiffness matrix for the axisymmetric triangular element. It accepts four parameters: **DOF** (predefined as 6); **YoungMod**, which represents Young's modulus; **Nu**, Poisson's ratio; and **vec_nodalcoordinates**, a vector of nodal coordinate in the form $(r_i, z_i, r_j, z_j, r_m, z_m)$.
AxisymmetricT_ExpandedElement_Matrix(TDOF,eMatrix,i,j,m) This function forms the expanded stiffness matrix for the CST element, and its parameters are of the same form as those defined in previous chapters: value of total degrees of freedom (**TDOF**), element matrix (**eMatrix**), and the nodal labels i, j and m.
AxisymmetricT_ReducedStiffnessMatrix(bigKmatrix,knownloadnodes) This function generates a reduced stiffness matrix for the element, and its parameters are of the same form as those defined in the previous chapters.
AxisymmetricT_ReducedLoadVector(loadvector) This function generates a reduced load vector needed to find the unknown nodal displacements. It accepts the parameter **loadvector** (a vector of known nodal forces) and transforms this vector to a column matrix.

TABLE 9.3

Functions for Considerations of Distributed Body and Surface Loads

AxisymmetricT_SF(DOF=6,pr,pz,case,vec_nodalcoordinates) This function generates and returns the column matrix for uniform surface loads p_r and p_z. It accepts five parameters: **DOF** (predefined as 6); **pr**, which represents the magnitude of a positive radial surface traction on edge; **pz**, which represents the magnitude of an axial surface traction on an element's edge; **case**, which is a number (1, 2, or 3) that denotes the edge supporting the loads (1, if load is on edge $j-m$, 2 if load is on edge $i-m$, and 3 if load is on edge $i-j$). The last parameter is the vector of nodal coordinates.
AxisymmetricT_BF(DOF=6,Rb,Zb,vec_nodalcoordinates) This function generates and returns the column matrix for the body force **Rb** and **Zb** in the radial and axial directions, respectively.
AxisymmetricT_ExpandedSFBF(TDOF,LoadColumnMatrix,i,j,m) This function forms the expanded column matrix for the body and surface forces, and its parameters are **TDOF** (total degrees of freedom of the system); **LoadColumnMatrix** (load column matrix, obtained in row 1 of this table); and **i, j, m** representing the *i*th, *j*th and *m*th nodes of the element with the surface traction.

TABLE 9.4

Functions for the Solution Phase (Axisymmetric Element)

`AxisymmetricT_NodalDisplacement(reducedmatrix,vec_reducedforce)` Given the parameters **`reducedmatrix`** (reduced stiffness matrix) and **`reducedforcevec`** (reduced force vector), this function solves for the unknown nodal displacements.
`AxisymmetricT_GlobalForces(bigKmatrix,vec_globalnodaldisp)` This function calculates the global forces (and global reaction forces) developed in the axisymmetric model. Its parameters are **`bigKmatrix`** (the global stiffness matrix) and **`vec_globalnodaldisp`** (a vector of the global nodal displacement).
`AxisymmetricT_Centroid(vec_nodalcoordinates)` This function calculates the centroid of an axisymmetric triangular element using an element's nodal coordinates.
`AxisymmetricT_Stresses(YoungMod,Nu,vec_nodalcoord,vec_globalnodaldisp, i,j,m)` This function returns the stress developed in an axisymmetric triangular element. Its parameters are **`YoungMod,Nu`**, Poisson's ratio; **`vec_nodalcoord`**, a vector of nodal coordinate in the form; **`vec_globalnodaldisp`** (a vector of the global nodal displacement); and **`i,j,m`** representing the *i*th, *j*th and *m*th nodes of the element.

9.4 Examples

9.4.1 Example 9.1

Figure 9.3 shows a long thick-walled cylindrical pressure vessel with inner and outer diameters of 80 and 160 mm, respectively. If the vessel is subjected to inner pressure of 2 MPa, $E = 200$ GPa and $v = 0.3$:

1. Establish the global stiffness matrix for the structure using the 2-element axisymmetric model in Figure 9.3b.
2. Determine the radial displacement of the inner surface of the vessel due to the applied pressure.

9.4.1.1 Solution

The above-mentioned example is motivated by a similar axisymmetric problem solved in Reference 4. Following the axisymmetric model in Figure 9.3b, we can form the nodal coordinates of the elements as shown in Table 9.5.

Step 1: We specify the nodal coordinates (base on the connectivity table) and the properties of each element.

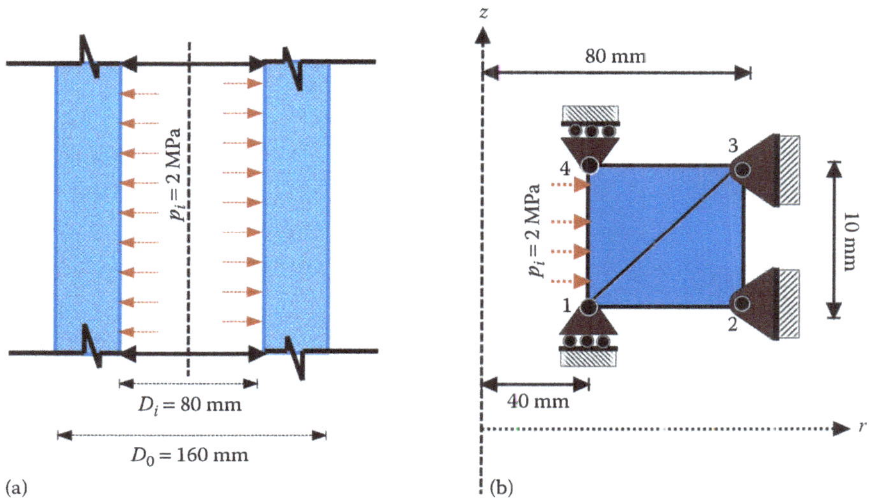

FIGURE 9.3
(a) An internally pressurized long cylinder and (b) a planar axisymmetric approximation.

TABLE 9.5
Connectivity Table from the Discretized System

Elements	Nodes (i,j,m)	Coordinates $(r_i,z_i,r_j,z_j,r_m,z_m)$
1	1,2,3	(0.04,0,0.08,0,0.08,0.01)
2	3,4,1	(0.08,0.01,0.04,0.01,0.04,0)

```
e1coord=c(0.04,0,0.08,0,0.08,0.01)
e2coord=c(0.08,0.01,0.04,0.01,0.04,0)
E1=E2=200e9;Nu=0.3
```

All parameters are in SI units.

Step 2: Write the stiffness matrix for each element using the coordinates and properties specified in Step 1 as arguments here.

```
k1=AxisymmetricT_Element_Matrix(6,E1,Nu,e1coord);k1
k2=AxisymmetricT_Element_Matrix(6,E2,Nu,e2coord);k2
```

The stiffness matrices for the two elements are computed, and they are as shown in the next block of code.

```
> k1=AxisymmetricT_Element_Matrix(6,E1,Nu,e1coord);k1
            [,1]        [,2]        [,3]         [,4]        [,5]         [,6]
[1,]  12244155983           0 -13533014508  19332877868   -644429262 -19332877868
[2,]            0  4027682889  16110731557  -4027682889 -16110731557            0
[3,] -13533014508 16110731557  81520301678 -45110048359 -62670745756  28999316802
[4,]  19332877868 -4027682889 -45110048359 229577924685  11277512090 -225550241796
[5,]   -644429262 -16110731557 -62670745756  11277512090 65006801832   4833219467
[6,] -19332877868           0  28999316802 -225550241796  4833219467 225550241796
> k2=AxisymmetricT_Element_Matrix(6,E2,Nu,e2coord);k2
            [,1]        [,2]        [,3]         [,4]        [,5]         [,6]
[1,]  14398966329           0 -10572667584  24166097335  1913149372 -24166097335
[2,]            0  3222146311  12888585245  -3222146311 -12888585245            0
[3,] -10572667584 12888585245  61120087844 -27388243647 -52057801343  14499658401
[4,]  24166097335 -3222146311 -27388243647 183662339748  17721804713 -180440193437
[5,]   1913149372 -12888585245 -52057801343  17721804713 52259185488  -4833219467
[6,] -24166097335           0  14499658401 -180440193437 -4833219467 180440193437
```

Element 2 carries a radial pressure (a surface force) of magnitude 2 MPa. We determine the equivalent nodal loads for this element by using the function **AxisymmetricT_SF()**. As noted in Table 8.3, this function accepts five parameters:

- The first is the degree of freedom of the element (predefined as 6).
- The second and third parameters represent the magnitude of radial and axial surface forces, respectively. For this problem, **pr** is replaced by 2 MPa and **pz** is 0 due to the absence of axial pressure.
- The fourth parameter, **case**, is used to indicate the edge on which the load is applied, which in the present example is 1, because the pressure is on the edge $j - m$.
- The last parameter is the vector of nodal coordinates.

By using these information, we can now determine the equivalent nodal loads from the surface load on element 2 as

```
k2sf=AxisymmetricT_SF(6,2e6,0,1,e2coord);k2sf
```

The output of this then reads

```
> k2sf=AxisymmetricT_SF(6,2e6,0,1,e2coord);k2sf
          [,1]
[1,]     0.000
[2,]     0.000
[3,]  2513.274
[4,]     0.000
[5,]  2513.274
[6,]     0.000
```

Step 3: We determine the global degree of freedom of the continuum and expand the matrix equation for each triangular element (from Step 2). The triangulated continuum has four nodes, and each node has two degrees of freedom. We thus have a total of 8 degrees of freedom for the discretized model.

The expanded matrices for each element are obtained via **AxisymmetricT_ExpandedElement_Matrix()**, which requires **TDOF** (the total degrees of freedom), **ematrix** (the element stiffness matrix) and **i,j,m** (the respective nodal numbers of the element).

```
n_nodes=4;Total_dof=n_nodes*2;
K1=AxisymmetricT_ExpandedElement_Matrix(Total_
dof,k1,1,2,3);K1
K2=AxisymmetricT_ExpandedElement_Matrix(Total_
dof,k2,3,4,1);K2
K2SF=AxisymmetricT_ExpandedSFBF(Total_
dof,k2sf,3,4,1);K2SF
```

Notice that the last line of the above-shown block of code deals with expansion of the vector of equivalent loads acting on element 2, which has been done via the function **AxisymmetricT_ExpandedSFBF()**.

Step 4: The global stiffness matrix is determined by adding the expanded matrices of the two elements as shown next.

```
GlobalK=K1+K2;GlobalK
```

The output of this gives the 8×8 global stiffness matrix $[K]$ displayed next.

```
> GlobalK=K1+K2;GlobalK
             [,1]          [,2]         [,3]          [,4]         [,5]          [,6]         [,7]          [,8]
[1,]  64503341471   -4833219467 -13533014508   19332877868   1268720110  -32221463114 -52057801343   17721804713
[2,]  -4833219467  184467876326  16110731557   -4027682889 -40276828892             0  14499658401 -180440193437
[3,] -13533014508   16110731557  81520301678  -45110048359 -62670745756   28999316802            0             0
[4,]  19332877868   -4027682889 -45110048359  229577924685  11277512090 -225550241796            0             0
[5,]   1268720110  -40276828892 -62670745756   11277512090  79405768161    4833219467 -10572667584   24166097335
[6,] -32221463114             0  28999316802 -225550241796   4833219467  228772388108  12888585245   -3222146311
[7,] -52057801343   14499658401            0             0 -10572667584   12888585245  61120087844  -27388243647
[8,]  17721804713 -180440193437            0             0  24166097335   -3222146311 -27388243647  183662339748
```

A reformatted form of the global stiffness matrix K_{global} is provided as follows:

$$10^6 \begin{bmatrix} 64503.341 & -4833.219 & -13533.015 & 19332.878 & 1268.72 & -32221.463 & -52057.801 & 17221.805 \\ -4833.219 & 184467.876 & 16110.732 & -4027.683 & -40276.829 & 0 & 14499.658 & -180440.193 \\ -13533.015 & 16110.732 & 81520.302 & -45110.048 & -62670.746 & 28999.317 & 0 & 0 \\ 19332.878 & -4027.683 & -45110.048 & 229577.925 & 11277.512 & -225550.242 & 0 & 0 \\ 1268.72 & -40276.829 & -62670.746 & 11277.512 & 79405.768 & 4833.219 & -10572.668 & 24166.097 \\ -32221.463 & 0 & 28999.317 & -225550.242 & 4833.219 & 22877.388 & 12888.585 & -3222.146 \\ -52057.801 & 14499.658 & 0 & 0 & -10572.668 & 12888.585 & 61120.088 & -27388.244 \\ 17721.805 & -180440.193 & 0 & 0 & 24166.097 & -3222.146 & -27388.244 & 183662.34 \end{bmatrix}$$

Step 5: Identify the boundary condition(s) and obtain the reduced stiffness matrix and the reduced force vector.

By using the global stiffness matrix obtained, we can write the global matrix equation for the idealized structure as

$$\begin{Bmatrix} F_{1r} \\ F_{1z} \\ F_{2r} \\ F_{2z} \\ F_{3r} \\ F_{3z} \\ F_{4r} \\ F_{4z} \end{Bmatrix} = \left[K_{\text{global}}\right] \begin{Bmatrix} U_{1r} \\ U_{1z} \\ U_{2r} \\ U_{2z} \\ U_{3r} \\ U_{3z} \\ U_{4r} \\ U_{4z} \end{Bmatrix} \tag{9.28}$$

For the boundary conditions, we note that the model is pinned at nodes 2 and 3 (translating to zero horizontal and vertical displacements at these nodes), and axially restrained roller supports are placed at nodes 1 and 4. Accordingly, the identified displacement boundary conditions will be

$$U_{1z} = U_{4z} = 0; = U_{2r} = U_{2z} = U_{3r} = U_{3z} = 0; \tag{9.29}$$

The set of known concentrated nodal loads corresponding to the 2 unknown nodal degrees of freedom is

$$F_{1r}^{c} = 0; \; F_{4r}^{c} = 0 \tag{9.30}$$

In addition, the equivalent nodal loads from the surface force acting on element 2 contributes to the nodal forces at nodes 1 and 4, and this has been obtained from Step 2 as

$$F_{1r}^{eq} = 2513.274; \; F_{4r}^{eq} = 2513.274 \tag{9.31}$$

By combining Equations 9.30 and 9.31, we estimate that nodes 1 and 4 must each carry a radial load of 2513.274 N. Furthermore, F_{1r} and F_{4r} belong to rows 1 and 7 of Equation 9.28, and this information is required to obtain the reduced stiffness matrix via the function **AxisymmetricT_ReducedStiffnessMatrix()**. The following code is used for this purpose:

```
vec_rows_of_knownloads=c(1,7);
ReducedK=AxisymmetricT_ReducedStiffnessMatrix(GlobalK,
vec_rows_of_knownloads);
ReducedK
```

The reduced stiffness matrix then reads

```
> ReducedK
              [,1]         [,2]
[1,]  64503341471 -52057801343
[2,] -52057801343  61120087844
```

Thus:

$$[K_{\text{reduced}}]=\begin{bmatrix} 64503341471 & -52057801343 \\ -52057801343 & 61120087844 \end{bmatrix} \tag{9.32}$$

In the next code, the set of known radial loads is transformed to establish the reduced load vector.

```
vec_values_of_knownloads=c(0,0)+K2SF[vec_rows_of_knownloads]
Reducedloadvector=AxisymmetricT_ReducedLoadVector(vec_
values_of_knownloads);
Reducedloadvector
```

where the first line of the preceding code snippet has two parts: The first part is the vector of the known concentrated radial loads, whereas the second part represents the equivalent nodal loads. The output of the preceding code reads

```
> Reducedloadvector
         [,1]
[1,] 2513.274
[2,] 2513.274
```

Combining the reduced stiffness matrix and the reduced vector of applied loads, one obtains the reduced matrix equation in the form

$$\begin{Bmatrix} 2513.274 \\ 2513.274 \end{Bmatrix} = \begin{bmatrix} 64503341471 & -52057801343 \\ -52057801343 & 61120087844 \end{bmatrix} \begin{Bmatrix} U_{3r} \\ U_{4r} \end{Bmatrix} \tag{9.33}$$

Step 6: Determine the unknown nodal displacements by using the function **`AxisymmetricT_NodalDisplacement()`**.

```
UnknwonNodalDisp=AxisymmetricT_NodalDisplacement(ReducedK,
Reducedloadvector)
UnknwonNodalDisp
```

The determined radial displacements (in metres) read

```
> UnknwonNodalDisp
                  [,1]
[1,] 0.0000002308008
[2,] 0.0000002377002
```

From this output, we have found the radial displacements at the inner surface of the vessel due to the applied pressure to be about 2.3 μm.

9.4.2 Example 9.2

Consider the sectioned view, shown in Figure 9.4a, of a flywheel rotating with a constant speed of 5000 rev/min. Given that the flywheel has geometric dimensions $a = 60$ mm, $b = 190$ mm, $h = 10$ mm, use the approximate axisymmetric model provided in Figure 9.4b to determine the unknown nodal displacements and the stresses developed within the elements. Weight density = 7860 kN/m^3, $E = 200$ GPa, and $v = 0.3$.

9.4.2.1 Solution

Following Figure 9.4b, we can form the nodal coordinates of the four axisymmetric elements as shown in Table 9.6. We then use the information from the connectivity table in the subsequent steps of the analyses.

Step 1: Specify the nodal coordinates (base on the connectivity table) and the properties of the triangular elements.

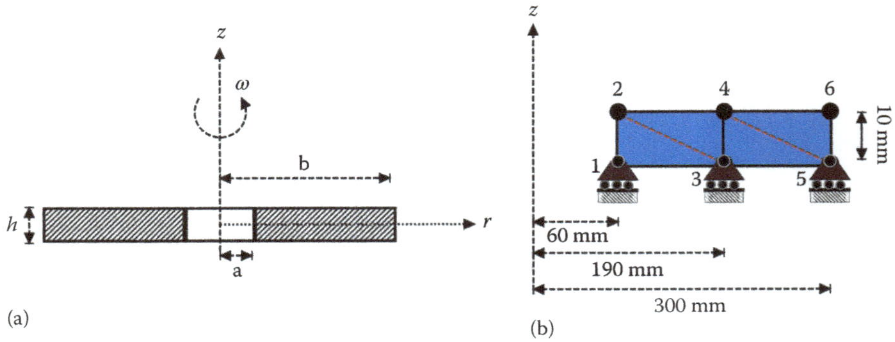

FIGURE 9.4
(a) A sectioned view of a rotating flywheel and (b) an axisymmetric finite element model.

TABLE 9.6
Connectivity Table from the Discretized System

Elements	Nodes (i,j,m)	Coordinates $(r_i,z_i,r_j,z_j,r_m,z_m)$
1	1,3,2	$(0.06,0,0.19,0,0.06,0.01)$
2	2,3,4	$(0.06,0.01,0.19,0,0.19,0.01)$
3	4,3,5	$(0.19,0.01,0.19,0,0.3,0)$
4	4,5,6	$(0.19,0.01,0.3,0,0.3,0.01)$

```
e1coord=c(0.06,0,0.19,0,0.06,0.01)
e2coord=c(0.06,0.01,0.19,0,0.19,0.01)
e3coord=c(0.19,0.01,0.19,0,0.3,0)
e4coord=c(0.19,0.01,0.3,0,0.3,0.01)
YoungMod=200e9;Nu=0.3
```

All coordinate values have been converted into metres.

Step 2: The stiffness matrices for the four elements are obtained via the function **AxisymmetricT_Element_Matrix()**.

```
k1=AxisymmetricT_Element_Matrix(6,YoungMod,Nu,e1coord);k1
k2=AxisymmetricT_Element_Matrix(6,YoungMod,Nu,e2coord);k2
k3=AxisymmetricT_Element_Matrix(6,YoungMod,Nu,e3coord);k3
k4=AxisymmetricT_Element_Matrix(6,YoungMod,Nu,e4coord);k4
```

Which produces:

```
> k1=AxisymmetricT_Element_Matrix(6,YoungMod,Nu,e1coord);k1
             [,1]          [,2]         [,3]         [,4]          [,5]           [,6]
[1,]  330120083161   46721121515  -5540812392 -24971633913 -324657225922   -21749487602
[2,]   46721121515 1138130237965 -53165414138  -1920894916  -40679597181 -1136209343048
[3,]   -5540812392  -53165414138  10322061757            0    2390624683    53165414138
[4,]  -24971633913   -1920894916            0   1920894916   24971633913              0
[5,] -324657225922  -40679597181   2390624683  24971633913  325813560687    15707963268
[6,]  -21749487602 -1136209343048  53165414138            0   15707963268  1136209343048
> k2=AxisymmetricT_Element_Matrix(6,YoungMod,Nu,e2coord);k2
            [,1]         [,2]          [,3]           [,4]          [,5]           [,6]
[1,]  7958898548            0    -375306815    37457450870   -8709512178   -37457450870
[2,]           0   2726431494   35443609425              0  -35443609425    -2726431494
[3,]  -375306815  35443609425  461599920579   -15707963268 -458725619608   -19735646157
[4,] 37457450870            0  -15707963268  1612684228843  -68873377406 -1612684228843
[5,] -8709512178 -35443609425 -458725619608   -68873377406  473559040542   104316986831
[6,] -37457450870 -2726431494  -19735646157 -1612684228843  104316986831  1615410660337
> k3=AxisymmetricT_Element_Matrix(6,YoungMod,Nu,e3coord);k3
             [,1]          [,2]          [,3]           [,4]          [,5]          [,6]
[1,]  602997436084   13291353534 -603293589237   -68067840828    1664380723   54776487293
[2,]   13291353534 2108894760794  -68873377406 -2108894760794   95456084474             0
[3,] -603293589237  -68873377406  618009708671   123649864699  -16972806464  -54776487293
[4,]  -68067840828 -2108894760794 123649864699  2113874441457  -95456084474   -4979680663
[5,]    1664380723   95456084474  -16972806464   -95456084474   20301567911             0
[6,]   54776487293             0  -54776487293    -4979680663             0    4979680663
> k4=AxisymmetricT_Element_Matrix(6,YoungMod,Nu,e4coord);k4
            [,1]         [,2]          [,3]           [,4]          [,5]           [,6]
[1,] 18224222233            0    -815733243    82164730940  -19855688720   -82164730940
[2,]           0   5785217241   63637389650              0  -63637389650    -5785217241
[3,]  -815733243  63637389650  700403857769   -13291353534 -698410409656   -50346036115
[4,] 82164730940            0  -13291353534  2450039501511 -108747438009 -2450039501511
[5,] -19855688720 -63637389650 -698410409656 -108747438009  723068727846   172384827659
[6,] -82164730940 -5785217241  -50346036115 -2450039501511  172384827659  2455824718752
```

Step 3: Next we determine the body forces acting on the axisymmetric triangular elements. As expressed in Section 9.2.4, the equivalent body force, in the case of a rotating body, is couched in the load vector:

$$f_{body}^{eq} = \frac{2\pi\bar{r}A}{3}\begin{Bmatrix} \bar{R}_b \\ \bar{Z}_b \\ \bar{R}_b \\ \bar{Z}_b \\ \bar{R}_b \\ \bar{Z}_b \end{Bmatrix} \tag{9.34}$$

where $\bar{Z}_b = 0$ (if we neglect the effect of gravity) and $\bar{r}$ is the centroid of each element. By using the data provided, we can estimate the radial body force as

$$\bar{R}_b = \omega^2\rho\bar{r} = (2\pi f)^2\,\rho\bar{r} = \left(2\pi \times \frac{5000}{60}\right)^2 \times \left(\frac{7861}{9.8}\right) \tag{9.35}$$

In the block of code that follows, we determine the centroid of the triangular elements using the function **AxisymmetricT_Centroid()**.

```
rbar1=AxisymmetricT_Centroid(e1coord)[1];
rbar2=AxisymmetricT_Centroid(e2coord)[1];
rbar3=AxisymmetricT_Centroid(e3coord)[1];
rbar4=AxisymmetricT_Centroid(e4coord)[1];
```

Notice that in the preceding lines of code, we extracted the first component of the vector returned by the function **AxisymmetricT_Centroid()**, as this function returns two components ($\bar{r}$ and $\bar{z}$).

Next, we calculate the radial body forces using the function **AxisymmetricT_BF()**.

```
ForcePerArea=(2*pi*5000/60)^2*(7861/9.8);
k1bf=AxisymmetricT_BF(6,ForcePerArea*rbar1,0,e1coord);k1bf
k2bf=AxisymmetricT_BF(6,ForcePerArea*rbar2,0,e2coord);k2bf
k3bf=AxisymmetricT_BF(6,ForcePerArea*rbar3,0,e3coord);k3bf
k4bf=AxisymmetricT_BF(6,ForcePerArea*rbar4,0,e4coord);k4bf
```

The vectors of body forces for each element are then obtained as

```
> k1bf=AxisymmetricT_BF(6,ForcePerArea*rbar1,0,e1coord);k1bf
       [,1]
[1,] 3196.7
[2,]    0.0
[3,] 3196.7
[4,]    0.0
[5,] 3196.7
[6,]    0.0
> k2bf=AxisymmetricT_BF(6,ForcePerArea*rbar2,0,e2coord);k2bf
         [,1]
[1,] 6439.969
[2,]    0.000
[3,] 6439.969
[4,]    0.000
[5,] 6439.969
[6,]    0.000
> k3bf=AxisymmetricT_BF(6,ForcePerArea*rbar3,0,e3coord);k3bf
         [,1]
[1,] 13015.04
[2,]     0.00
[3,] 13015.04
```

```
[4,]      0.00
[5,] 13015.04
[6,]      0.00
> k4bf=AxisymmetricT_BF(6,ForcePerArea*rbar4,0,e4coord);k4bf
         [,1]
[1,] 17566.37
[2,]     0.00
[3,] 17566.37
[4,]     0.00
[5,] 17566.37
[6,]     0.00
```

Step 4: Determine the global degree of freedom of the triangulated domain and expand the matrix equation for each triangular element. For the element matrices, the expansion is done by using **AxisymmetricT_ExpandedElement_Matrix()**.

Besides, the expanded vector of equivalent nodal loads of each element is obtained via **AxisymmetricT_ExpandedSFBF()** as done in the following:

```
n_nodes=6;Total_dof=n_nodes*2;
K1=AxisymmetricT_ExpandedElement_Matrix(Total_dof,
k1,1,3,2);K1
K2=AxisymmetricT_ExpandedElement_Matrix(Total_dof,
k2,2,3,4);K2
K3=AxisymmetricT_ExpandedElement_Matrix(Total_dof,
k3,4,3,5);K3
K4=AxisymmetricT_ExpandedElement_Matrix(Total_dof,
k4,4,5,6);K4

K1BF=AxisymmetricT_ExpandedSFBF(Total_dof,k1bf,1,3,2);K1BF
K2BF=AxisymmetricT_ExpandedSFBF(Total_dof,k2bf,2,3,4);K2BF
K3BF=AxisymmetricT_ExpandedSFBF(Total_dof,k3bf,4,3,5);K3BF
K4BF=AxisymmetricT_ExpandedSFBF(Total_dof,k4bf,4,5,6);K4BF
```

Step 5: The global stiffness matrix and the global body force vectors are determined as shown next.

```
GlobalK=K1+K2+K3+K4;GlobalK
GlobalBF=K1BF+K2BF+K3BF+K4BF;GlobalBF
```

The first line of the preceding code produces the 12×12 global stiffness matrix, which reads

$$
10^9 \begin{bmatrix}
330.1201 & 46.7211 & -324.6572 & -21.7495 & -5.5408 & -24.9716 & 0. & 0. & 0. & 0. & 0. & 0. \\
46.7211 & 1138.1302 & -40.6796 & -1136.2093 & -53.1654 & -1.9209 & 0. & 0. & 0. & 0. & 0. & 0. \\
-324.6572 & -40.6796 & 333.7725 & 15.708 & 2.0153 & 62.4291 & -8.7095 & -37.4575 & 0. & 0. & 0. & 0. \\
-21.7495 & -1136.2093 & 15.708 & 1138.9358 & 88.609 & 0. & -35.4436 & -2.7264 & 0. & 0. & 0. & 0. \\
-5.5408 & -53.1654 & 2.0153 & 88.609 & 1089.9358 & 107.9419 & -1062.0192 & -88.609 & -16.9728 & -54.7765 & 0. & 0. \\
-24.9716 & -1.9209 & 62.4291 & 0. & 107.9419 & 3728.4796 & -136.9412 & -3721.579 & -95.4561 & -4.9797 & 0. & 0. \\
0. & 0. & -8.7095 & -35.4436 & -1062.0192 & -136.9412 & 1094.7807 & 117.6083 & 0.8486 & 136.9412 & -19.8557 & -82.1647 \\
0. & 0. & -37.45 & -2.7264 & -88.609 & -3721.579 & 117.6083 & 3730.0906 & 159.0935 & 0. & -63.6374 & -5.7862 \\
0. & 0. & 0. & 0. & -16.9728 & -95.4561 & 0.8486 & 159.0935 & 720.7054 & -13.2914 & -698.4104 & -50.346 \\
0. & 0. & 0. & 0. & -54.7765 & -4.9797 & 136.9412 & 0 & -13.2914 & 2455.0192 & -108.7474 & -2450.0395 \\
0. & 0. & 0. & 0. & 0. & 0. & -19.8557 & -63.4104 & -108.7474 & -108.7474 & 723.0687 & 172.3848 \\
0. & 0. & 0. & 0. & 0. & 0. & -82.1647 & -5.7852 & -50.346 & 2450.0395 & 172.3848 & 2455.8247
\end{bmatrix}
$$

And the second line yields the vector of global body forces as

```
> GlobalBF=K1BF+K2BF+K3BF+K4BF;GlobalBF
              [,1]
 [1,]   3196.700
 [2,]      0.000
 [3,]   9636.669
 [4,]      0.000
 [5,]  22651.711
 [6,]      0.000
 [7,]  37021.378
 [8,]      0.000
 [9,]  30581.409
[10,]      0.000
[11,]  17566.367
[12,]      0.000
```

Step 6: We now identify the boundary condition(s) and obtain the reduced stiffness matrix and the reduced force vector.

From the foregoing discussions, the global matrix equation for the system will be

$$\begin{Bmatrix} F_{1r} \\ F_{1z} \\ F_{2r} \\ F_{2z} \\ F_{3r} \\ F_{3z} \\ F_{4r} \\ F_{4z} \\ F_{5r} \\ F_{5z} \\ F_{6r} \\ F_{6z} \end{Bmatrix}^{c} + \begin{Bmatrix} F_{1r} \\ F_{1z} \\ F_{2r} \\ F_{2z} \\ F_{3r} \\ F_{3z} \\ F_{4r} \\ F_{4z} \\ F_{5r} \\ F_{5z} \\ F_{6r} \\ F_{6z} \end{Bmatrix}^{eq} = [K_{\text{global}}] \begin{Bmatrix} U_{1r} \\ U_{1z} \\ U_{2r} \\ U_{2z} \\ U_{3r} \\ U_{3z} \\ U_{4r} \\ U_{4z} \\ U_{5r} \\ U_{5z} \\ U_{6r} \\ U_{6z} \end{Bmatrix} \tag{9.36}$$

where the right-hand side of Equation 9.36 represents the combination of the concentrated nodal loads (the vector with superscript c) and the equivalent loads from the body forces (the vector with superscript eq).

For the displacement boundary conditions, we noticed that the finite element model has axially restrained roller supports at nodes 1, 2, and 3 (translating to zero vertical displacements at these nodes). For this reason, the identified displacement boundary conditions take the form

$$U_{1z} = U_{2z} = U_{3z} = 0 \tag{9.37}$$

The set of known concentrated nodal loads corresponding to the remaining nine unknown nodal degrees of freedom is

$$F_{1r}^{c} = 0;\ F_{2r}^{c} = F_{3r}^{c} = 0;\ F_{4r}^{c} = F_{4z}^{c} = F_{5r}^{c} = F_{5z}^{c} = F_{6r}^{c} = F_{6z}^{c} = 0 \tag{9.38}$$

Furthermore, the equivalent nodal loads from the radial body force (as obtained in Step 5) are as follows:

$$\begin{aligned} &F_{1r}^{eq} = 3196.700;\ F_{2r}^{eq} = 9636.669;\ F_{3r}^{eq} = 22651.711; \\ &F_{4r}^{c} = 37021.378;\ F_{5r}^{eq} = 30581.409;\ F_{6r}^{eq} = 17566.367 \end{aligned} \tag{9.39}$$

Notice that the forces in Equation 9.38 belong to rows 1, 3, 4, 5, 7, 8, 9, 11, and 12 (see the full matrix of Equation 9.36), so using this information along with the function **AxisymmetricT_ReducedStiffnessMatrix()** we establish the reduced stiffness matrix as

```
vec_rows_of_knownloads=c(1,3,4,5,7,8,9,11,12);
ReducedK=AxisymmetricT_ReducedStiffnessMatrix(GlobalK,
vec_rows_of_knownloads);
ReducedK
```

The output reads

```
> ReducedK
             [,1]          [,2]          [,3]          [,4]          [,5]          [,6]          [,7]          [,8]          [,9]
[1,]  330120083161 -324657225922  -21749487602 -5.540812e+09  0.000000e+00             0             0             0             0
[2,] -324657225922  333772459235   15707963268  2.015318e+09 -8.709512e+09  -37457450870             0             0             0
[3,]  -21749487602   15707963268 1138935774543  8.860902e+10 -3.544361e+10   -2726431494             0             0             0
[4,]   -5540812392    2015317868   88609023563  1.089932e+12 -1.062019e+12  -88609023563  -16972806464             0             0
[5,]             0   -8709512178  -35443609425 -1.062019e+12  1.094781e+12  117608340365     848647480  -19855688720  -82164730940
[6,]             0  -37457450870   -2726431494 -8.860902e+10  1.176083e+11 3730090638372  159093474124  -63637389650   -5785217241
[7,]             0             0             0 -1.697281e+10  8.486475e+08  159093474124  720705425680 -698410409656  -50346036115
[8,]             0             0             0  0.000000e+00 -1.985569e+10  -63637389650 -698410409656  723068727846  172384827659
[9,]             0             0             0  0.000000e+00 -8.216473e+10   -5785217241  -50346036115  172384827659 2455824718752
```

Or:

$$[K_{reduced}] = 10^9 \begin{bmatrix} 330.1201 & -324.6572 & -21.7495 & -5.5408 & 0. & 0. & 0. & 0. & 0. \\ -324.6572 & 333.7725 & 15.708 & 2.0153 & -8.7095 & -37.4575 & 0. & 0. & 0. \\ -21.7495 & 15.708 & 1138.9358 & 88.609 & -35.4436 & -2.7264 & 0. & 0. & 0. \\ -5.5408 & 2.0153 & 88.609 & 1089.9358 & -1062.0192 & -88.609 & -16.9728 & 0. & 0. \\ 0. & -8.7095 & -35.4436 & -1062.0192 & 1094.7807 & 117.6083 & 0.8486 & -19.8557 & -82.1647 \\ 0. & -37.4575 & -2.7264 & -88.609 & 117.6083 & 3730.0906 & 159.0935 & -63.6374 & -5.7852 \\ 0. & 0. & 0. & -16.9728 & 0.8486 & 159.0935 & 720.7054 & -698.4104 & -50.346 \\ 0. & 0. & 0. & 0. & -19.8557 & -63.6374 & -698.4104 & 723.3848 & 172.3848 \\ 0. & 0. & 0. & 0. & -82.1647 & -5.7852 & -50.346 & 172.3848 & 2455.8247 \end{bmatrix} \quad (9.40)$$

In the next code snippet, the reduced vector of known loads is formulated.

```
vec_values_of_knownloads=rep(0,9)+GlobalKBF[
vec_rows_of_knownloads]
Reducedloadvector=AxisymmetricT_ReducedLoadVector(vec_
values_of_knownloads);
Reducedloadvector
```

Again, the first line of the preceding code snippet considers both the vector of the known concentrated radial loads and vector of the equivalent nodal loads. If executed, the code produces:

```
> Reducedloadvector
          [,1]
 [1,]  3196.700
 [2,]  9636.669
 [3,]     0.000
 [4,] 22651.711
 [5,] 37021.378
 [6,]     0.000
 [7,] 30581.409
 [8,] 17566.367
 [9,]     0.000
```

Combining the reduced stiffness matrix and the reduced vector of applied loads, one obtains the reduced matrix equation.

$$\begin{Bmatrix} 3196.700 \\ 9636.669 \\ 0 \\ 22651.711 \\ 37021.378 \\ 0 \\ 30581.409 \\ 17566.367 \\ 0 \end{Bmatrix} = \left[K_{\text{reduced}}\right] \begin{Bmatrix} U_{1r} \\ U_{2r} \\ U_{3r} \\ U_{4r} \\ U_{4z} \\ U_{5r} \\ U_{5z} \\ U_{6r} \\ U_{6z} \end{Bmatrix} \tag{9.41}$$

Step 7: Determine the unknown nodal displacements by using **AxisymmetricT_NodalDisplacement()**.

```
UnknwonNodalDisp=AxisymmetricT_NodalDisplacement(ReducedK,
Reducedloadvector)
UnknwonNodalDisp*1000
```

The determined radial displacements (in *millimetres*) are then obtained as

```
> UnknwonNodalDisp*1000
                [,1]
 [1,]  0.0052005066
 [2,]  0.0051968198
 [3,] -0.0002386942
 [4,]  0.0057033749
 [5,]  0.0057114337
 [6,] -0.0001458310
 [7,]  0.0059751330
 [8,]  0.0059648060
 [9,] -0.0001054567
```

The previous box shows that the nodal displacements are approximately

$$\begin{Bmatrix} U_{1r} \\ U_{2r} \\ U_{3r} \\ U_{4r} \\ U_{4z} \\ U_{5r} \\ U_{5z} \\ U_{6r} \\ U_{6z} \end{Bmatrix} = \begin{Bmatrix} 0.00520 \\ 0.00520 \\ -0.00024 \\ 0.00570 \\ 0.00571 \\ -0.00015 \\ 0.00598 \\ 0.00596 \\ -0.00011 \end{Bmatrix} \text{mm} \tag{9.42}$$

At this juncture, it is necessary to combine the determined displacements and the boundary conditions (from Equation 9.37).

```
allglobalNodalDisp=matrix(rep(0,Total_dof),byrow=T);
allglobalNodalDisp[vec_rows_of_knownloads]=UnknwonNodalDisp
allglobalNodalDisp*1000
```

which produces

```
> allglobalNodalDisp*1000
              [,1]
 [1,]  0.0052005066
 [2,]  0.
 [3,]  0.0051968198
 [4,] -0.0002386942
 [5,]  0.0057033749
 [6,]  0.
 [7,]  0.0057114337
 [8,] -0.0001458310
 [9,]  0.0059751330
[10,]  0.
[11,]  0.0059648060
[12,] -0.0001054567
```

Finally, the stresses in elements 1, 2, 3, and 4 are computed by using **AxisymmetricT_Stresses()** as follows.

```
Element1_Stress=AxisymmetricT_
Stresses(YoungMod,Nu,e1coord,allglobalNodalDisp,1,3,2)
Element1_Stress/10^3
Element2_Stress=AxisymmetricT_
Stresses(YoungMod,Nu,e2coord,allglobalNodalDisp,2,3,4)
Element2_Stress/10^3
Element3_Stress=AxisymmetricT_
Stresses(YoungMod,Nu,e3coord,allglobalNodalDisp,4,3,5)
Element3_Stress/10^3
Element4_Stress=AxisymmetricT_
Stresses(YoungMod,Nu,e4coord,allglobalNodalDisp,4,5,6)
Element4_Stress/10^3
```

Accordingly, the output returns the stress σ_{rr},σ_{zz},$\sigma_{\theta\theta}$ and τ_{rz} (in kPa) for elements 1, 2, 3, and 4 as

```
> Element1_Stress/10^3
             [,1]
[1,]  4280.09634
[2,]    12.76755
[3,] 11675.40840
[4,]   -28.35958
> Element2_Stress=AxisymmetricT_Stresses(YoungMod,Nu,
e2coord,allglobalNodalDisp,2,3,4)
> Element2_Stress/10^3
           [,1]
[1,] 3739.2987
[2,]  886.7339
[3,] 8938.5500
[4,]  116.9391
> Element3_Stress=AxisymmetricT_Stresses(YoungMod,Nu,
e3coord,allglobalNodalDisp,4,3,5)
> Element3_Stress/10^3
            [,1]
[1,] 1933.25832
[2,] -690.37732
[3,] 5487.55299
[4,]   61.99049
> Element4_Stress=AxisymmetricT_Stresses(YoungMod,Nu,
e4coord,allglobalNodalDisp,4,5,6)
> Element4_Stress/10^3
             [,1]
[1,] 1981.431935
[2,]    4.654931
[3,] 5064.528003
[4,]  -51.204242
```

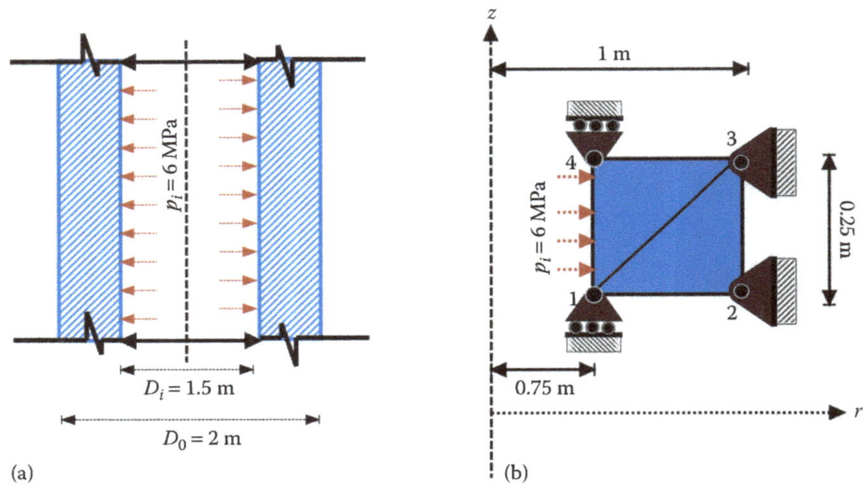

FIGURE 9.5
(a and b) Exercise 9Q1.

9.5 Exercises

1. Figure 9.5a shows a long thick-walled cylindrical pressure vessel with inner and outer diameter of 1.5 and 2 m, respectively. If the vessel is subjected to inner pressure of 6 MPa, determine the radial displacement of the inner surface of the vessel by using the two-element axisymmetric model in Figure 9.5b. Take $E = 210$ GPa and $v = 0.3$.
2. Determine the percentage change in radial and axial displacements for the flywheel problem solved in Example 9.2 if the effect of gravity is included in the analysis.

References

1. S. S. Rao, *The Finite Element Method in Engineering*, Elsevier Science, Burlington, NJ, 2010.
2. D. L. Logan, *First Course in the Finite Element Method*, Thomson, Toronto, ON, 2007.

3. D. W. A. Rees, *Mechanics of Solids and Structures*, 2nd ed., World Scientific Publishing Company Imperial College Press, London, UK, 2016.
4. T. R. Chandrupatla, A. D. Belegundu, *Introduction to Finite Elements in Engineering*, Prentice Hall, Upper Saddle River, NJ, 1997.
5. W. Hayes, L. Swenson, D. Schurman, Axisymmetric finite element analysis of the lateral tibial plateau, *Journal of Biomechanics*, 11: 21–33, 1978.

10

Finite Element Formulations for Thermal Stress Problems

10.1 Introduction

The thermal stress analysis of bar and truss elements is the objective of this chapter. Solutions to thermal stress problems are necessary for applications in the design of components for high-temperature environments (e.g., turbine, nuclear reactor, thermal protection systems, etc.) or systems in which temperature rise or drop can lead to unintended changes in the geometric properties of load-supporting members (e.g., bridge, microelectronics packaging, and microelectromechanical systems) [1,2].

The theoretical basis for thermal stress considerations in the design of mechanical components is best acquired from specialized sources such as References 3,4. That said, in this chapter, we will acquaint ourselves with the finite element matrix equations for thermal stresses developed in bars and trusses in the presence of arbitrary change in temperature. As done in the previous chapters, we employed the mathematical convenience proffered by the principle of minimum potential energy to arrive at the elements' equations. The relevant R functions and example problems dealing with thermal stress problems in simple structures are presented in the rest of the chapters.

10.2 Finite Element Equations for Thermal Stress Problems

10.2.1 Stress–Strain–Temperature Relationship

Consider the bar in Figure 10.1 having length L and cross-sectional area A. When the temperature of the bar changes from an initial temperature T_1 to a final temperature T_2, the developed strain is generally a function of stress and temperature [5]:

$$\varepsilon = \varepsilon(\sigma, T) \tag{10.1}$$

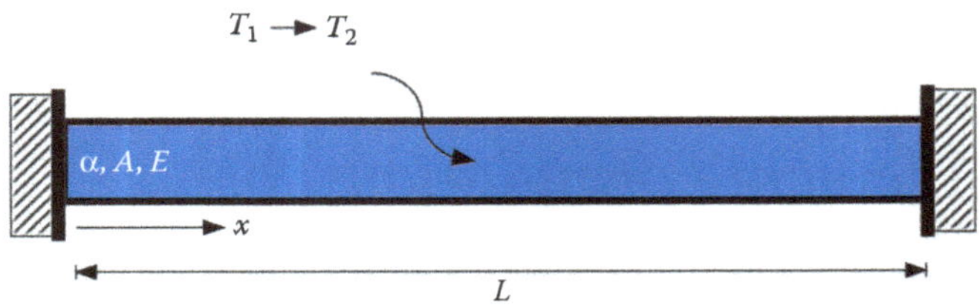

FIGURE 10.1
A statically indeterminate bar under temperature change.

where the stress dependence arises from the bar being statically indeterminate. Given this, a differential strain can be computed as

$$d\varepsilon = \left(\frac{\partial \varepsilon}{\partial \sigma}\right)_T d\sigma + \left(\frac{\partial \varepsilon}{\partial T}\right)_\sigma dT \tag{10.2}$$

One quickly recognizes that $(\partial\varepsilon/\partial\sigma)_T$ represents the inverse of the bar's Young's modulus, and $(\partial\varepsilon/\partial T)_\sigma$ represents the coefficient of thermal expansion. Consequently, we can recast Equation 10.2 as

$$d\varepsilon = \frac{1}{E} d\sigma + \alpha\, dT \tag{10.3}$$

By integrating Equation 10.3 within the temperature limits T_0 and T_1, one retrieves the stress–strain–temperature relationship for simple one-dimensional systems as

$$\varepsilon = \frac{\sigma}{E} + \alpha\left(T_1 - T_0\right) \tag{10.4}$$

The first part of Equation 10.4 is the mechanically induced strain, whereas the second part relates to the thermally induced strain. A simple manipulation of Equation 10.4 produces the stress–strain–temperature relationship that emerges when there is a simultaneous application of an internal force and the temperature change in the bar as

$$\sigma = E\varepsilon - E\alpha\nabla T = E\left(\varepsilon - \alpha\nabla T\right) = E\left(\varepsilon_m - \varepsilon_T\right) \tag{10.5}$$

where ε_m and ε_T are the mechanical and thermally induced strain, respectively. For purpose of generality, Equation 10.5 can be rewritten in terms of the material property matrix $[D]$ as

$$\{\sigma\} = [D]\left(\{\varepsilon_m\} - \{\varepsilon_T\}\right) \tag{10.6}$$

10.2.2 Minimization of the Total Potential Energy for Thermal Stress Problems in Bar Elements

For a bar of cross-sectional area A, the total strain energy (U) remains

$$U = \frac{1}{2}\iiint_{\text{vol}} \{\sigma\}^T \{\varepsilon\} dV = \frac{A}{2}\int_0^L \{\sigma\}^T \{\varepsilon\} dx \tag{10.7}$$

Considering the thermal stress effect, we recast Equation 10.7 as

$$U = \frac{A}{2}\int_0^L \left(\{\varepsilon_m\} - \{\varepsilon_T\}\right)^T [D]\left(\{\varepsilon_m\} - \{\varepsilon_T\}\right) dx \tag{10.8}$$

From the earlier chapters, we know that the strain can be mapped to the displacement field using the $[B]$ matrix:

$$\{\varepsilon_m\} = [B]\{u_e\} \tag{10.9}$$

Plugging of Equation 10.9 into Equations 10.6 through 10.8 transforms the strain-energy expression to

$$U = \frac{A}{2}\int_0^L \left([B]\{u_e\} - \{\varepsilon_T\}\right)^T [D]\left([B]\{u_e\} - \{\varepsilon_T\}\right) dx \tag{10.10}$$

The work done by externally applied concentrated nodal forces (W) is of the form

$$W = \{u_e\}^T \begin{Bmatrix} f_{1x} \\ f_{2x} \end{Bmatrix} \tag{10.11}$$

Consequently, the total potential energy can be written as

$$P_t = U - W = \frac{A}{2}\int_0^L \left([B]\{u_e\} - \{\varepsilon_T\}\right)^T [D]\left([B]\{u_e\} - \{\varepsilon_T\}\right) dx - \{u_e\}^T \begin{Bmatrix} f_{1x} \\ f_{2x} \end{Bmatrix} \tag{10.12}$$

To obtain the stiffness matrix, we minimize Equation 10.12 with respect to u_e to obtain

$$\frac{\partial}{\partial \{u_e\}^T} P_t = \frac{\partial}{\partial \{u_e\}^T} U_1 + \frac{\partial}{\partial \{u_e\}^T} U_2 + \frac{\partial}{\partial \{u_e\}^T} U_3 - \{f_e\} \tag{10.13}$$

where

$$U_1 = \frac{A}{2}\int_0^L ([B]\{u_e\})^T [D]([B]\{u_e\})dx \tag{10.14}$$

$$U_2 = -\frac{A}{2}\int_0^L 2([B]\{u_e\})^T [D]\{\varepsilon_T\}dx \tag{10.15}$$

$$U_3 = \frac{A}{2}\int_0^L \{\varepsilon_T\}^T [D]\{\varepsilon_T\}dx \tag{10.16}$$

Carrying out the integration of Equation 10.13, we obtain

$$\left(A\int_0^L [B]^T [D][B]dx\right)\{u_e\} - \left(A\int_0^L [B][D]\{\varepsilon_T\}dx\right) = \{f_e\} \tag{10.17}$$

For a horizontally positioned bar element, we have

$$[B] = \begin{bmatrix} -\frac{1}{L} & \frac{1}{L} \end{bmatrix}; [D] = E \tag{10.18}$$

Therefore, the element matrix equation for a statically indeterminate bar experiencing a temperature change takes the form

$$\frac{AE}{L}\begin{bmatrix} 1 & -1 \\ -1 & 1 \end{bmatrix}\begin{Bmatrix} u_{1x} \\ u_{2x} \end{Bmatrix} - \begin{Bmatrix} -AE\nabla T\alpha \\ AE\nabla T\alpha \end{Bmatrix} = \begin{Bmatrix} f_{1x} \\ f_{2x} \end{Bmatrix} \tag{10.19}$$

10.2.3 Element Equations for Thermal Stress Problems in Planar Trusses

In Chapter 5, we derived the vector of mechanical strains for the planar truss element as

$$\varepsilon_{xx} = \frac{1}{L_e}\begin{bmatrix} -\cos\theta & -\sin\theta & \cos\theta & \sin\theta \end{bmatrix}\begin{Bmatrix} u_{1X} \\ u_{1Y} \\ u_{2X} \\ u_{2Y} \end{Bmatrix} \tag{10.20}$$

We also obtained $[B]$ for the truss (in Chapter 5) as

$$[B]=\frac{1}{L_e}\begin{bmatrix}-\cos\theta & -\sin\theta & \cos\theta & \sin\theta\end{bmatrix} \tag{10.21}$$

By substituting Equations 10.20 and 10.21 into Equation 10.17, the matrix equation for a planar truss element experiencing temperature change turns out to be

$$\begin{Bmatrix} f_{1x} \\ f_{1y} \\ f_{2x} \\ f_{2y} \end{Bmatrix} = [k]\begin{Bmatrix} u_{1x} \\ u_{1y} \\ u_{2x} \\ u_{2y} \end{Bmatrix} - AE\begin{Bmatrix} -\nabla T\alpha\cos\theta \\ -\nabla T\alpha\sin\theta \\ \nabla T\alpha\cos\theta \\ \nabla T\alpha\sin\theta \end{Bmatrix} \tag{10.22}$$

where

$$[k]=\frac{A_eE}{L_e}\begin{bmatrix} \cos^2\theta & \cos\theta\sin\theta & -\cos^2\theta & -\cos\theta\sin\theta \\ \cos\theta\sin\theta & \sin^2\theta & -\cos\theta\sin\theta & -\sin^2\theta \\ -\cos^2\theta & -\cos\theta\sin\theta & \cos^2\theta & \cos\theta\sin\theta \\ -\cos\theta\sin\theta & -\sin^2\theta & \cos\theta\sin\theta & \sin^2\theta \end{bmatrix} \tag{10.23}$$

10.3 Computer Implementation

10.3.1 Systematic Procedure for Analyses of Thermal Stress in Bars and Trusses

10.3.1.1 Precomputation Phase

- Establish a global coordinate system.
- Discretize the structure into finite elements.
- Label the nodes and elements to form the connectivity table.

10.3.1.2 Computation Phase

1. Specify the material and geometric properties of each bar element.
2. Establish the finite element matrix equation for each element by using Equation 10.19 for a bar element or Equation 10.22 for a truss element. Or simply, form the stiffness matrix for each element.

3. For elements with thermal loads, determine the column matrix of the equivalent thermal nodal loads.
4. Determine the global degree of freedom of the connected bars (or truss), and carry out the expansion of
   a. The matrix equation for each element (from Step 1).
   b. The vector of thermal nodal loads (for any element with a thermal load).
5. Form the global stiffness matrix and the global matrix equation.
6. Identify the essential boundary condition(s) and apply the boundary conditions to the global matrix equation to obtain a reduced matrix equation. Programmatically, this step involves the following two actions:
   a. Obtain the reduced load vector, which will be equal to the reduced load vector from concentrated applied loads (if they exist) *plus* the reduced load vector from the expanded vector of equivalent thermal nodal loads.
   b. Use the identified boundary conditions to obtain the reduced stiffness matrix.
7. Determine the unknown global nodal degree of freedom (displacements and slopes) by using the reduced matrix equation. By using the determined displacements, obtain other variables such as stress and so on.

10.3.2 Implemented R Functions for the Thermal Stress Analysis (Bar Elements)

```
#*******************************************************#
# R functions for thermal stress analysis of     #
#     linear bar elements                         #
#                                                 #
#       KB Mustapha, 2017                         #
#*******************************************************#

ThermalBar_Element_Matrix = function(DOF=2,stiffbar)
{
  ematrix=stiffbar*matrix(c(1,-1,-1,1),nrow=DOF,byrow=T);
  return (ematrix)
}
```

```
ThermalBar_BF = function(DOF=2,YoungMod,area,alpha,
tempchange)
{
  nodalthermal=area*alpha*tempchange*YoungMod
  equivalentbodyload=matrix(c(-nodalthermal,nodalthermal),
nrow=DOF,byrow=T)
  return (equivalentbodyload)
}

ThermalBar_ExpandedBF = function(TDOF,LoadColumn
Matrix,i,j)
{
  bigColumnMatrix=matrix(vector(l=TDOF),nrow=TDOF,
  byrow=T);
  bigColumnMatrix[c(i,j)]=LoadColumnMatrix;
  return (bigColumnMatrix)
}

ThermalBar_Element_ExpMatrix = function(TDOF)
{
  expmatrix=matrix(vector(l=TDOF*TDOF),nrow=TDOF,
  byrow=T);
  return (expmatrix)
}

ThermalBar_ExpandedElement_Matrix = function
(TDOF,eMatrix,i,j)
{
  bigMatrix = matrix(vector(l = TDOF * TDOF),nrow = TDOF,
  byrow = T);
  bigMatrix[c(i,j),c(i,j)] = eMatrix;
  return (bigMatrix)
}

ThermalBar_Element_ReducedStiffnessMatrix = function
(bigKmatrix,knownforcenodes)
{
  reducedsmatrix=bigKmatrix[c(knownforcenodes),
  (knownforcenodes)];
  return(reducedsmatrix)
}
```

```
ThermalBar_Element_ReducedForceVector = function
(forcevector)
{
  reducedforcevector = matrix(forcevector,ncol = 1)
  return(reducedforcevector)
}

ThermalBar_Element_NodalDisplacement = function
(reducedmatrix,reducedforcevec)
{
    return(solve(reducedmatrix,reducedforcevec))
}

ThermalBar_Element_Forces = function(DOF=2,stiffbar,nodaldisp)
{
  stiffnessmatrix=stiffbar*matrix(c(1,-1,-1,1),
  nrow=DOF,byrow=T)nodal_forces=stiffnessmatrix %*%
  matrix(nodaldisp,ncol=1)
  return(nodal_forces)
}

ThermalBar_Global_Forces = function(bigKmatrix,totalnodal
disp)
{
  global_forces = bigKmatrix %*% totalnodaldisp
  return(round(global_forces))
}

ThermalBar_Stresses = function(YoungMod,alpha,tempchange,
Length,vec_globalnodaldisp,i,j)
{
  r1=i; r2=j
  BMatrix=matrix(c(-1/Length,1/Length),nrow=1,byrow=T);
  localstresses=(YoungMod%*%BMatrix%*%vec_globalnodaldisp
  [c(r1,r2)])-(YoungMod*alpha*tempchange)
  return(localstresses)
}
```

10.3.3 Implemented R Functions for Thermal Stress Analysis (Truss Elements)

```
#*********************************************#
# R functions for thermal stress analysis of  #
#        plane truss element                  #
#       with thermal loads                    #
#          KB Mustapha                        #
#*********************************************#

PlaneTruss_LengthOrientation = function(vec_nodal
coordinates)
{
  ysquared=(vec_nodalcoordinates[4]-vec_nodal
  coordinates[2])^2
  xsquared=(vec_nodalcoordinates[3]-vec_nodal
  coordinates[1])^2
  length_of_element=sqrt(ysquared+xsquared)
  ydiff=(vec_nodalcoordinates[4]-vec_nodalcoordinates[2])
  xdiff=(vec_nodalcoordinates[3]-vec_nodalcoordinates[1])
  orientation_of_element=(180/pi)*atan(ydiff/xdiff)
  return(round(c(length_of_element,orientation_of_
  element),3))

}

ThermalPlaneTruss_Element_Matrix = function(DCF=4,
stifftruss,theta)
{
  m=cos(theta*pi/180);
  n=sin(theta*pi/180);
  stiffnesscoefficients=c(m^2,m*n,-m^2,-m*n,m*n,n^2,
  -m*n,-n^2,-m^2,-m*n,m^2,m*n,-m*n,-n^2,m*n,n^2)
  ematrix=stifftruss*matrix(stiffnesscoefficients,
  nrow=DOF,byrow=T)
}

ThermalPlaneTruss_BF = function(DOF=4,YoungMod,area,alpha,
tempchange,theta)
{
  m=cos(theta*pi/180);
  n=sin(theta*pi/180);
  nodalthermal=area*alpha*tempchange*YoungMod
  equivalentbodyload=matrix(c(-nodalthermal*m,-nodal
  thermal*n,nodalthermal*m,nodalthermal*n),
                                nrow=DOF,byrow=T)
  return (equivalentbodyload)
}
```

```
ThermalPlaneTruss_ExpandedBF = function(TDOF,LoadColumn
Matrix,i,j)
{
  r1=(i-1)+i;r2=(i-1)+(i+1);r3=(j-2)+(j+1);
  r4=(j-2)+(j+2);
  bigColumnMatrix=matrix(vector(l=TDOF),nrow=TDOF,
  byrow=T);
  bigColumnMatrix[c(r1,r2,r3,r4)]=LoadColumnMatrix;
  return (bigColumnMatrix)
}


ThermalPlaneTruss_ExpandedElement_Matrix = function
(TDOF,eMatrix,i,j)
{

  r1=(i-1)+i;r2=(i-1)+(i+1);r3=(j-2)+(j+1);
  r4=(j-2)+(j+2);
  bigMatrix=matrix(vector(l=TDOF*TDOF),nrow=TDOF,
  byrow=T);
  bigMatrix[c(r1,r2,r3,r4),c(r1,r2,r3,r4)]=eMatrix;
  return (bigMatrix)
}


ThermalPlaneTruss_ReducedStiffnessMatrix = function(bigK
matrix,knownforcenodes)
{
  reducedstiffnessmatrix = bigKmatrix[c(knownforcenodes),
  (knownforcenodes)]
  return(reducedstiffnessmatrix)
}

ThermalPlaneTruss_ReducedForceVector = function
(forcevector)
{
  reducedforcevector = matrix(forcevector,ncol = 1)
  return(reducedforcevector)
}
ThermalPlaneTruss_NodalDisplacement = function
(reducedmatrix,vec_reducedforce)
{
  return(solve(reducedmatrix,vec_reducedforce))
```

```
}

ThermalPlaneTruss_Global_Forces = function(bigKmatrix,
vec_globalnodaldisp)
{
  global_forces = bigKmatrix %*% vec_globalnodaldisp
  return(round(global_forces))
}

ThermalPlaneTruss_Local_Forces = function(stifftruss,
theta,vec_globalnodaldisp)
{
  m=cos(theta*pi/180);
  n=sin(theta*pi/180);
  localstiffnessmatrix=matrix(c(1,-1,-1,1),
  byrow=T,nrow=2);
  transformationmatrix=matrix(c(m,n,0,0,0,0,m,n),
  byrow=T,nrow=2);
  local_forces = (stifftruss)*localstiffnessmatrix %*%
  transformationmatrix %*%
    vec_globalnodaldisp
  return(round(local_forces))
}

ThermalPlaneTruss_AxialStress = function(YoungMod,elem_
len,theta,alpha,tempchange,vec_globalnodaldisp,i,j)
{
  m=cos(theta*pi/180);
  n=sin(theta*pi/180);
  r1=(i-1)+i;r2=(i-1)+(i+1);r3=(j-2)+(j+1);r4=(j-2)+
  (j+2);
  localdisp=vec_globalnodaldisp[c(r1,r2,r3,r4)];
  disp_vec=matrix(localdisp,nrow=length(localdisp),
  byrow=T,ncol = 1);
  transformationmatrix=matrix(c(-m,-n,m,n),nrow=1,
  byrow=T);
  local_stress=((YoungMod/elem_len)*transformation
  matrix%*%disp_vec)-(YoungMod*alpha*tempchange);
  return(local_stress)
}
```

The usage of these functions will be illustrated in the examples that follow.

10.4 Examples

10.4.1 Example 10.1

Each member of the bar assembly shown in Figure 10.2 has the material and diameters as indicated. The assembly is constrained between two supports at an initial temperature $T_0 = 20°$. If the temperature reaches $T_1 = 45°$ during the operational use of the system, employ the bar element developed in this chapter, to determine the

1. Displacements of nodes 2 and 3;
2. The reaction forces at the nodes;
3. The normal stress developed in each element.

10.4.1.1 Solution

This problem is a variant of a problem exercise solved in Reference 6 using the analytical method. We will solve the same problem with the code snippet provided in Section 10.3.2. For the solution, we start by designating the aluminum, bronze, and stainless-steel bars as elements 1, 2, and 3, respectively. The material properties for each element are then listed in Table 10.1.

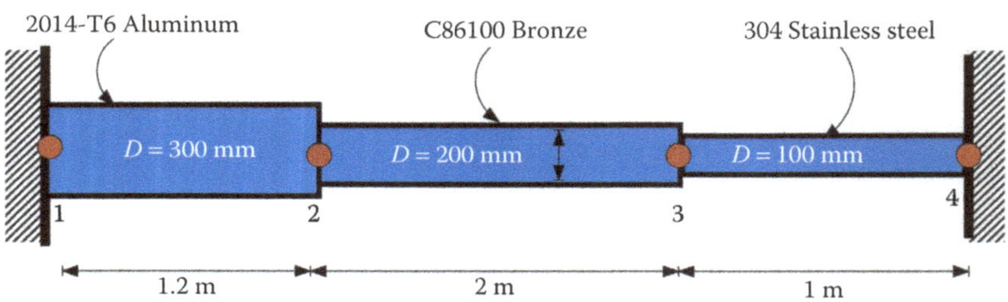

FIGURE 10.2
A three-element assembly under temperature change.

TABLE 10.1
Material Properties

Element	Young's Modulus (*E*) (GPa)	Coefficient of Thermal Expansion (*α*) (/°C)
1	73.1	23×10^{-6}
2	103	17×10^{-6}
3	193	17×10^{-6}

TABLE 10.2

Elements Connectivity and Axial Stiffness

Elements	Nodes (*i*, *j*)	Axial Stiffness (*AE*/*L*) (MN/m)
1	1,2	4305.945
2	2,3	1617.920
3	3,4	1515.818

Furthermore, we form Table 10.2, which reveals the connectivity of the element.

By using these two tables, the rest of the steps is carried out using the R functions.

Step 1: Specify the elements' properties.

```
E1=73.1e9;E2=103e9;E3=193e9;
A1=pi*(0.3^2)/4
A2=pi*(0.2^2)/4
A3=pi*(0.1^2)/4;
L1=1.2;L2=2;L3=1;
tempchange=20;
alpha1=23e-6;alpha2=17e-6;alpha3=17e-6;
```

Running the above-mentioned code, we have the output for the axial stiffness of each bar as

```
> Axialstiffness1=(E1*A1)/L1;Axialstiffness1
[1] 4305945431
> Axialstiffness2=(E2*A2)/L2;Axialstiffness2
[1] 1617920217
> Axialstiffness3=(E3*A3)/L3;Axialstiffness3
[1] 1515818455
```

Step 2: Write the stiffness matrix for each element using the function **ThermalBar_Element_Matrix()**.

```
k1=ThermalBar_Element_Matrix(2,Axialstiffness1);k1
k2=ThermalBar_Element_Matrix(2,Axialstiffness2);k2
k3=ThermalBar_Element_Matrix(2,Axialstiffness3);k3
```

The stiffness matrices are obtained and shown in the following block:

```
> k1=ThermalBar_Element_Matrix(2,Axialstiffness1);k1
            [,1]        [,2]
[1,]  4305945431 -4305945431
[2,] -4305945431  4305945431
> k2=ThermalBar_Element_Matrix(2,Axialstiffness2);k2
            [,1]        [,2]
[1,]  1617920217 -1617920217
[2,] -1617920217  1617920217
> k3=ThermalBar_Element_Matrix(2,Axialstiffness3);k3
            [,1]        [,2]
[1,]  1515818455 -1515818455
[2,] -1515818455  1515818455
```

As all the elements experience a temperature rise of 20°C, the distributed thermal load needs to be determined for all as well. We do this by calling on the function **`ThermalBar_BF()`** as shown next:

```
k1ThermalLoad=ThermalBar_BF(2,E1,A1,alpha1,tempchange);k1
ThermalLoad
k2ThermalLoad=ThermalBar_BF(2,E2,A2,alpha2,tempchange);k2
ThermalLoad
k3ThermalLoad=ThermalBar_BF(2,E3,A3,alpha3,tempchange);k3
ThermalLoad
```

The output of which reads as

```
> k1ThermalLoad=ThermalBar_BF(2,E1,A1,alpha1,tempchange);
k1ThermalLoad
         [,1]
[1,] -2376882
[2,]  2376882
> k2ThermalLoad=ThermalBar_BF(2,E2,A2,alpha2,tempchange);
k2ThermalLoad
         [,1]
[1,] -1100186
[2,]  1100186
> k3ThermalLoad=ThermalBar_BF(2,E3,A3,alpha3,tempchange);
k3ThermalLoad
          [,1]
[1,] -515378.3
[2,]  515378.3
```

Step 3: Determine the global degree of freedom of the structure and expand the matrix equation for each element.

We note that the structure has 4 nodes, and each node has 1 degree of freedom (being a bar). We thus have a total of 4 degrees of freedom for this system. Consequently, the modified elements' matrices are obtained: **ThermalBar_ExpandedElement_Matrix()**, as done in the following:

```
n_nodes=4;Total_dof=n_nodes*1;
K1=ThermalBar_ExpandedElement_Matrix(Total_dof,k1,1,2);K1
K2=ThermalBar_ExpandedElement_Matrix(Total_dof,k2,2,3);K2
K3=ThermalBar_ExpandedElement_Matrix(Total_dof,k3,3,4);K3
```

For consistency, the vector of thermal loads must also be expanded accordingly using the function **ThermalBar_ExpandedBF()**.

```
K1Thermal=ThermalBar_ExpandedBF(Total_dof,k1ThermalLoad,
1,2);K1Thermal
K2Thermal=ThermalBar_ExpandedBF(Total_dof,k2ThermalLoad,
2,3);K2Thermal
K3Thermal=ThermalBar_ExpandedBF(Total_dof,k3ThermalLoad,
3,4);K3Thermal
```

Step 4: Establish the global stiffness matrix by adding the expanded matrices of the elements as shown next.

As we have only three elements, the global stiffness matrix is computed from the expanded matrix equations for these elements. This takes the form

```
GlobalK=K1+K2+K3;
GlobalK
```

The output of this gives the 4×4 global stiffness matrix $[K]$ produced in the following:

```
> GlobalK
             [,1]        [,2]        [,3]        [,4]
[1,]  4305945431 -4305945431           0           0
[2,] -4305945431  5923865647 -1617920217           0
[3,]           0 -1617920217  3133738672 -1515818455
[4,]           0           0 -1515818455  1515818455
```

To consider the global thermal body force, we must carry out the addition of the expanded nodal thermal loads as well:

```
GlobalThermal=K1Thermal+K2Thermal+K3Thermal;
GlobalThermal
```

Which produces

```
> GlobalThermal
          [,1]
[1,] -2376881.9
[2,]  1276696.1
[3,]   584807.5
[4,]   515378.3
```

Step 5: Identify the boundary condition(s) and obtain the reduced stiffness matrix and reduced force vector.

Having obtained the global stiffness matrix in Step 3, we may write the global matrix equation as

$$\begin{Bmatrix} F_{1x} \\ F_{2x} \\ F_{3x} \\ F_{4x} \end{Bmatrix} = \begin{bmatrix} K_{\text{global}} \end{bmatrix} \begin{Bmatrix} U_{1x} \\ U_{2x} \\ U_{3x} \\ U_{4x} \end{Bmatrix} - \begin{Bmatrix} F_{T1} \\ F_{T2} \\ F_{T3} \\ F_{T4} \end{Bmatrix} \tag{10.24}$$

From Figure 10.2, the assembly's boundary conditions are identified as

$$U_{1x} = U_{4x} = 0 \tag{10.25}$$

Furthermore, we can identify the known nodal forces as

$$F_{2x} = F_{3x} = 0 \tag{10.26}$$

We are now able to extract the reduced stiffness matrix by using **ThermalBar_Element_ReducedStiffnessMatrix()**, which of course requires the vector of row(s) of the known nodal forces as indicated in the next block of code.

```
vec_rows_of_knownloads=c(2,3);
ReducedK=ThermalBar_Element_ReducedStiffnessMatrix(Global
K,vec_rows_of_knownloads);
ReducedK
```

The output of the above-mentioned box reads

```
> ReducedK
              [,1]        [,2]
[1,]  5923865647 -1617920217
[2,] -1617920217  3133738672
```

In the next code snippet, the set of known loads identified in Equation 10.26 along with the relevant thermal loads are combined to form the column matrix of reduced loads.

```
vec_values_of_knownloads=c(0,0)+GlobalThermal[c(2,3)];
ReducedLoad
=ThermalBar_Element_ReducedForceVector(vec_values_of_
knownloads);
ReducedLoad
```

With the result being

```
> ReducedLoad
          [,1]
[1,] 1276696.1
[2,]  584807.5
```

Combining the reduced stiffness matrix and the reduced vector of load leads to the reduced matrix equation.

$$\begin{Bmatrix} 1276696.1 \\ 584807.5 \end{Bmatrix} = \begin{bmatrix} 5923865647 & -1617920217 \\ -1617920217 & 3133738672 \end{bmatrix} \begin{Bmatrix} U_{2x} \\ U_{3x} \end{Bmatrix} \tag{10.27}$$

Step 6: Determine the unknown nodal displacements and other variables of interest.

```
UnknwonNodalDisp=ThermalBar_Element_NodalDisplacement(Red
ucedK,ReducedLoad);
UnknwonNodalDisp
```

The determined displacements read

```
> UnknwonNodalDisp
              [,1]
[1,] 0.0003102313
[2,] 0.0003467861
```

The earlier box shows that the displacements of nodes 2 and 3 in the x-direction are

$$\begin{Bmatrix} U_{2x} \\ U_{3x} \end{Bmatrix} = \begin{Bmatrix} 0.3102313 \\ 0.3467861 \end{Bmatrix} \text{mm} \tag{10.28}$$

As has been the practice, the determined displacements and the boundary conditions are combined in a single variable **allglobalNodalDisp**.

```
allglobalNodalDisp=matrix(rep(0,Total_dof),byrow=T);
allglobalNodalDisp[vec_rows_of_
knownloads]=UnknwonNodalDisp
allglobalNodalDisp
```

Next, we determine the reaction forces. For this purpose, we must consider both the thermal and mechanical load effects.

```
MechLoadsEffect=ThermalBar_Global_Forces(GlobalK,allglobal
NodalDisp)
ThermalLoadEffect=GlobalThermal
Globalloads=MechLoadsEffect-ThermalLoadEffect
Globalloads
```

It should be noted that in the first line of the above-mentioned code, we called upon the function **ThermalBar_Global_Forces()** to determine the mechanical load effect. As apparent from the third line of the code snippet, the actual reaction forces are determined by subtracting the vector of global thermal loads from the vector of mechanical loads.

The computed reaction forces produced from running the forgoing lines of code reads

```
> Globalloads
                  [,1]
[1,]   1041042.8778162
[2,]        -0.1305291
[3,]        -0.4724657
[4,] -1041043.2748214
```

Based on the preceding output, the reaction forces at the nodes are as follows:

$$\begin{Bmatrix} F_{1x} \\ F_{2x} \\ F_{3x} \\ F_{4x} \end{Bmatrix} = \begin{Bmatrix} 1041042.877\,N \\ -0.1305291\,N \\ -0.4724657\,N \\ -1041043.274\,N \end{Bmatrix} \tag{10.29}$$

To find the normal axial stress developed in each element, we employ the function **ThermalBar_Stresses()** as shown next.

```
Element1Stress=ThermalBar_Stresses(E1,alpha1,tempchange,
L1,allglobalNodalDisp,1,2);
Element1Stress/10^6
Element2Stress=ThermalBar_Stresses(E2,alpha2,tempchange,
L2,allglobalNodalDisp,2,3);
Element2Stress/10^6
Element3Stress=ThermalBar_Stresses(E3,alpha3,tempchange,
L3,allglobalNodalDisp,3,4);
Element3Stress/10^6
```

Which gives

```
> Element1Stress/10^6
          [,1]
[1,] -14.72775
> Element2Stress=ThermalBar_Stresses(E2,alpha2,tempchange,
L2,allglobalNodalDisp,2,3);
> Element2Stress/10^6
          [,1]
```

```
[1,] -33.13743
> Element3Stress=ThermalBar_Stresses(E3,alpha3,tempchange,
L3,allglobalNodalDisp,3,4);
> Element3Stress/10^6
         [,1]
[1,] -132.5497
```

The stress developed in the aluminum, bronze, and stainless-steel component of the assembly are therefore approximately −14.72 MPa,−33.13 MPa and −132.54 MPa, respectively. These results are very consistent with the one reported in Reference 6.

10.4.2 Example 10.2

For the truss framework shown in Figure 10.3, the middle element (AC) is subjected to a uniform temperature drop of $T_0 = 20°C$. Given that $E = 70\,GPa$, $A = 4 \times 10^{-2}\,m^2$, and $\alpha = 23 \times 10^{-6}(mm/mm)/°C$. Determine the

1. Displacement of node 1;
2. Reaction forces at nodes 2, 3, and 4;
3. Axial stress developed in each element.

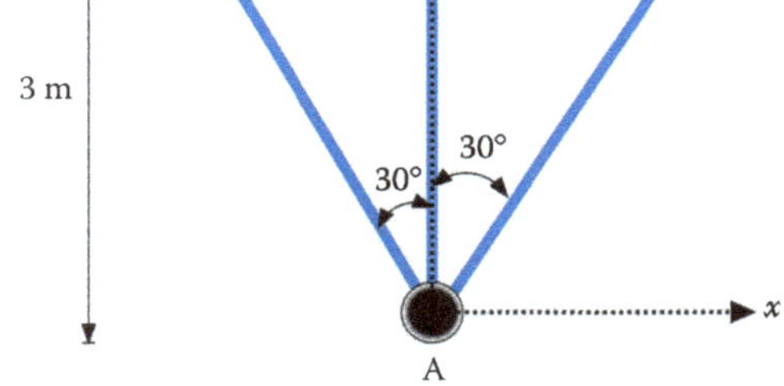

FIGURE 10.3
A three-member planar truss with an element under temperature change.

TABLE 10.3

Connectivity Table from the Discretized System

Elements	Nodes (i, j)	Coordinates (x_i, y_i, x_j, y_j)
1	1,2	(0,0,1.73,3)
2	1,3	(0,0,0,3)
3	1,4	(0,0,−1.73,3)

10.4.2.1 Solution

We will treat each points A,B,C, and D as node 1, 2, 3, and 4, respectively. This results in three elements: AB, AC, and AD. Furthermore, we locate the position of each node via the xy coordinate system placed at node 1. The ensuing connectivity of the elements then follows the form in Table 10.3.

The other steps of the solution are presented next. All the functions used will be those implemented for the planar truss element provided in Section 10.3.3.

Step 1: Specify and obtain the properties of each element (material, geometric lengths, and orientation angles).

```
E1=E2=E3=70e9;
A1=A2=A3=4e-2;
alpha1=alpha2= alpha3=23e-6;
tempchange=-20
L1T1=PlaneTruss_LengthOrientation(vec_nodalcoordinates=
c(0,0,1.73,3));
L1T1
L2T2=PlaneTruss_LengthOrientation(vec_nodalcoordinates=
c(0,0,0,3));
L2T2
L3T3=PlaneTruss_LengthOrientation(vec_nodalcoordinates=
c(0,0,-1.73,3));
L3T3
```

The variables **L1T1,L2T2,and L3T3** return the geometric lengths and orientations of the members as

```
> L1T1=PlaneTruss_LengthOrientation(vec_nodalcoordinates=
c(0,0,1.73,3));L1T1
[1]  3.463 60.029
```

```
> L2T2=PlaneTruss_LengthOrientation(vec_nodalcoordinates=
c(0,0,0,3));L2T2
[1]  3 90
> L3T3=PlaneTruss_LengthOrientation(vec_nodalcoordinates=
c(0,0,-1.73,3));L3T3
[1]   3.463 -60.029
```

Which means

$$L_1 = 3.463\,\text{m}, \theta_1 = 60°; L_2 = 3\,\text{m}, \theta_2 = 90°; L_3 = 3.463\,\text{m}, \theta_3 = -60°$$

Step 2: Write the stiffness matrix for each element with the function **ThermalPlaneTruss_Element_Matrix().**

```
stiffness1=E1*A1/L1T1[1];stiffness1
stiffness2=E2*A2/L2T2[1];stiffness2
stiffness3=E3*A3/L3T3[1];stiffness3
k1=ThermalPlaneTruss_Element_Matrix(4,stiffness1,L1T1[2]);
round(k1)
k2=ThermalPlaneTruss_Element_Matrix(4,stiffness2,L2T2[2]);
round(k2)
k3=ThermalPlaneTruss_Element_Matrix(4,stiffness3,L3T3[2]);
round(k3)
```

The element's stiffness matrices are obtained and shown in the following block:

```
> k1=ThermalPlaneTruss_Element_Matrix(4,stiffness1,L1T1[2]);
round(k1)
           [,1]       [,2]       [,3]       [,4]
[1,]  201782565  349906538 -201782565 -349906538
[2,]  349906538  606764938 -349906538 -606764938
[3,] -201782565 -349906538  201782565  349906538
[4,] -349906538 -606764938  349906538  606764938
> k2=ThermalPlaneTruss_Element_Matrix(4,stiffness2,L2T2[2]);
round(k2)
     [,1]       [,2] [,3]       [,4]
[1,]    0          0    0          0
[2,]    0  933333333    0 -933333333
[3,]    0          0    0          0
[4,]    0 -933333333    0  933333333
```

```
> k3=ThermalPlaneTruss_Element_Matrix(4,stiffness3,L3T3[2]);
round(k3)
             [,1]       [,2]       [,3]       [,4]
[1,]  201782565 -349906538 -201782565  349906538
[2,] -349906538  606764938  349906538 -606764938
[3,] -201782565  349906538  201782565 -349906538
[4,]  349906538 -606764938 -349906538  606764938
```

Step 3: Determine the column matrix of the thermal load.

Only element 2 (member AC) supports a thermal load, and we need to determine the thermal nodal loads for this element by using the function **ThermalPlaneTruss_BF()**.

```
k2ThermalLoad=ThermalPlaneTruss_BF(4,E2,A2,alpha2,temp
change,L2T2[2]);
round(k2ThermalLoad)
```

The output of this box reads

```
k2ThermalLoad=ThermalPlaneTruss_BF(4,E2,A2,alpha2,temp
change,L2T2[2]);
round(k2ThermalLoad)
          [,1]
[1,]         0
[2,]   1288000
[3,]         0
[4,]  -1288000
```

Step 4: Determine the global degree of freedom of the truss and expand the stiffness matrix and the vector of thermal loads.

For the stiffness matrix expansion, we will be using the function **ThermalPlaneTruss_ExpandedElement_Matrix()** to obtain the expanded stiffness matrix for each element as

```
n_nodes=4;Total_dof=n_nodes*2;
K1=ThermalPlaneTruss_ExpandedElement_Matrix(Total_
dof,k1,1,2);
K2=ThermalPlaneTruss_ExpandedElement_Matrix(Total_
dof,k2,1,3);
K3=ThermalPlaneTruss_ExpandedElement_Matrix(Total_
dof,k3,1,4);
```

```
K2Thermal=ThermalPlaneTruss_ExpandedBF(Total_dof,
k2ThermalLoad,1,3);
round(K2Thermal)
```

In a similar manner, the thermal load vector of each element is expanded using the function **ThermalPlaneTruss_ExpandedBF()** as

```
K2Thermal=ThermalPlaneTruss_ExpandedBF(Total_dof,
k2ThermalLoad,1,3);
round(K2Thermal)
```

Notice that the numbers 1 and 3 supplied as parameters to the function **ThermalPlaneTruss_ExpandedBF()** in the preceding code reflect the nodes for element 2.

Step 5: Establish the global stiffness matrix and the vector of global thermal nodal loads by adding the expanded matrices of the elements as shown next.

```
GlobalK=K1+K2+K3; round(GlobalK)
```

The output of this gives the 8×8 global stiffness matrix $[K]$ as

```
> GlobalK=K1+K2+K3;round(GlobalK)
           [,1]       [,2]       [,3]       [,4] [,5]       [,6]       [,7]       [,8]
[1,]  403565129          0 -201782565 -349906538    0          0 -201782565  349906538
[2,]          0 2146863208 -349906538 -606764938    0 -933333333  349906538 -606764938
[3,] -201782565 -349906538  201782565  349906538    0          0          0          0
[4,] -349906538 -606764938  349906538  606764938    0          0          0          0
[5,]          0          0          0          0    0          0          0          0
[6,]          0 -933333333          0          0    0  933333333          0          0
[7,] -201782565  349906538          0          0    0          0  201782565 -349906538
[8,]  349906538 -606764938          0          0    0          0 -349906538  606764938
```

For readability, the global stiffness matrix for the planar truss system is reproduced in the following:

$$[K_{\text{global}}] = 10^6 \begin{bmatrix} 403.565129 & 0. & -201.782565 & -349.906538 & 0. & 0. & -201.782565 & 349.906538 \\ 0. & 2146.863208 & -349.906538 & -606.764938 & 0. & -933.333333 & 349.906538 & -606.764938 \\ -201.782565 & -349.906538 & 201.782565 & 349.906538 & 0. & 0. & 0. & 0. \\ -349.906538 & -606.764938 & 349.906538 & 606.764938 & 0. & 0. & 0. & 0. \\ 0. & 0. & 0. & 0. & 0. & 0. & 0. & 0. \\ 0. & -933.333333 & 0. & 0. & 0. & 933.333333 & 0. & 0. \\ -201.782565 & 349.906538 & 0. & 0. & 0. & 0. & 201.782565 & -349.906538 \\ 349.906538 & -606.764938 & 0. & 0. & 0. & 0. & -349.906538 & 606.764938 \end{bmatrix}$$

Meanwhile, the vector of global thermal load is formed as

```
GlobalThermal=K2Thermal; round(GlobalThermal)
```

With the output being

```
> GlobalThermal=K2Thermal;round(GlobalThermal)
          [,1]
[1,]         0
[2,]   1288000
[3,]         0
[4,]         0
[5,]         0
[6,]  -1288000
[7,]         0
[8,]         0
```

Step 6: Identify the boundary condition(s) and obtain the reduced stiffness matrix along with the reduced force vector.

Noting the global stiffness matrix and vector of thermal load found in Step 5, we can write the global matrix equation as

$$\begin{Bmatrix} F_{1x} \\ F_{1y} \\ F_{2x} \\ F_{2x} \\ F_{3x} \\ F_{3y} \\ F_{4x} \\ F_{4y} \end{Bmatrix} = [K_{\text{global}}] \begin{Bmatrix} U_{1x} \\ U_{1y} \\ U_{2x} \\ U_{2y} \\ U_{3x} \\ U_{3y} \\ U_{4x} \\ U_{4y} \end{Bmatrix} - \begin{Bmatrix} F_{T1x} \\ F_{T1y} \\ F_{T2x} \\ F_{T2y} \\ F_{T3x} \\ F_{T3y} \\ F_{T4x} \\ F_{T4y} \end{Bmatrix} \tag{10.30}$$

The boundary conditions to be applied on the above-mentioned system of equations are derived from the knowledge that the truss is fixed at nodes 2, 3, and 4, leading to zero horizontal and vertical displacements at each of these nodes. Therefore, the displacement boundary conditions will be

$$U_{2x} = U_{2y} = U_{3x} = U_{3y} = U_{4x} = U_{4y} = 0 \tag{10.31}$$

To obtain the reduced stiffness matrix, we need to identify the known nodal loads. Meanwhile, for the six known nodal degrees of freedom listed in Equation 10.31, the corresponding loads will not be known. As a result, we are left with node 1, in which we see that no force is applied in both x and y directions: Or that

$$F_{1x} = F_{1y} = 0 \tag{10.32}$$

Apparently, from Equation 10.30, it is observed that F_{1x} and F_{1y} belong to rows 1 and 2, respectively. Therefore, the reduced stiffness matrix depends on rows 1 and 2 as reflected in the following block of code:

```
vec_rows_of_knownloads=c(1,2);
ReducedK=ThermalPlaneTruss_ReducedStiffnessMatrix
(GlobalK,vec_rows_of_knownloads);
round(ReducedK)
```

The output of which reads as

```
> round(ReducedK)
          [,1]       [,2]
[1,] 403565129          0
[2,]         0 2146863208
```

In the next lines code, we extract the corresponding reduced force vector that will match the stiffness matrix.

```
vec_values_of_knownloads=c(0,0)+GlobalThermal[c(1,2)];
ReducedForceVector=ThermalPlaneTr
uss_ReducedForceVector(vec_values_of_knownloads);
round(ReducedForceVector)
```

In the preceding code, the variable **vec_values_of_knownloads()** has two parts. The first part (i.e., **c(0,0)**) is a vector of the forces listed in Equation 10.32, whereas the second part (i.e., **GlobalThermal[c(1,2)]**) refers to the thermal loads that correspond to rows 1 and 2 (which of course belong to node 1).

```
> round(ReducedForceVector)
        [,1]
[1,]       0
[2,] 1288000
```

Step 7: Determine the unknown nodal displacements and nodal global forces by using the **ThermalPlaneTruss_Nodal Displacement()** and **ThermalPlaneTruss_Global_Forces ()**.

For the determination of the displacements, the function **Thermal PlaneTruss_NodalDisplacement()** is used.

```
UnknwonNodalDisp=ThermalPlaneTruss_NodalDisplacement(Redu
cedK,ReducedForceVector)
round(UnknwonNodalDisp,6)
```

And the computed displacements are obtained as

```
> round(UnknwonNodalDisp,6)
       [,1]
[1,] 0.0000
[2,] 0.0006
```

The previous box shows that the displacements U_{1x} and U_{1y} are as follows:

$$\begin{Bmatrix} U_{1x} \\ U_{1y} \end{Bmatrix} = \begin{Bmatrix} 0 \\ 0.0006 \end{Bmatrix} \tag{10.33}$$

Meanwhile, for the determination of the other variables, it is logical to combine the determined displacements and the boundary condition in a single variable as done in the following.

```
allglobalNodalDisp=matrix(rep(0,Total_dof),byrow=T);
allglobalNodalDisp[vec_rows_of_knownloads]=Unknwon
NodalDisp;
```

Next, we determine the global reaction forces and the stress developed within each element. For the global reaction force, we first determine the reaction forces from mechanical loads by using the function **ThermalPlaneTruss_ Global_Forces()**. This is then combined with the effect of the global thermal load (contained in the variable **GlobalThermal** that was determined in Step 6). The codes for these calculations are shown next:

```
MechLoadsEffect=ThermalPlaneTruss_Global_Forces(GlobalK,
allglobalNodalDisp)
ThermalLoadEffect=GlobalThermal
Globalloads=MechLoadsEffect-ThermalLoadEffect
round(Globalloads)
```

Running the preceding block of code returns the forces (in N) at the nodes as

```
> round(Globalloads)
        [,1]
[1,]       0
[2,]       0
[3,] -209925
[4,] -364026
[5,]       0
[6,]  728051
[7,]  209925
[8,] -364026
```

From the previous box, the reaction forces are as follows:

$$\begin{Bmatrix} F_{1x} \\ F_{1y} \\ F_{2x} \\ F_{2x} \\ F_{3x} \\ F_{3y} \\ F_{4x} \\ F_{4y} \end{Bmatrix} = \begin{Bmatrix} 0 \\ 0 \\ -209925 \\ -364026 \\ 0 \\ 728051 \\ 209925 \\ -364026 \end{Bmatrix} N \tag{10.34}$$

Finally, we find the stress in each element using the function **ThermalPlaneTruss_AxialStress()**, as shown next.

```
Element1Stress=ThermalPlaneTruss_AxialStress(E1,L1T1[1],L1T1[2],alpha1,
                                              0,allglobalNodalDisp,1,2);
Element1Stress

Element2Stress=ThermalPlaneTruss_AxialStress(E2,L2T2[1],L2T2[2],alpha2,

tempchange,allglobalNodalDisp,1,3);
Element2Stress
Element3Stress=ThermalPlaneTruss_AxialStress(E3,L3T3[1],L3T3[2],alpha3,
                                              0,allglobalNodalDisp,1,4);
Element3Stress
```

After running the preceding block of code, we obtained the following stress values:

```
> Element1Stress=ThermalPlaneTruss_AxialStress(E1,L1T1[1],L1T1[2],alpha1,
+
0,allglobalNodalDisp,1,2);
> Element1Stress
          [,1]
[1,] -10505446
> Element2Stress=ThermalPlaneTruss_AxialStress(E2,
L2T2[1],L2T2[2],alpha1,
+
tempchange,allglobalNodalDisp,1,3);
> Element2Stress
         [,1]
[1,] 18201282
> Element3Stress=ThermalPlaneTruss_AxialStress(E3,L3T3[1],L3T3[2],alpha1,
+
0,allglobalNodalDisp,1,4);
> Element3Stress
         [,1]
[1,] 10505446
```

From the previous box, we have $\sigma_1 = -10.5\,\text{MPa}$; $\sigma_2 = 18.20\,\text{MPa}$; $\sigma_3 = 10.5\,\text{MPa}$. The results obtained from this problem match that of a similar problem contained in Reference 7.

10.5 Exercises

1. For the system of bars shown in Figure 10.4. The length for each element is as shown. Bar 1 is made of aluminum ($A = 10 \times 10^{-4} \text{m}^2$, $E = 70$ GPa, $\alpha = 23 \times 10^{-6}/°\text{C}$), bars 2 and 3 are made of brass ($A = 5 \times 10^{-4} \text{m}^2$, $E = 100$ GPa, $\alpha = 20 \times 10^{-6}/°\text{C}$). If the aluminum bar experiences a temperature rise of 35°C, find the (a) displacement of node 2; (b) the reaction force at the fixed ends; and (c) the axial stress developed in each element.
2. A two-member pin-connected truss structure shown in Figure 10.5 is loaded by a vertical force of 10 kN at node 1. The cross-sectional area for elements 1 (nodes 1–2) and 2 (nodes 2–3) are $A_1 = 300 \text{ mm}^2$

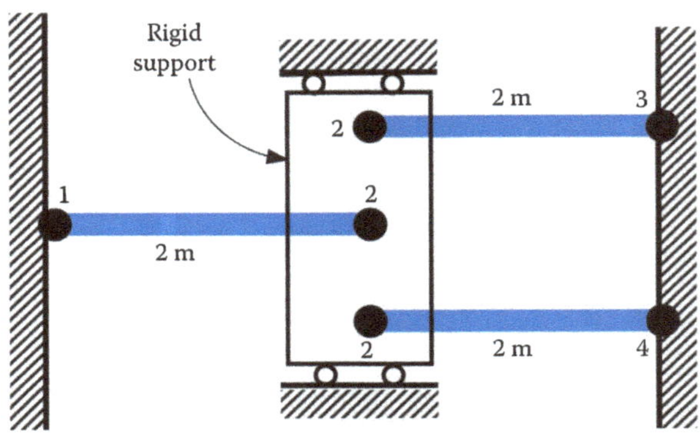

FIGURE 10.4
Exercise 10Q1.

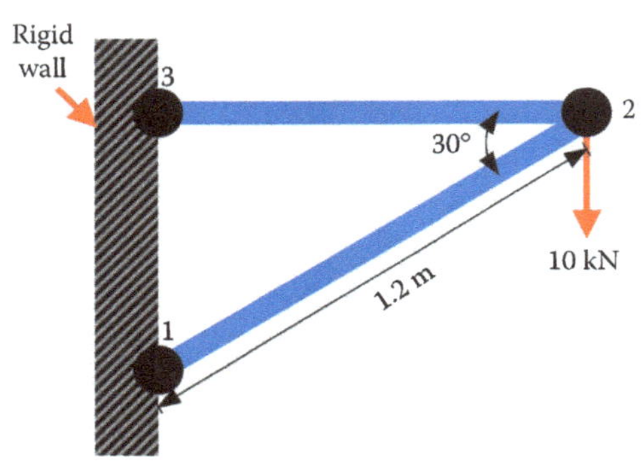

FIGURE 10.5
Exercise 10Q2.

and $A_2 = 60\ \text{mm}^2$, respectively. Both elements are made of the same material with Young's modulus, $E = 200$ GPa and thermal expansion coefficient, $\alpha = 11\times10^{-6}/^\circ\text{C}$. Under operating conditions, elements 1 and 2 experience a temperature rise of 0°C and 60°C, respectively. Determine the vertical and horizontal deflections of joint 2 and the stress developed in each element.

References

1. Y. Wang, M. Kong, Y.-C. Lee, Thermal stress in MEMS. In *Encyclopedia of Thermal Stresses*, R. B. Hetnarski (Ed.), Dordrecht, the Netherlands: Springer, 2014, pp. 5237–5248.
2. N. Noda, *Thermal Stresses*, CRC Press, Boca Raton, FL, 2002.
3. D. J. Johns, B. G. Neal, *Thermal Stress Analyses*, Elsevier Science, Burlington, NJ, 2013.
4. R. B. Hetnarski, M. R. Eslami, G. Gladwell, *Thermal Stresses: Advanced Theory and Applications*, volume 158, Springer, Dordrecht, the Netherlands, 2009.
5. R. F. Barron, B. R. Barron, *Design for Thermal Stresses*, John Wiley & Sons, Hoboken, N.J., 2011.
6. R. C. Hibbeler, *Mechanics of Materials SI edition*, 9th ed., Pearson Education South Asia, Singapore, 2014.
7. D. L. Logan, *First Course in the Finite Element Method*: Thomson, Toronto, ON, 2007.

11

General Methods for Formulations of Element Matrices

11.1 Introduction

In Chapters 3 through 10, we used the principle of *minimum potential energy* to derive the finite element mathematical models for different types of element. As demonstrated, this method allows the formulation of the equations for a wide range of problems in mechanics. A main weakness of the minimum potential energy method is that one faces a considerable difficulty in its application to higher order elements for which the potential energy expression is not known *a priori*. In such situations, other methods come to the rescue. In a broader context, the method of minimum potential energy belongs to a larger pool of procedures called the variational methods. In this chapter, we will formally introduce, even if briefly, the principle of variational calculus as it relates to the finite element method. We shall also familiarize ourselves with the Rayleigh–Ritz method and the method of weighted residual (both of which are powerful techniques in the family of variational methods).

11.2 Variational Methods

Among the throng of competing techniques for dealing with formidable physical problems that involve finding extrema (minimum, maximum, or optimal solutions), a group of potent approaches comes under the umbrella of the variational methods. Procedurally, the variational methods rely on the calculus of variations, which is a branch of mathematics that stipulates rules for finding the extrema of *functionals*. In mathematical parlance, a functional is defined as *a function of functions* [1]. A common incarnation of a functional is a definite integral, the integrand of which is a collection of unknown functions and their derivatives.

Widely used in many fields (ranging through elasticity, optics, electromagnetism, quantum and particle mechanics, etc.), the variational methods have benefited from the contributions of many notable minds: the Bernoulli brothers, Newton, Euler, and so on [2].

In this text, we do not aim to provide an exhaustive exploration of the topic; instead, our principal concern is to provide a cursory account of how the method can be used to obtain approximate solutions in the context of the finite element method.

11.2.1 The Variation Operator and Its Properties

In the realm of calculus of variation, the delta operator (δ) is used to denote the variation of functions or functionals. We will follow the exposition in Reference 3 and demonstrate what we mean by this with some examples. First, let us start with the variation of functions by considering a function $y(x)$ defined as

$$y(x) = a_0 + a_1 x + a_2 x^2 \tag{11.1}$$

Now, assuming we have a small change in the coefficients (but not in x) of the function, one may generate a related function $\overline{y}(x)$ as

$$\overline{y}(x) = \left(a_0 + \delta a_0\right) + \left(a_1 + \delta a_1\right)x + \left(a_2 + \delta a_2\right)x^2 \tag{11.2}$$

The difference between $\overline{y}(x)$ and $y(x)$ quantifies the variation of $y(x)$, and this is denoted as δy. That is,

$$\delta y = \overline{y}(x) - y(x) \text{ or } \overline{y}(x) = y(x) + \delta y \tag{11.3}$$

The variation of similar other algebraic functions with higher order power follows the same approach. Next, we examine a functional J, which is a function of x and $y(x)$ defined as

$$J(y(x)) = \left[y(x)\right]^2 \tag{11.4}$$

The variation of J takes the form

$$\delta J = \left[\overline{y}(x)\right]^2 - \left[y(x)\right]^2 \tag{11.5}$$

By using Equations 11.1 and 11.2, the variation of the functional J can be rewritten as

$$\delta J = \left[\left(a_0 + \delta a_0\right) + \left(a_1 + \delta a_1\right)x + \left(a_2 + \delta a_2\right)x^2\right]^2 - \left[a_0 + a_1 x + a_2 x^2\right]^2 \tag{11.6}$$

Carrying out the expansion of each term on the right-hand side leads to

$$\delta J = 2a_0\delta a_0 + 2xa_1\delta a_0 + 2x^2a_2\delta a_0 + \underline{\delta a_0^2} + 2xa_0\delta a_1 + 2x^2a_1\delta a_1 + 2x^3a_2\delta a_1 + \underline{2x\delta a_0\delta a_1}$$
$$+ \underline{x^2\delta a_1^2} + 2x^2a_0\delta a_2 + 2x^3a_1\delta a_2 + 2x^4a_2\delta a_2 + \underline{2x^2\delta a_0\delta a_2} + \underline{2x^3\delta a_1\delta a_2} + \underline{x^4\delta a_2^2} \tag{11.7}$$

Given that the underlined terms are infinitesimally small, they can be omitted to arrive at a clean expression for δJ as

$$\delta J = 2a_0\delta a_0 + 2xa_1\delta a_0 + 2x^2a_2\delta a_0 + 2xa_0\delta a_1 + 2x^2a_1\delta a_1$$
$$+ 2x^3a_2\delta a_1 + 2x^2a_0\delta a_2 + 2x^3a_1\delta a_2 + 2x^4a_2\delta a_2 \tag{11.8}$$

In principle, the functional does not have to be simple algebraic equations such as Equation 11.5. For instance, consider functionals of the form

$$G(y(x)) = \frac{dy(x)}{dx} \tag{11.9}$$

$$H(y(x)) = \int_0^2 y(x)\,dx \tag{11.10}$$

where G and H are differential and integral functionals, respectively. Recalling Equations 11.1 and 11.2, the variation of $G(y(x))$ and $H(y(x))$ takes the form

$$\delta G = \delta\left(\frac{dy}{dx}\right) = \frac{d\bar{y}}{dx} - \frac{dy}{dx} = \frac{d}{dx}\left[(a_0 + \delta a_0) + (a_1 + \delta a_1)x + (a_2 + \delta a_2)x^2\right]$$
$$- \frac{d}{dx}\left[a_0 + a_1x + a_2x^2\right] \tag{11.11}$$

$$\delta H(y(x)) = \delta\left(\int_0^0 y\,dx\right) = \left(\int_0^2 \bar{y}\,dx\right) - \left(\int_0^2 y\,dx\right)$$
$$= \left[x(a_0 + \delta a_0) + x^2\left(\frac{a_1}{2} + \frac{\delta a_1}{2}\right) + x^3\left(\frac{a_2}{3} + \frac{\delta a_2}{3}\right)\right]_0^2 - \left[xa_0 + \frac{x^2a_1}{2} + \frac{x^3a_2}{3}\right]_0^2 \tag{11.12}$$

Both of which simplifies to

$$\delta G = \delta a_1 + 2x\delta a_2 \tag{11.13}$$

$$\delta H = 2\delta a_0 + 2\delta a_1 + \frac{8\delta a_2}{3} \tag{11.14}$$

Significantly, we could have obtained Equations 11.13 and 11.14 by simply interchanging the variational operator with the differential and integral operators as

$$\delta G = \delta\left(\frac{dy}{dx}\right) = \frac{d}{dx}(\delta y) = \frac{d}{dx}\left(\overline{y} - y\right) = \delta a_1 + 2x\delta a_2$$

$$\delta H = \delta\left(\int_0^2 y\,dx\right) = \left(\int_0^2 \delta y\,dx\right) = 2\delta a_0 + 2\delta a_1 + \frac{8\delta a_2}{3}$$

The preceding steps lead to some essential properties in the calculus of variations [3,4]:

1. The variational operator is commutative with respect to both differentiation and integration. Which means, the variation of the differentiation (or integration) of a function is equal to the differentiation (or integration) of the variation of the same function.
2. The variational operator deals with the variation of the function (say, $y(t)$) for some fixed values of the independent variable (say, t). As a result, the variation of the independent variable is zero (i.e., $\delta t = 0$ in the case where t is the independent variable).
3. In situations where a functional is a function of several variables, the variational operator could be construed as a total derivative (with respect to the parameters of the function upon which the functional depends).

To demonstrate the last point, we consider the functional $J(x,y,y')$. The variation of J, following the interpretation of the variation as a total derivative, is obtained as

$$\delta J = \frac{\partial J}{\partial x}\underbrace{\delta x}_{\delta x=0} + \frac{\partial J}{\partial y}\delta y + \frac{\partial J}{\partial y'}\delta y' = \frac{\partial J}{\partial y}\delta y + \frac{\partial J}{\partial y'}\delta y' \tag{11.15}$$

Other properties include the following:

- The variation of the sum (or subtraction) of two functionals (e.g., φ and ϕ) is the sum (or subtraction) of the variation of each functional. That is, $\delta(\varphi \pm \phi) = \delta\varphi \pm \delta\phi$.
- The variation of the product of two functionals obeys the chain rule of differentiation. That is, $\delta(\varphi\phi) = (\delta\varphi)\phi + \varphi(\delta\phi)$.

- The variation of the functionals with exponents is analogous to the differentiation rule of functions with exponents. For instance,
  - $\delta F^n = nF^{n-1}\delta F$
  - $\delta\left(F/G\right) = \delta\left(FG^{-1}\right) = G^{-1}(\delta F) + F\left(\delta G^{-1}\right) = G^{-1}(\delta F) + F\left(-G^{-2}\delta G\right)$
    $= \dfrac{\delta F}{G} - \dfrac{F(\delta G)}{G^2} = \dfrac{G(\delta F) - F(\delta G)}{G^2}$

11.2.2 Extremization of Functionals and the Euler–Lagrange Differential Equation

Extremization of functionals is about finding the family of functions that make a functional to assume an extreme value. As a rule, in elementary calculus, the procedure for the extremization of a function of a single variable is hinged on the necessary condition that the first derivative of the function be zero.

In a similar vein, in the calculus of variation, a necessary condition for the extremization of a functional is that its first variation be zero. This concept comes up repeatedly in a host of applications of the variational approach in mechanics. To demonstrate the procedure for the extremization of functionals, we restrict ourselves to the problems of fixed end-point variational problems, and we consider a known function $F(x,y,y',y'')$, which is a function of $x, y(x), y'(x)$, and $y''(x)$. Our goal is to establish the extremum (maximum or minimum) of the functional I defined as

$$I(y) = \int_{x_1}^{x_2} F\left(x,y,y',y''\right)dx \tag{11.16}$$

subject to the constrain

$$y(x_2) = y_2; y(x_1) = y_1 \tag{11.17}$$

where x is the independent variable, y is the primary dependent variable assumed to be continuous and at least twice differentiable on $[x_1, x_2]$, whereas y' and y'' are the first and second derivatives of y, respectively.[1]

To solve the postulated problem associated with Equation 11.16, we presuppose that there exists a path $y(x)$, the values of which at the boundaries

[1] Notice that in Equation 11.16, F is itself a functional, and it has physical meaning in most applications (e.g., as the total potential energy of elastic systems, or as the distance between two points on a plane along some path etc.).

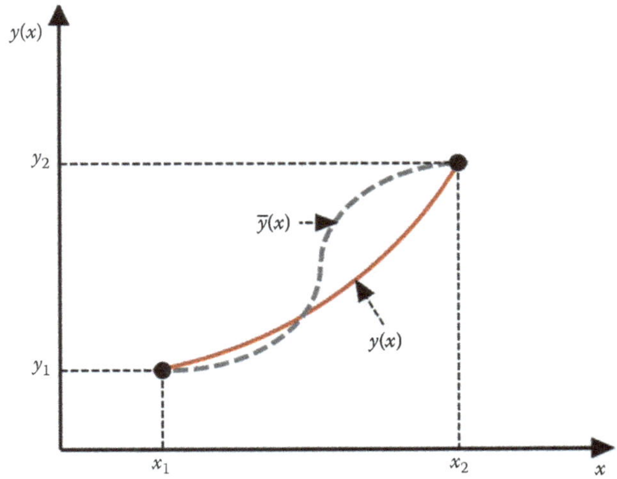

FIGURE 11.1
The desired function $y(x)$ and a neighborhood function $\overline{y}(x)$.

(i.e., y_1 and y_2) are known. Furthermore, let us assume that this function $y(x)$ extremizes the functional I as shown in Figure 11.1.

Noticeably, there exists a series of other non-optimal prospective paths in the vicinity of $y(x)$. For convenience, we will refer to these other non-optimal varied paths as *comparison functions* (some other authors use the term *trial functions* for the varied paths) [5]. Conveniently, we designate the family of varied paths as $\overline{y}(x)$. A single-parameter family of trial function is then written as

$$\overline{y}(x) = y(x) + \epsilon\eta(x) \tag{11.18}$$

where ϵ is a small parameter. Besides, $\eta(x)$ is a differentiable auxiliary function of x that vanishes at the boundary as

$$\eta(x_1) = \eta(x_2) = 0 \tag{11.19}$$

Without the loss of generality, one may also write Equation 11.18 as

$$\delta y = \overline{y}(x) - y(x) = \epsilon\eta(x) \tag{11.20}$$

By using the comparison function, we rewrite Equation 11.16 as

$$\overline{I}(y,\epsilon) = \int_{x_1}^{x_2} F(x,\overline{y},\overline{y}',\overline{y}'')dx = \int_{x_1}^{x_2} F(x, y+\epsilon\eta, y'+\epsilon\eta', y''+\epsilon\eta'')dx \tag{11.21}$$

Apparently, in the case when $\epsilon = 0$, the new functional with the comparison function is undistinguishable from the original.

Following the point mentioned earlier, a necessary condition for extremizing the functional $\bar{I}$ is to have

$$\left.\frac{d\bar{I}}{d\epsilon}\right|_{\epsilon=0} = 0 \tag{11.22}$$

Performing the differentiation of Equation 11.22 with the chain rule, bearing in mind Equation 11.21, one obtains

$$\left.\frac{d\bar{I}}{d\epsilon}\right|_{\epsilon=0} = \int_{x_1}^{x_2}\left[\frac{\partial F}{\partial \bar{y}}\left(\frac{d\bar{y}}{d\epsilon}\right)+\frac{\partial F}{\partial \bar{y}'}\left(\frac{d\bar{y}'}{d\epsilon}\right)+\frac{\partial F}{\partial \bar{y}''}\left(\frac{d\bar{y}''}{d\epsilon}\right)\right]dx \tag{11.23}$$

We know that: $\frac{d\bar{y}}{d\epsilon}=\eta;\frac{d\bar{y}'}{d\epsilon}=\eta';\frac{d\bar{y}''}{d\epsilon}=\eta''$, thus,

$$\left.\frac{d\bar{I}}{d\epsilon}\right|_{\epsilon=0} = \left\{\int_{x_1}^{x_2}\left[\frac{\partial F}{\partial \bar{y}}\eta+\frac{\partial F}{\partial \bar{y}'}\eta'+\frac{\partial F}{\partial \bar{y}''}\eta''\right]dx\right\}_{\epsilon=0} \tag{11.24}$$

When evaluated at $\epsilon = 0$:

$$\bar{y}=y;\ \bar{y}'=y';\ \bar{y}''=y'' \tag{11.25}$$

Therefore,

$$\left.\frac{d\bar{I}}{d\epsilon}\right|_{\epsilon=0} = \int_{x_1}^{x_2}\left[\frac{\partial F}{\partial y}\eta+\underbrace{\frac{\partial F}{\partial y'}\eta'}_{T_1}+\underbrace{\frac{\partial F}{\partial y''}\eta''}_{T_2}\right]dx \tag{11.26}$$

We will integrate the underlined terms (T_1 and T_2) of Equation 11.26, by using integration by part as follows:

$$\int_{x_1}^{x_2}\left[\frac{\partial F}{\partial y'}\eta'\right]dx = \left[\left.\frac{\partial F}{\partial y'}\eta\right|_{x_1}^{x_2} - \int_{x_1}^{x_2}\frac{d}{dx}\frac{\partial F}{\partial y'}\eta dx\right] \tag{11.27}$$

$$\begin{aligned}\int_{x_1}^{x_2}\left[\frac{\partial F}{\partial y''}\eta''\right]dx &= \left[\left.\frac{\partial F}{\partial y''}\eta'\right|_{x_1}^{x_2} - \int_{x_1}^{x_2}\frac{d}{dx}\left(\frac{\partial F}{\partial y''}\right)\eta' dx\right]\\ &= \left[\left.\frac{\partial F}{\partial y''}\eta'\right|_{x_1}^{x_2} - \left(\left.\frac{d}{dx}\frac{\partial F}{\partial y''}\eta\right|_{x_1}^{x_2} - \int_{x_1}^{x_2}\frac{d^2}{dx^2}\left(\frac{\partial F}{\partial y''}\right)\eta dx\right)\right]\end{aligned} \tag{11.28}$$

By using Equations 11.27 and 11.28, we rewrite Equation 11.26 as

$$\left.\frac{d\bar{I}}{d\epsilon}\right|_{\epsilon=0} = \int_{x_1}^{x_2} \left[\frac{\partial F}{\partial y} - \frac{d}{dx}\frac{\partial F}{\partial y'} + \frac{d^2}{dx^2}\left(\frac{\partial F}{\partial y''}\right) \right] \eta dx + \left[\underline{\left.\frac{\partial F}{\partial y''}\eta'\right|_{x_1}^{x_2}} + \underline{\left.\left(\frac{\partial F}{\partial y'} - \frac{d}{dx}\frac{\partial F}{\partial y''}\right)\eta\right|_{x_1}^{x_2}} \right] \tag{11.29}$$

For the first part of Equation 11.29, we noticed that η is known at the boundaries, but it is arbitrary between the endpoints. Therefore, the term within the integral must vanish to yield the *Euler–Lagrange* equation:

$$\left[\frac{\partial F}{\partial y} - \frac{d}{dx}\frac{\partial F}{\partial y'} + \frac{d^2}{dx^2}\left(\frac{\partial F}{\partial y''}\right) \right] = 0 \tag{11.30}$$

Typically, the *Euler–Lagrange* equation is an ordinary differential equation of order $2n$. Accordingly, its solution requires $2n$ boundary conditions, n being the highest order derivative contained in the functional.

For the underlined parts of Equation 11.29, assuming that the function $y(x)$ and its derivatives are specified at the endpoints (x_1 and x_2), then η and η' will be zero at the endpoints (x_1 and x_2), and hence,

$$\left.\left(\frac{\partial F}{\partial y'} - \frac{d}{dx}\frac{\partial F}{\partial y''} \right)\right|_{x_1}^{x_2} = 0 \tag{11.31}$$

$$\left.\frac{\partial F}{\partial y''}\right|_{x_1}^{x_2} = 0 \tag{11.32}$$

Equations 11.31 and 11.32 are called the natural boundary conditions. One may handle, in a similar manner, functionals with higher order derivatives and those with several dependent variables. However, interested readers may consult the following references for a deeper dive into this delightful topic [1,3,6].

11.3 Finite Element Equations via the Rayleigh–Ritz Method

The Rayleigh–Ritz method is one of the variational-based methods suitable for formulating the finite element equations of structural problems for which the potential energy expression is known. In structural mechanics, the following steps are used in obtaining the finite element equations using the Rayleigh–Ritz method:

- Identify the expression for the potential energy functional, the extremization of which is required. For instance, assuming that we are interested in the extremization of a functional I over a domain V and boundary surface S, we will start with

$$I = \iiint_v F(y, y', \ldots) dV + \iint_s g(x, y', \ldots) ds \tag{11.33}$$

- Discretize the domain into smaller elements and construct an admissible assumed solution. As an example, the field variable of interest (e.g., y in Equation 11.33) is assumed to vary in each element as

$$y = [N]\phi^e \tag{11.34}$$

where $[N]$ and ϕ^e denote the vector of shape functions and nodal values of y associated with the element.
- Set up element equations by employing the extremization of the functional. As an example, with respect to I set up in Equation 11.33, this involves invoking the necessary condition for the minimum of I with respect to the parameters ϕ associated with the entire domain. If for instance, there exist N_e number of elements, then the minimization results, for the overall domain, in a system of equations as

$$\frac{\partial I}{\partial \phi_i} = \sum_{j=1}^{N_e} \frac{\partial I^j}{\partial \phi_i} = 0, i = 1, 2, 3, \ldots, N_e \tag{11.35}$$

It is customary to decompose Equation 11.35 to the form we have used in previous chapters. That is,

$$\frac{\partial I}{\partial \phi_i} = [K]\{\phi\} - \{Q\} = 0 \tag{11.36}$$

where $[K]$ and $\{Q\}$ are the matrix of the global stiffness properties and the vector of nodal actions, respectively.
- Obtain approximate solutions of the system of equations. In general, this requires solving Equation 11.36 after applying boundary conditions to find the unknown nodal parameters ϕ_i.
- The field variable y in each element is obtained by substitution of ϕ_i into Equation 11.34.

It is necessary to point out that the quality of a solution via the highlighted steps mentioned above depends on two important factors: (1) the number of elements N_e as shown in Equation 11.36; and (2) the nature of shape functions used in Equation 11.34. An elaborate discussion of these points is given in Reference 7.

11.4 The Method of Weighted Residuals

The method of weighted residuals (MWR) comprises a group of general mathematical tools developed to obtain approximate solution to linear and nonlinear differential equations. Unlike the variational method of Section 11.3, MWR do not assume the existence of an integral functional. This makes the MWR suitable for dealing with a wider range of models of physical processes in which the governing equation is in the form of differential equations or the energy functional is not known *a priori* or difficult to obtain (e.g., in mass transport phenomenon). MWR is the very basis of various standard numerical techniques such as the [8]

- Galerkin method
- Least-square method
- Collocation method
- Subdomain method

At the very basic, the starting point for all the MWR is an estimate of the solution of the differential equation. This is then followed by a subsequent request that the weighted average error is minimized. Among all the listed MWR, the Galerkin method is the most general, and a short description of its procedure is provided in the following.

11.4.1 The Galerkin Method

To explore the main features of the Galerkin method, we start with the following linear differential equation:

$$L(u(\boldsymbol{x})) = f \tag{11.37}$$

which governs the behavior of a physical problem in a domain Ω, where L is a differential operator, u is the field variable, and f is the forcing term. Equation 11.37 is supplemented with the corresponding boundary conditions:

$$S(u) = 0 \tag{11.38}$$

on the boundary of Ω. In reality, Equation 11.37 could have been for a one-, two-, or three-dimensional problem. For simplicity, assuming that Equation 11.37 models the behavior of a one-dimensional problem, a representative example of the differential equation could be a simple boundary value problem of the form

$$\frac{d^2u}{dx^2} + p(x) = 0, \qquad 0 \le x \le 1 \tag{11.39}$$

with the corresponding boundary conditions being

$$u\big|_{x=0} = 0; \ \frac{du}{dx}\bigg|_{x=1} = 0 \tag{11.40}$$

The Galerkin method requires that we seek the solution of Equation 11.39 by representing u with an approximate function such as

$$\tilde{u} = \sum_{k=1}^{N} c_k \phi_k(x) \tag{11.41}$$

where ϕ_k's are known analytic functions (also called trial functions) and the c_k's are unknown coefficients. For accurate solutions, the trial functions ϕ_k's must be chosen to satisfy the essential boundary conditions of the problem and must be differentiable at least up to the highest derivative in the governing equation.

Plugging of the trial function $\tilde{u}$ into Equation 11.39 yields a non-zero residual. That is,

$$\frac{d^2\tilde{u}}{dx^2} + p(x) = R(x) \tag{11.42}$$

The Galerkin method requires that the residual $R(x)$ be minimized. To achieve this, we multiply the terms of Equation 11.42 by the trial functions, integrate over the element, and equate the total weighted residual to zero:

$$\int_0^1 R(x)\phi_k \, dx = 0, \ \ k = 1,2,3,\ldots N \tag{11.43}$$

When evaluated, Equation 11.43 produces a set of N number of equations:

$$\int_0^1 \left(\underline{\frac{d^2\tilde{u}}{dx^2}\phi_k} + p(x)\phi_k \right) dx = 0, \ \ k = 1,2,3,\ldots N \tag{11.44}$$

A common step in MWR is to reduce all higher order derivatives in the residuals by using integration-by-parts. Beyond the reduction of higher order terms, however, integration-by-parts facilitates the introduction of the boundary conditions in the weighted residual equations. For the problem at hand, we apply integration-by-parts to the underlined second-order derivative in Equation 11.44 and obtain

$$\int_0^1 \frac{d^2\tilde{u}}{dx^2}\phi_k \, dx = \phi_k \frac{d\tilde{u}}{dx}\bigg|_0^1 - \int_0^1 \frac{d\tilde{u}}{dx}\frac{d\phi_k}{dx} dx \tag{11.45}$$

With the aid of Equation 11.45, we rewrite Equation 11.44 as

$$\phi_k \frac{d\tilde{u}}{dx}\bigg|_0^1 - \int_0^1 \frac{d\tilde{u}}{dx}\frac{d\phi_k}{dx}dx = -\int_0^1 p(x)\phi_k dx \tag{11.46}$$

Recalling that the trial function $\tilde{u}$ is defined in terms of the unknown function in Equation 11.41, one can rewrite Equation 11.46 as

$$\int_0^1 \frac{d\phi_k}{dx}\sum_{k=1}^{N} c_k \frac{d\phi_k}{dx}dx = \int_0^1 p(x)\phi_k dx + \underline{\left(\phi_k \frac{du}{dx}\right)\bigg|_0^1} \tag{11.47}$$

Notice that we have not used the trial function for the boundary condition (the underlined term in Equation 11.47); instead, we retained the original function as contained in the boundary condition of Equation 11.40. The problem now is to find coefficients c_k, and for this we rewrite Equation 11.47 in a compact matrix form as

$$[K]\{c\} = \{Q\} \tag{11.48}$$

where

$$K_{ik} = \int_0^1 \frac{d\phi_i}{dx}\frac{d\phi_k}{dx}dx \qquad k = 1,2,3,\ldots N \tag{11.49}$$

$$\{Q\} = \int_0^1 p(x)\phi_k dx + \underline{\left(\phi_k \frac{du}{dx}\right)\bigg|_0^1} \tag{11.50}$$

For a known value of p and an N set of ϕ_k's, Equation 11.48 gives a set of N simultaneous equations to be solved to determine the unknown coefficients c_k's.

11.5 Finite Element Equations via the Galerkin Method

Consider the governing differential equation:

$$L(u) = f \tag{11.51}$$

In a domain Ω with boundary condition:

$$S(u) = 0 \tag{11.52}$$

Following the procedure of Section 11.4, we assume a trial function:

$$\tilde{u} = \sum_{k=1}^{N} c_k \phi_k(x) \tag{11.53}$$

and formulate the weighted residual integral that is valid over the whole problem domain:

$$\int_{v} \left(L(\tilde{u}) - f\right)\phi_k \, dV = 0 \qquad k = 1,2,3,\ldots N \tag{11.54}$$

However, in the formulation of a finite element matrix, the weighted integral is formulated so as to be valid for an element. As a result, the trial functions are taken to be the shape functions.

$$\tilde{u} = \sum_{k=1}^{N} N_k \varphi_k = \left[N^e\right]\left\{\varphi_k^e\right\} \tag{11.55}$$

And the equations for an element are then derived as

$$\int_{V^e} \left(L\left(\tilde{u}^e\right) - f^e\right)\varphi_k^e \, dV^e = 0 \qquad k = 1,2,3,\ldots N \tag{11.56}$$

Unlike the trial functions of Section 11.4.1, the shape functions of Equation 11.55 do not need to satisfy all the boundary conditions, but they are required to satisfy the continuity conditions. In Chapter 12, we shall deploy the Galerkin method for the derivation of finite element equations for vibration analyses of bars and beams.

References

1. L. Komzsik, *Applied Calculus of Variations for Engineers*, CRC Press, Boca Raton, FL, 2008.
2. B. van Brunt, *The Calculus of Variations*, Springer, New York, 2003.
3. M. A. Bhatti, *Fundamental Finite Element Analysis and Applications: with Mathematica and Matlab Computations*, Wiley, Hoboken, NJ, 2005.
4. S. S. Rao, *Vibration of Continuous Systems*, John Wiley & Sons, Hoboken, NJ, 2007.
5. C. L. Dym, I. H. Shames, Introduction to the calculus of variations. In *Solid Mechanics*, Springer, New York, 2013, pp. 71–116.
6. C. L. Dym, I. H. Shames, Variational principles of elasticity. In *Solid Mechanics*, Springer, New York, 2013, pp. 117–186.
7. J. N. Reddy, *Energy Principles and Variational Methods in Applied Mechanics*, Wiley, New York, 2002.
8. S. S. Rao, *The Finite Element Method in Engineering*, Elsevier Science, Burlington, NJ, 2010.

12

Finite Element Formulations for Vibration Problems

12.1 Introduction

Vibration analyses occupy a strategic importance in the design-against-failure of mechanical systems under the effect of time-dependent loads. In terms of coverage, books on the application of the finite element methods (FEM) are characteristically geared more to statics than structural dynamics and vibration studies. A comprehensive treatise on the application of the FEM for structural dynamics is offered by Petyt [1]. Gentle introductions to the topic of FEM for vibration studies can also be found in the selective chapters of References 2–4.

Given the introductory nature of this book, we will only study the finite element vibration analyses of bars, truss structures, and the Euler beams. In contrast to the other chapters, the Galerkin method of weighted residuals is employed to derive the finite element matrix equations (presented in Sections 12.2 and 12.3) for the vibration analyses of these elements. The implemented R functions and three illustrative examples demonstrating the use of the functions are contained in Sections 12.4 and 12.5, respectively.

12.2 Finite Element Equations for Vibration Problems

The governing equations for the dynamics of structural continua can be derived using several methods [5]. Biased toward conciseness, we will present the governing equations for the vibration of bars and beams in this section in a ready-made form (without derivations). The finite element formulations then follow in a systematic manner thereafter.

12.2.1 Finite Element Matrix Equation for Axial Vibration of Bars

For a homogeneous bar with Young's modulus (E), mass density (ρ), and cross-sectional area (A), the differential equation that characterizes its undamped vibration behavior, in the presence of an external system of force f, is of the form

$$-EA\frac{\partial^2 u(x,t)}{\partial x^2}+\rho A\frac{\partial^2 u(x,t)}{\partial t^2}=f(x,t);\quad 0\le x\le L \tag{12.1}$$

subject to

$$u(0,t)=0;\ \frac{\partial u(L,t)}{\partial x}=0 \tag{12.2}$$

where u describes the axial displacement of the bar. The field variable u depends on both space and time. In principle, Equation 12.1 is applicable for both forced response and free vibration analyses of the bar. Of course, for free vibration analyses, the force term (f) vanishes, and Equation 12.1 becomes

$$EA\frac{\partial^2 u}{\partial x^2}-\rho A\frac{\partial^2 u}{\partial t^2}=0;\quad 0\le x\le L \tag{12.3}$$

Equation 12.3 holds for every point in the entire domain of the bar. Clearly, analytical solutions do exist for this simple differential equation, but here we seek the finite element solution. Specifically, we will use the Galerkin weighted residual method explored in Chapter 11. For this reason, we assume an approximate solution for u in the form u_a for an element. By replacing u with u_a in Equation 12.3, one obtains a residual equation:

$$R=EA\frac{\partial^2 u_a}{\partial x^2}-\rho A\frac{\partial^2 u_a}{\partial t^2} \tag{12.4}$$

We further require that a weighted value of the residual R be a minimum over the whole region. By using a weighting function W, the integral of the inner product of the weighting function and the residual can then be written as

$$\int_0^L RW\,dx=\int_0^L W\left(EA\frac{\partial^2 u_a}{\partial x^2}-\rho A\frac{\partial^2 u_a}{\partial t^2}\right)dx=0 \tag{12.5}$$

By the Galerkin criterion, the weighting functions are chosen to be the same as the spatial shape functions **N**, which is a vector. Replacing W by **N** yields

$$\int_0^L \mathbf{N}\left(EA\frac{\partial^2 u_a}{\partial x^2}-\rho A\frac{\partial^2 u_a}{\partial t^2}\right)dx=0 \tag{12.6}$$

By applying integration by part to Equation 12.6, one obtains

$$\left[\mathbf{N}\left(EA\frac{\partial u_a}{\partial x}\right)\right]_0^L - \int_0^L \frac{d\mathbf{N}}{dx}\left(EA\frac{\partial u_a}{\partial x}\right)dx - \int_0^L \mathbf{N}\left(\rho A\frac{\partial^2 u_c}{\partial t^2}\right)dx = 0 \quad (12.7)$$

Or:

$$\left[\mathbf{N}\left(EA\frac{\partial u_a}{\partial x}\right)\right]_0^L = \int_0^L \frac{d\mathbf{N}}{dx}\left(EA\frac{\partial u_a}{\partial x}\right)dx + \int_0^L \mathbf{N}\left(\rho A\frac{\partial^2 u_a}{\partial t^2}\right) \quad (12.8)$$

We will assume the approximate solution u_a is valid for a typical bar element with two nodes. Consequently, u_a may be represented by the nodal values $u_j^e\,(j=1,2)$ and the interpolation functions $N_j\,(j=1,2)$, resulting in

$$u_a^e = N_1(x)u_1^e(t) + N_2(x)u_2^e(t) = \left[N_j(x)\right]\left\{u_j^e(t)\right\} = \mathbf{N}^T\boldsymbol{u}_e \quad (12.9)$$

Substituting Equation 12.9 into Equation 12.8 leads to a system of simultaneous equations, which, if written in the customary form of the equation of motion in structural dynamics, takes the form

$$\boldsymbol{M}\left(\frac{d^2\boldsymbol{u}_e(t)}{dt^2}\right) + \boldsymbol{K}\boldsymbol{u}_e(t) = \boldsymbol{f}(t) \quad (12.10)$$

where

$$\boldsymbol{M} = \int_0^L \rho A\,\mathbf{N}^T\mathbf{N}\,dx \quad (12.11)$$

$$\boldsymbol{K} = \int_0^L \left(EA\frac{d\mathbf{N}^T}{dx}\right)\frac{d\mathbf{N}}{dx}dx \quad (12.12)$$

$$\boldsymbol{f} = \left[\mathbf{N}\left(EA\frac{\partial u_a}{\partial x}\right)\right]_0^L \quad (12.13)$$

Notice that all partial differential operators have been converted into ordinary differential operators because of the format of the approximate function stated in Equation 12.9. To evaluate each of the integral, the full form of the shape function for a bar is presented as

$$\mathbf{N} = \left[\left(1-\frac{x}{L}\right) \quad \left(\frac{x}{L}\right)\right] \quad (12.14)$$

We substitute Equation 12.14 into Equations 12.11 through 12.13 to obtain the following matrix equation for the bar element:

$$\underbrace{\frac{\rho AL}{6}\begin{bmatrix} 2 & 1 \\ 1 & 2 \end{bmatrix}}_{\text{mass matrix}}\begin{bmatrix} \ddot{u}_{1x} \\ \ddot{u}_{2x} \end{bmatrix} + \underbrace{\frac{EA}{L}\begin{bmatrix} 1 & -1 \\ -1 & 1 \end{bmatrix}}_{\text{stiffness matrix}}\begin{bmatrix} u_{1x} \\ u_{2x} \end{bmatrix} = \begin{bmatrix} F_{1x}(t) \\ F_{2x}(t) \end{bmatrix} \tag{12.15}$$

Notice that the stiffness matrix of the bar is the same as that obtained in Chapter 4.

12.2.2 Finite Element Matrix Equation for Vibration of Truss Structures

For a truss element, the governing differential equation for its axial vibration (with respect to the local xy coordinate system) is the same as that of the bar. Hence, the integral expressions for the mass and stiffness matrices are analogous to that of the bar. That is,

$$\boldsymbol{M}_{\text{truss}} = \int_0^L \rho A \boldsymbol{N}^T \boldsymbol{N} dx\,;\ \boldsymbol{K}_{\text{truss}} = \int_0^L \left(EA \frac{d\boldsymbol{N}^T}{dx} \right) \frac{d\boldsymbol{N}}{dx} dx \tag{12.16}$$

However, a point of departure occurs if the nodal degree of freedom of the truss is to be characterized in the global coordinate XY, for instance. In this case, the axial displacement u must be expressed in terms of u_X and u_Y, denoting the horizontal and vertical displacements in the global coordinate system (XY), respectively. Consequently, the vector of shape functions for the interpolation of the axial displacement takes the form presented in Chapter 5 as

$$\boldsymbol{N} = \begin{bmatrix} \left(1 - \dfrac{x}{L}\right) & 0 & \dfrac{x}{L} & 0 \\ 0 & \left(1 - \dfrac{x}{L}\right) & 0 & \dfrac{x}{L} \end{bmatrix} \tag{12.17}$$

On integrating Equation 12.16, bearing in mind Equation 12.17, one obtains the mass matrix for the truss element:

$$\boldsymbol{M}_{\text{truss}} = \frac{\rho AL}{6}\begin{bmatrix} 2 & 0 & 1 & 0 \\ 0 & 2 & 0 & 1 \\ 1 & 0 & 2 & 0 \\ 0 & 1 & 0 & 2 \end{bmatrix} \tag{12.18}$$

The stiffness matrix for the truss element was established in Chapter 5 as

$$K_{\text{truss}} = \frac{A_e E}{L_e}\begin{bmatrix} C^2 & CS & -C^2 & -CS \\ CS & S^2 & -CS & -S^2 \\ -C^2 & -CS & C^2 & CS \\ -CS & -S^2 & CS & S^2 \end{bmatrix} \tag{12.19}$$

where $C = \cos\theta$ and $S = \sin\theta$

Hence, the system of equations that hold for the dynamic behavior of a truss element is therefore,

$$M_{\text{truss}}\begin{bmatrix} \ddot{u}_{1X} \\ \ddot{u}_{1Y} \\ \ddot{u}_{2X} \\ \ddot{u}_{2Y} \end{bmatrix} + K_{\text{truss}}\begin{bmatrix} u_{1X} \\ u_{1Y} \\ u_{2X} \\ u_{2Y} \end{bmatrix} = \begin{bmatrix} F_{1X}(t) \\ F_{1Y}(t) \\ F_{2X}(t) \\ F_{2Y}(t) \end{bmatrix} \tag{12.20}$$

12.2.3 Finite Element Matrix Equation for Bending Vibration of the Euler–Bernoulli Beams

Consider an Euler–Bernoulli beam with a mass per unit length ρA, Young's modulus E and moment of inertia I. The strong form of the equation that governs its free vibration behaviour is [5]

$$EI\frac{\partial^4 v(x,t)}{\partial x^4} + \rho A\frac{\partial^2 v(x,t)}{\partial t^2} = 0; \quad 0 \le x \le L \tag{12.21}$$

By using the Galerkin weighted residual method, we again assume an approximate solution for v in the form of v_a. Substituting this into Equation 12.21 yields a residual equation:

$$R = EI\frac{\partial^4 v_a}{\partial x^4} + \rho A\frac{\partial^2 v_a}{\partial t^2} \tag{12.22}$$

Consonant with the Galerkin approach, we employ a weighting function W and then enforce

$$\int_0^L RW\,dx = \int_0^L W\left(EI\frac{\partial^4 v_a}{\partial x^4} + \rho A\frac{\partial^2 v_a}{\partial t^2}\right)dx = 0 \tag{12.23}$$

Replacing the weighting functions with the spatial shape functions **N** gives

$$\int_0^L \mathbf{N}\left(EI\frac{\partial^4 v_a}{\partial x^4} + \rho A\frac{\partial^2 v_a}{\partial t^2}\right)dx = 0 \tag{12.24}$$

By applying integration by part to Equation 12.24, one obtains

$$\left[\mathbf{N}\left(EI\frac{\partial^3 v_a}{\partial x^3}\right)\right]_0^L - \left[\frac{d\mathbf{N}}{d\mathbf{x}}\left(EI\frac{\partial^2 v_a}{\partial x^2}\right)\right]_0^L + \int_0^L \frac{d^2\mathbf{N}}{dx^2}\left(EI\frac{\partial^2 v_a}{\partial x^2}\right)dx$$
$$+\int_0^L \mathbf{N}\left(\rho A\frac{\partial^2 v_a}{\partial t^2}\right)dx = 0 \tag{12.25}$$

We will assume the approximate solution v_a is valid for a typical beam element with two nodes. Consequently, v_a may be represented by the nodal degrees of freedom $\boldsymbol{v}_e$ and the interpolation functions **N**, resulting in

$$v_a^e = \mathbf{N}^T \boldsymbol{v}_e \tag{12.26}$$

where

$$\boldsymbol{v}_e = \begin{bmatrix} v_1 & \theta_1 & v_2 & \theta_2 \end{bmatrix}^T \tag{12.27}$$

With Equation 12.26, Equation 12.25 is rewritten as

$$\underbrace{\left(\int_0^L \left(EI\frac{d^2\mathbf{N}^T}{dx^2}\right)\frac{d^2\mathbf{N}}{dx^2}dx\right)}_{\text{Stiffness}}\boldsymbol{v}_e + \underbrace{\left(\rho A\int_0^L \mathbf{N}^T\mathbf{N}dx\right)}_{\text{mass}}\frac{\partial^2 \boldsymbol{v}_e}{\partial t^2} = \{\boldsymbol{F}\} \tag{12.28}$$

where

$$\{\boldsymbol{F}\} = \left[\left(EI\frac{d^3\mathbf{N}^T}{dx^3}\right)\mathbf{N}\boldsymbol{v}_e\right]_0^L - \left[\left(EI\frac{\partial^2\mathbf{N}^T}{\partial x^2}\right)\frac{d\mathbf{N}}{dx}\boldsymbol{v}_e\right]_0^L \tag{12.29}$$

The vector of shape functions that interpolate the deflection and rotation along the length of the beam has been presented in Chapter 6 as

$$\mathbf{N} = \left[\left(1-\frac{3x^2}{L^2}+\frac{2x^3}{L^3}\right) \quad \left(x-\frac{2x^2}{L}+\frac{x^3}{L^2}\right) \quad \left(\frac{3x^2}{L^2}-\frac{2x^3}{L^3}\right) \quad \left(-\frac{x^2}{L}+\frac{x^3}{L^2}\right)\right] \tag{12.30}$$

Employing these shape functions, we can evaluate the mass matrix of the Euler–Bernoulli beam $\boldsymbol{M}_{EB}$ as

$$M_{EB} = \left(\int_0^L \rho A \mathbf{N}^T \mathbf{N} dx \right) = \frac{\rho AL}{420} \begin{bmatrix} 156 & 22L & 54 & -13L \\ 22L & 4L^2 & 13L & -3L^2 \\ 54 & 13L & 156 & -22L \\ -13L & -3L^2 & -22L & 4L^2 \end{bmatrix} \tag{12.31}$$

The stiffness matrix is also established as

$$K_{EB} = \int_0^L \left(EI \frac{d^2\mathbf{N}^T}{dx^2} \right) \frac{d^2\mathbf{N}}{dx^2} dx = \frac{EI}{L^3} \begin{bmatrix} 12 & 6L & -12 & 6L \\ 6L & 4L^2 & -6L & 2L^2 \\ -12 & -6L & 12 & -6L \\ 6L & 2L^2 & -6L & 4L^2 \end{bmatrix} \tag{12.32}$$

The following matrix equations then hold for the dynamic behavior of the Euler–Bernoulli beam element:

$$M_{EB} \begin{bmatrix} \ddot{v}_1 \\ \ddot{\theta}_1 \\ \ddot{v}_2 \\ \ddot{\theta}_2 \end{bmatrix} + K_{EB} \begin{bmatrix} v_1 \\ \theta_1 \\ v_2 \\ \theta_2 \end{bmatrix} = \begin{bmatrix} F_{1y}(t) \\ M_{1z}(t) \\ F_{2y}(t) \\ M_{2z}(t) \end{bmatrix} \tag{12.33}$$

12.3 Evaluation of Natural Frequencies and Eigenvalues

The generalized problem in free vibration analysis of structures is concerned with the determination of the natural frequencies. For the bar, truss, and beam elements, the natural frequencies are determined by solving Equations 12.15, 12.20, and 12.33, respectively, in the absence of the time-dependent force $\boldsymbol{F}(\boldsymbol{t})$. Given the absence of the time-dependent force, free vibration problems involve solving the matrix equation involving only the mass matrix $\boldsymbol{M}$ and stiffness matrix $\boldsymbol{K}$. That is,

$$\boldsymbol{M}\ddot{\boldsymbol{u}}_e(t) + \boldsymbol{K}\boldsymbol{u}_e(t) = 0 \tag{12.34}$$

A common praxis in dynamic analyses is to assume harmonic motion for certain vibration problems. As a result, a generic vector of nodal acceleration will be represented by

$$\ddot{\boldsymbol{u}}_e(t) = u_e e^{i\omega t} \tag{12.35}$$

An eigenvalue problem emerges with the substitution of the exponential function in Equation 12.35 into Equation 12.34 as

$$\left[-\omega^2 \boldsymbol{M} + \boldsymbol{K}\right] u_e e^{i\omega t} = 0 \tag{12.36}$$

Seeking a nontrivial solution of the above leads to

$$\left|\boldsymbol{K} - \omega^2 \boldsymbol{M}\right| = 0 \tag{12.37}$$

The R functions that will be used to deal with Equation 12.37, contextualized for the three elements, are presented in Section 12.4.

12.4 Computer Implementations

12.4.1 Procedure for the Free Vibration Analyses of Bars

12.4.1.1 Precomputation Phase

- Discretize the structure into finite elements.
- Label the nodes and elements to form the connectivity table.

12.4.1.2 Computation Phase

1. Specify the material and geometric properties of each bar element (if more than one). In most cases, more than one element will be needed to get an accurate result for free vibration problems.
2. Establish the finite element matrix equation for each bar element by using Equation 12.15. For computational purposes, we only need to form the stiffness and mass matrices for each element.
3. Determine the global degrees of freedom of the connected bars and carry out the expansion of the mass and stiffness matrices to conform to the size of the global degree of freedom.
4. Form the global stiffness and mass matrices needed for the global matrix.
5. Identify the essential boundary condition(s) and apply these to the global stiffness and mass matrices to obtain a reduced set of matrices.
   a. Establish an eigenvalue problem from the reduced set of mass and stiffness matrices.
   b. Determine the natural frequencies of the system from the eigenvalue problem.

The procedure listed here for the bar element should act as a basis for the modification of the procedures for the analyses of trusses and beams, initially given in Chapters 5 and 6.

12.4.2 Implemented R Functions for the Free Vibration Analysis of Bars

```
#*********************************************#
# R functions for free vibration analysis of    #
#        bar elements                            #
#       KB Mustapha                              #
#*********************************************#
DynamicBar_StiffnessMatrix = function(DOF=2,YoungMod,Area,Length)
{
  stiffbar=YoungMod*Area/Length
  ematrix=stiffbar*matrix(c(1,-1,-1,1),nrow=DOF,byrow=T);
  return (ematrix)
}

DynamicBar_MassMatrix = function(DOF=2,Density,Area,Length)
{
massmatrix=(Density*Area*Length/6)*matrix(c(2,1,1,2),nrow=DOF,byrow=T);
  return (massmatrix)
}

DynamicBar_ExpandedElement_Matrix = function(TDOF,eMatrix,i,j)
{
  bigMatrix = matrix(vector(l = TDOF * TDOF),nrow = TDOF,byrow = T);
  bigMatrix[c(i,j),c(i,j)] = eMatrix;
  return (bigMatrix)
}

DynamicBar_ReducedK = function(bigKmatrix,knownforcenodes)
{
  reducedK=bigKmatrix[c(knownforcenodes),(knownforcenodes)];
  return(reducedK)
}

DynamicBar_ReducedM = function(bigMmatrix,knownforcenodes)
{
  reducedM=bigMmatrix[c(knownforcenodes),(knownforcenodes)];
  return(reducedM)
}

DynamicBar_NaturalFrequencies = function(reducedM,reducedK)
{
  massinv=solve(reducedM);
  productMK=massinv%*%reducedK
  syseigen=eigen(productMK)
  sortedfreq=sort(syseigen$values)
  return(sqrt(sortedfreq))
}
```

12.4.3 Implemented R Functions for the Free Vibration Analysis of Truss Structures

```
#**********************************************#
#      R functions for vibration analysis of     #
#              plane truss elements              #
#            KB Mustapha                         #
#**********************************************#
DynamicPlaneTruss_LengthOrientation = function(vec_nodalcoordinates)
{
  ysquared=(vec_nodalcoordinates[4]-vec_nodalcoordinates[2])^2
  xsquared=(vec_nodalcoordinates[3]-vec_nodalcoordinates[1])^2
  length_of_element=sqrt(ysquared+xsquared)
  ydiff=(vec_nodalcoordinates[4]-vec_nodalcoordinates[2])
  xdiff=(vec_nodalcoordinates[3]-vec_nodalcoordinates[1])
  orientation_of_element=(180/pi)*atan(ydiff/xdiff)
  return(round(c(length_of_element,orientation_of_element),3))
}

DynamicPlaneTruss_StiffnessMatrix = function(DOF = 4,stifftruss,theta)
{
  m=cos(theta*pi/180);
  n=sin(theta*pi/180);
  stiffnesscoefficients=c(m^2,m*n,-m^2,-m*n,
                          m*n,n^2,-m*n,-n^2,
                          -m^2,-m*n,m^2,m*n,
                          -m*n,-n^2,m*n,n^2)
  ematrix=stifftruss*matrix(stiffnesscoefficients,nrow=DOF,byrow=T);
  return(ematrix)
}

DynamicPlaneTruss_MassMatrix = function(DOF = 4,Density,Area,Length,theta)
{
  m=cos(theta*pi/180);
  n=sin(theta*pi/180);
  masscoefficients=c(2,0,1,0,
                     0,2,0,1,
                     1,0,2,0,
                     0,1,0,2)
  mmatrix=(Density*Area*Length/6)*matrix(masscoefficients,nrow=DOF,byrow=T);
  return(mmatrix)
}

DynamicPlaneTruss_ExpandedMatrix = function(TDOF,eMatrix,i,j)
{

  r1=(i-1)+i;r2=(i-1)+(i+1);r3=(j-2)+(j+1);r4=(j-2)+(j+2);
  bigMatrix=matrix(vector(l=TDOF*TDOF),nrow=TDOF,byrow=T);
  bigMatrix[c(r1,r2,r3,r4),c(r1,r2,r3,r4)]=eMatrix;
  return (bigMatrix)
}

DynamicPlaneTruss_ReducedK = function(bigKmatrix,knownforcenodes)
{
  reducedk = bigKmatrix[c(knownforcenodes),(knownforcenodes)];
  return(reducedk)
}
```

```
DynamicPlaneTruss_ReducedM = function(bigMmatrix,knownforcenodes)
{
  reducedm = bigMmatrix[c(knownforcenodes),(knownforcenodes)];
  return(reducedm)
}

DynamicPlaneTruss_NaturalFrequencies = function(reducedM,reducedK)
{
  massinv=solve(reducedM);
  productMK=massinv%*%reducedK;
  syseigen=eigen(productMK);
  sortedfreq=sort(syseigen$values);
  return(sqrt(sortedfreq))
}
```

12.4.4 Implemented R Functions for the Free Vibration Analysis of Euler–Bernoulli Beams

```
#**************************************************#
#     R functions for vibration analysis           #
#       of Euler Bernoulli beam elements           #
#            KB Mustapha                           #
#**************************************************#

DynamicEulerBeam_
StiffnessMatrix = function(DOF = 4,YoungMod,MomentI,Length)
{
  L=Length;
  stiffness=YoungMod*MomentI/Length^3;
  coefficients=c(12,6*L,-12,6*L,
                 6*L,4*L^2,-6*L,2*L^2,
                 -12,-6*L,12,-6*L,
                 6*L,2*L^2,-6*L,4*L^2);
  ematrix=stiffness*matrix(coefficients,nrow=DOF,byrow=T);
  return(ematrix)
}

DynamicEulerBeam_MassMatrix = function(DOF = 4,Density,Area,Length)
{

  L=Length;
  masscoeff=c(156,22*L,54,-13*L,
             22*L,4*L^2,13*L,-3*L^2,
             54,13*L,156,-22*L,
             -13*L,-3*L^2,-22*L,4*L^2);

massmatrix=(Density*Area*Length/420)*matrix(masscoeff,nrow=DOF,
byrow=T);
  return(massmatrix)
}
```

```
DynamicEulerBeam_ExpandedMatrix = function(TDOF,eMatrix,i,j)
{
  r1=(i-1)+i;r2=(i-1)+(i+1);r3=(j-2)+(j+1);r4=(j-2)+(j+2);
  bigMatrix=matrix(vector(l=TDOF*TDOF),nrow=TDOF,byrow=T);
  bigMatrix[c(r1,r2,r3,r4),c(r1,r2,r3,r4)]=eMatrix;
  return (bigMatrix)
}

DynamicEulerBeam_ReducedK = function(bigKmatrix,knownloadnodes)
{
  reducedK = bigKmatrix[c(knownloadnodes),(knownloadnodes)]
  return(reducedK)
}

DynamicEulerBeam_ReducedM = function(bigMmatrix,knownloadnodes)
{
  reducedM = bigMmatrix[c(knownloadnodes),(knownloadnodes)]
  return(reducedM)
}

DynamicEulerBeam_NaturalFrequencies = function(reducedM,reducedK)
{
  massinv=solve(reducedM);
  productMK=massinv%*%reducedK
  syseigen=eigen(productMK)
  sortedfreq=sort(syseigen$values)
  return(sqrt(sortedfreq))
}
```

The usage of the above-listed functions will be illustrated in the examples that follow.

12.5 Examples

12.5.1 Example 12.1

For the fixed–fixed and fixed–free bar shown in Figure 12.1, use four bar finite elements to determine the first three natural frequencies. Compare the results with the exact solutions. Assume $L = 2\,\text{m}$; $A = 0.004\,\text{m}$; $E = 200\,\text{GPa}$, and $\rho = 7800\ \text{kg/m}^3$.

12.5.1.1 Solution

For purpose of analysis, the two bars in Figure 12.1 are discretized into four bar elements, with each element having a length $L_e = L/4 = 0.5$ m.

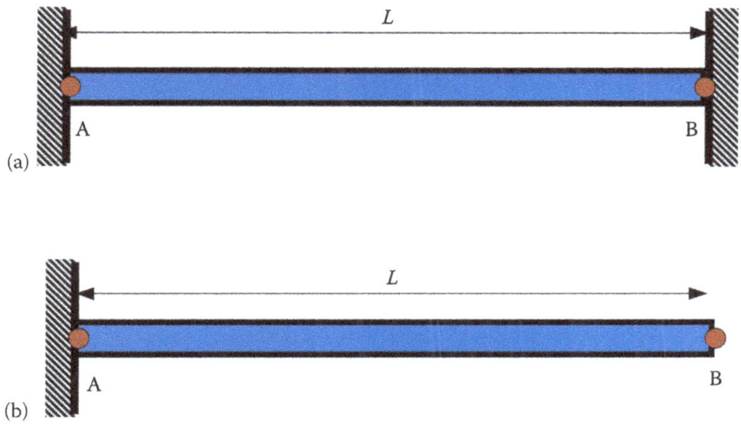

FIGURE 12.1
(a) A fixed–fixed bar and (b) a fixed–free bar.

TABLE 12.1
Element Connectivity and Properties for Figure 12.1

Element	Nodes (i, j)
1	1,2
2	2,3
3	3,4
4	4,5

The discretized structure is not shown, but the connectivity of the elements is contained in Table 12.1.

Based on the above-mentioned information, the rest of the analysis steps is carried out using the R functions.

Step 1: Specify the elements' properties.

```
YoungMod=200e9; area=0.004;L=2 Le=L/4;rho=7800;
```

Step 2: Write the stiffness and mass matrices for each element using **`DynamicBar_StiffnessMatrix()`** and **`DynamicBar_MassMatrix()`**.

```
k1=DynamicBar_StiffnessMatrix(2,YoungMod,area,Le);k1
k2=DynamicBar_StiffnessMatrix(2,YoungMod,area,Le);k2
k3=DynamicBar_StiffnessMatrix(2,YoungMod,area,Le);k3
k4=DynamicBar_StiffnessMatrix(2,YoungMod,area,Le);k4

m1=DynamicBar_MassMatrix(2,rho,area,Le);m1
m2=DynamicBar_MassMatrix(2,rho,area,Le);m2
m3=DynamicBar_MassMatrix(2,rho,area,Le);m3
m4=DynamicBar_MassMatrix(2,rho,area,Le);m4
```

The stiffness and mass matrices are obtained and shown as

```
> k1=DynamicBar_StiffnessMatrix(2,YoungMod,area,Le);k1
            [,1]        [,2]
[1,]  1600000000 -1600000000
[2,] -1600000000  1600000000
> k2=DynamicBar_StiffnessMatrix(2,YoungMod,area,Le);k2
            [,1]        [,2]
[1,]  1600000000 -1600000000
[2,] -1600000000  1600000000
> k3=DynamicBar_StiffnessMatrix(2,YoungMod,area,Le);k3
            [,1]        [,2]
[1,]  1600000000 -1600000000
[2,] -1600000000  1600000000
> k4=DynamicBar_StiffnessMatrix(2,YoungMod,area,Le);k4
            [,1]        [,2]
[1,]  1600000000 -1600000000
[2,] -1600000000  1600000000
>
>
> m1=DynamicBar_MassMatrix(2,rho,area,Le);m1
     [,1] [,2]
[1,]  5.2  2.6
[2,]  2.6  5.2
> m2=DynamicBar_MassMatrix(2,rho,area,Le);m2
     [,1] [,2]
[1,]  5.2  2.6
[2,]  2.6  5.2
> m3=DynamicBar_MassMatrix(2,rho,area,Le);m3
     [,1] [,2]
[1,]  5.2  2.6
[2,]  2.6  5.2
> m4=DynamicBar_MassMatrix(2,rho,area,Le);m4
     [,1] [,2]
[1,]  5.2  2.6
[2,]  2.6  5.2
```

Step 3: Determine the global degree of freedom of the structure and expand the stiffness and mass matrices for each element to conform to the global degree of freedom.

The expansion of the matrices is done with the help of the function **`DynamicBar_ExpandedElement_Matrix()`**.

```
n_nodes=5;Total_dof=n_nodes*1;
K1=DynamicBar_ExpandedElement_Matrix(Total_dof,k1,1,2);K1
K2=DynamicBar_ExpandedElement_Matrix(Total_dof,k2,2,3);K2
K3=DynamicBar_ExpandedElement_Matrix(Total_dof,k3,3,4);K3
K4=DynamicBar_ExpandedElement_Matrix(Total_dof,k4,4,5);K4

M1=DynamicBar_ExpandedElement_Matrix(Total_dof,m1,1,2);M1
M2=DynamicBar_ExpandedElement_Matrix(Total_dof,m2,2,3);M2
M3=DynamicBar_ExpandedElement_Matrix(Total_dof,m3,3,4);M3
M4=DynamicBar_ExpandedElement_Matrix(Total_dof,m4,4,5);M4
```

Step 4: Establish the global stiffness and mass matrices.

```
GlobalK=K1+K2+K3+K4;GlobalK
GlobalM=M1+M2+M3+M4;GlobalM
```

As we have just five nodes for the four elements, the output of this produces the 5×5 global stiffness matrix $[K]$ and the 5×5 mass matrix M as

```
> GlobalK=K1+K2+K3+K4;GlobalK
            [,1]        [,2]        [,3]        [,4]        [,5]
[1,]  1600000000 -1600000000           0           0           0
[2,] -1600000000  3200000000 -1600000000           0           0
[3,]           0 -1600000000  3200000000 -1600000000           0
[4,]           0           0 -1600000000  3200000000 -1600000000
[5,]           0           0           0 -1600000000  1600000000
> GlobalM=M1+M2+M3+M4;GlobalM
     [,1] [,2] [,3] [,4] [,5]
[1,]  5.2  2.6  0.0  0.0  0.0
[2,]  2.6 10.4  2.6  0.0  0.0
[3,]  0.0  2.6 10.4  2.6  0.0
[4,]  0.0  0.0  2.6 10.4  2.6
[5,]  0.0  0.0  0.0  2.6  5.2
```

Step 5: Identify the essential boundary condition(s) and apply these to the global stiffness and mass matrices to obtain a reduced set of matrices.

For the fixed–fixed bar, the ends of the bar (nodes 1 and 5) are restrained from movements, hence the boundary conditions are as follows:

$$U_{1x} = U_{5x} = 0 \tag{12.38}$$

However, for the fixed–free bar, only node 1 is fixed; therefore, the boundary condition is

$$U_{1x} = 0 \tag{12.39}$$

By using the boundary conditions specified in Equations 12.38 and 12.39, we can obtain the reduced stiffness and mass matrices for the two cases. These reduced matrices will then be used to find the natural frequencies.

For the fixed–fixed bar we have

```
vec_rows_of_knownloads=c(2,3,4);
ReducedK=DynamicBar_ReducedK(GlobalK,vec_rows_of_
knownloads);
ReducedK

ReducedM=DynamicBar_ReducedM(GlobalM,vec_rows_of_
knownloads);
ReducedM

naturalfrequencies=DynamicBar_NaturalFrequencies(ReducedM,
ReducedK);
cbind(naturalfrequencies)
```

Running code produces the following output:

```
> vec_rows_of_knownloads=c(2,3,4);
> ReducedK=DynamicBar_ReducedK(GlobalK,vec_rows_of_
knownloads);
> ReducedK
            [,1]        [,2]        [,3]
[1,]  3200000000 -1600000000           0
[2,] -1600000000  3200000000 -1600000000
[3,]           0 -1600000000  3200000000
>
> ReducedM=DynamicBar_ReducedM(GlobalM,vec_rows_of_
knownloads);
```

```
> ReducedM
      [,1] [,2] [,3]
[1,] 10.4  2.6  0.0
[2,]  2.6 10.4  2.6
[3,]  0.0  2.6 10.4
>
> naturalfrequencies=DynamicBar_NaturalFrequencies(Reduce
dM,ReducedK);
> cbind(naturalfrequencies)
     naturalfrequencies
[1,]            8159.72
[2,]           17541.16
[3,]           28505.08
```

On the basis of the preceding set of output, we can extract the natural frequencies as

$$\begin{Bmatrix} \omega_1 \\ \omega_2 \\ \omega_3 \end{Bmatrix}^{\text{FEM}} = \begin{Bmatrix} 8159.72 \\ 17541.16 \\ 28505.08 \end{Bmatrix} \text{rad/s} \tag{12.40}$$

Meanwhile, the exact solution for the natural frequency $\omega_n (n = 1,2,3,\ldots,N)$ of a fixed–fixed bar is [6]

$$\omega_n^{\text{exact}} = \frac{\lambda_n}{L}\sqrt{\frac{E}{\rho}};\ \lambda_n = n\pi \text{ where } n = 1,2,3,\ldots,N \tag{12.41}$$

Substituting the parameters (E,ρ,L) of the question into Equation 12.41 results in

$$\begin{Bmatrix} \omega_1 \\ \omega_2 \\ \omega_3 \end{Bmatrix}^{\text{Exact}} = \begin{Bmatrix} 7950.004 \\ 15900.008 \\ 23850.012 \end{Bmatrix} \text{rad/s} \tag{12.42}$$

As revealed, the first, second, and third natural frequencies determined from the finite element calculations (given in Equation 12.40) deviate from the exact solution by roughly 2%, 10%, and 20%, respectively. It is possible to get the deviation to be less than 1% for the lowest frequencies by using more elements.

For the fixed–free bar we have

```
vec_rows_of_knownloads=c(2,3,4,5);
ReducedK=DynamicBar_ReducedK(GlobalK,vec_rows_of_
knownloads);
ReducedK

ReducedM=DynamicBar_ReducedM(GlobalM,vec_rows_of_
knownloads);
ReducedM

naturalfrequencies=DynamicBar_NaturalFrequencies(ReducedM,
ReducedK);
cbind(naturalfrequencies)
```

Running the code produces

```
> vec_rows_of_knownloads=c(2,3,4,5);
> ReducedK=DynamicBar_ReducedK(GlobalK,vec_rows_of_
knownloads);
> ReducedK
            [,1]        [,2]        [,3]        [,4]
[1,]  3200000000 -1600000000           0           0
[2,] -1600000000  3200000000 -1600000000           0
[3,]           0 -1600000000  3200000000 -1600000000
[4,]           0           0 -1600000000  1600000000
>
> ReducedM=DynamicBar_ReducedM(GlobalM,vec_rows_of_
knownloads);
> ReducedM
     [,1] [,2] [,3] [,4]
[1,] 10.4  2.6  0.0  0.0
[2,]  2.6 10.4  2.6  0.0
[3,]  0.0  2.6 10.4  2.6
[4,]  0.0  0.0  2.6  5.2
>
> naturalfrequencies=DynamicBar_NaturalFrequencies(Reduce
dM,ReducedK);
> cbind(naturalfrequencies)
     naturalfrequencies
[1,]           4002.619
[2,]          12626.824
[3,]          22937.030
```

From the preceding output, the three lowest natural frequencies of the fixed–free bar are obtained as

$$\begin{Bmatrix} \omega_1 \\ \omega_2 \\ \omega_3 \end{Bmatrix}^{\text{FEM}} = \begin{Bmatrix} 4002.619 \\ 12626.824 \\ 22937.030 \end{Bmatrix} \text{rad/s} \tag{12.43}$$

The exact solution for the frequency $\omega_n \left(n = 1,2,3,\ldots,N\right)$ of a fixed–fixed bar is [6]

$$\omega_n^{\text{exact}} = \frac{\lambda_n}{L}\sqrt{\frac{E}{\rho}}; \lambda_n = \frac{(2n-1)\pi}{2} \text{ where } n = 1,2,3,\ldots,N \tag{12.44}$$

Plugging the parameters (E,ρ,L) of the question into Equation 12.44 gives

$$\begin{Bmatrix} \omega_1 \\ \omega_2 \\ \omega_3 \end{Bmatrix}^{\text{Exact}} = \begin{Bmatrix} 3977.018 \\ 11931.055 \\ 19885.091 \end{Bmatrix} \text{rad/s} \tag{12.45}$$

Comparing the predictions shows that the first, second, and third natural frequencies determined from the finite element calculations for the fixed–free bar deviate from the estimate of the exact solution by roughly 0.6%, 5%, and 15%, respectively. Again, we can use more elements to reduce the deviations.

12.5.2 Example 12.2

By using the functions for the truss element, estimate the two lowest frequencies of the plane truss shown in Figure 12.2. Assume $A = 0.06\,\text{in}^2$; $E = 30\times10^6\,\text{lbf/in}^2$; and $\rho = 7.35\times10^{-4}\,\text{lbf}\,\text{s}^2/\text{in}^4$.[1]

12.5.2.1 Solution

Our solution approach will be to treat points A, B, C, and D as nodes 1, 2, 3, and 4, respectively. With this in mind, members AB, CD, and DB become elements 1, 2, and 3, respectively. With this information, the elements' connectivity is shown in Table 12.2.

The rest of the solution steps follows, and we employ the functions from Section 12.4.3.

[1] To allow for comparison, the units of the variables in this example have been intentionally left as given in Ref [7] C. T. Ross, *Finite Element Programs for Structural Vibrations*, Springer Science & Business Media, 2012.

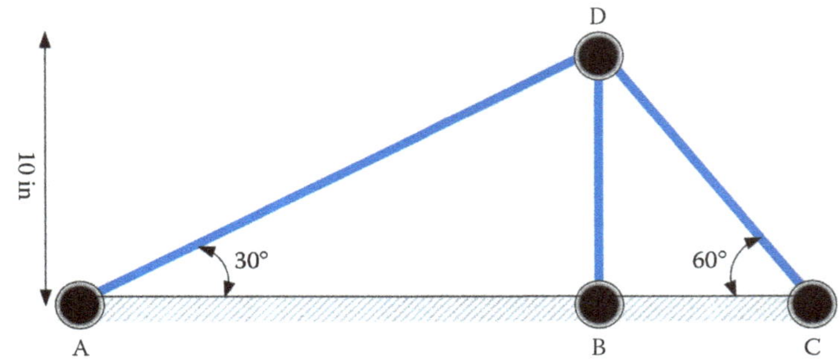

FIGURE 12.2
A three-member plane truss.

TABLE 12.2
Element Connectivity and Properties for Figure 12.2

Elements	Nodes (i,j)	Coordinates (x_i,y_i,x_j,y_j)
1	1,4	(0,0,17.32,10)
2	2,4	(17.32,10,17.32,0)
3	3,4	(0,0,0,6)

Step 1: Specify elements' properties (material and the geometric lengths and orientation angles of each).

```
e1coord=c(0,0,17.32,10);
e2coord=c(17.32,0,17.32,10);
e3coord=c(23.09,0,17.32,10);
E1=E2=E3=30e6;
Density=7.35e-4;
A1=A2=A3=0.06;

L1T1=DynamicPlaneTruss_LengthOrientation(e1coord);L1T1
L2T2=DynamicPlaneTruss_LengthOrientation(e2coord);L2T2
L3T3=DynamicPlaneTruss_LengthOrientation(e3coord);L3T3

stiffness1=E1*A1/L1T1[1];stiffness1
stiffness2=E2*A2/L2T2[1];stiffness2
stiffness3=E3*A3/L3T3[1];stiffness3
```

In the above-mentioned block of code, the first six lines supply information about the nodal coordinates of each element, their Young's modulus, density, and area. In the next three lines, we used the function **DynamicPlaneTruss_LengthOrientation()** to obtain the length and orientation angles of each element. The last three lines compute the axial stiffness for the elements. By running the above-mentioned code, we generate the following output:

```
> L1T1=DynamicPlaneTruss_LengthOrientation(e1coord);L1T1
[1] 20.000 30.001
> L2T2=DynamicPlaneTruss_LengthOrientation(e2coord);L2T2
[1] 10 90
> L3T3=DynamicPlaneTruss_LengthOrientation(e3coord);L3T3
[1]  11.545 -60.015

> E1=E2=E3=30e6;
> Density=7.35e-4;
> A1=A2=A3=0.06;
> stiffness1=E1*A1/L1T1[1];stiffness1
[1] 90000
> stiffness2=E2*A2/L2T2[1];stiffness2
[1] 180000
> stiffness3=E3*A3/L3T3[1];stiffness3
[1] 155911.7
```

As shown earlier, it should be noticed that the function **DynamicPlaneTruss_LengthOrientation()** returns a vector of two numbers (the length and orientation angles of the element). We will extract each of these two numbers in the next line of code.

Step 2: Write the stiffness and mass matrices for each element.

```
k1=DynamicPlaneTruss_StiffnessMatrix(4,stiffness1,L1T1[2]);
k2=DynamicPlaneTruss_StiffnessMatrix(4,stiffness2,L2T2[2]);
k3=DynamicPlaneTruss_StiffnessMatrix(4,stiffness3,L3T3[2]);

m1=DynamicPlaneTruss_MassMatrix(4,Density,A1,L1T1[1],
L1T1[2]);
m2=DynamicPlaneTruss_MassMatrix(4,Density,A2,L2T2[1],
L2T2[2]);
m3=DynamicPlaneTruss_MassMatrix(4,Density,A3,L3T3[1],
L3T3[2]);
```

where we have used the function **DynamicPlaneTruss_Stiffness Matrix()** for the expansion of the matrices, and the variables **L1T1**, **L2T2**, **L3T3** have been computed in the preceding block of code.

Step 3: Determine the global degree of freedom of the truss and expand the mass and stiffness matrices for each element to conform to the global degree of freedom.

```
n_nodes=4;Total_dof=n_nodes*2;
K1=DynamicPlaneTruss_ExpandedMatrix(Total_dof,k1,1,4);
K2=DynamicPlaneTruss_ExpandedMatrix(Total_dof,k2,2,4);
K3=DynamicPlaneTruss_ExpandedMatrix(Total_dof,k3,3,4);

M1=DynamicPlaneTruss_ExpandedMatrix(Total_dof,m1,1,4);
M2=DynamicPlaneTruss_ExpandedMatrix(Total_dof,m2,2,4);
M3=DynamicPlaneTruss_ExpandedMatrix(Total_dof,m3,3,4);
```

Step 4: Establish the global stiffness matrix by adding the expanded matrices of the elements as shown next.

```
GlobalK=K1+K2+K3;round(GlobalK,4)
GlobalM=M1+M2+M3;round(GlobalM,7)
```

The output of the preceding code produces the global stiffness and mass stiffness matrices $\boldsymbol{K}$ and $\boldsymbol{M}$ as

```
> GlobalK=K1+K2+K3;round(GlobalK,4)
          [,1]       [,2] [,3]    [,4]      [,5]       [,6]      [,7]       [,8]
[1,]  67498.64   38971.93    0       0      0.00       0.00 -67498.64  -38971.93
[2,]  38971.93   22501.36    0       0      0.00       0.00 -38971.93  -22501.36
[3,]      0.00       0.00    0       0      0.00       0.00      0.00       0.00
[4,]      0.00       0.00    0  180000      0.00       0.00      0.00 -180000.00
[5,]      0.00       0.00    0       0  38942.57  -67491.31 -38942.57   67491.31
[6,]      0.00       0.00    0       0 -67491.31  116969.08  67491.31 -116969.08
[7,] -67498.64  -38971.93    0       0 -38942.57   67491.31 106441.21  -28519.38
[8,] -38971.93  -22501.36    0 -180000  67491.31 -116969.08 -28519.38  319470.44
> GlobalM=M1+M2+M3;round(GlobalM,7)
         [,1]     [,2]      [,3]      [,4]      [,5]      [,6]      [,7]      [,8]
[1,] 0.000294 0.000000 0.0000000 0.0000000 0.0000000 0.0000000 0.0001470 0.0000000
[2,] 0.000000 0.000294 0.0000000 0.0000000 0.0000000 0.0000000 0.0000000 0.0001470
[3,] 0.000000 0.000000 0.0001470 0.0000000 0.0000000 0.0000000 0.0000735 0.0000000
[4,] 0.000000 0.000000 0.0000000 0.0001470 0.0000000 0.0000000 0.0000000 0.0000735
[5,] 0.000000 0.000000 0.0000000 0.0000000 0.0001697 0.0000000 0.0000849 0.0000000
[6,] 0.000000 0.000000 0.0000000 0.0000000 0.0000000 0.0001697 0.0000000 0.0000849
[7,] 0.000147 0.000000 0.0000735 0.0000000 0.0000849 0.0000000 0.0006107 0.0000000
[8,] 0.000000 0.000147 0.0000000 0.0000735 0.0000000 0.0000849 0.0000000 0.0006107
```

For readability, we reproduce the 8×8 global stiffness and mass matrices in the following.

$$[K] = \begin{bmatrix} 67498.64 & 38971.93 & 0. & 0. & 0. & 0. & -67498.64 & -38971.93 \\ 38971.93 & 22501.36 & 0. & 0. & 0. & 0. & -38971.93 & -22501.36 \\ 0. & 0. & 0. & 0. & 0. & 0. & 0. & 0. \\ 0. & 0. & 0. & 180000. & 0. & 0. & 0. & 180000. \\ 0. & 0. & 0. & 0. & 38942.57 & -67491.31 & -38942.57 & 67491.31 \\ 0. & 0. & 0. & 0. & -67491.31 & 116969.08 & 67491.31 & -116969.08 \\ -67948.64 & -38971.93 & 0. & 0. & -38942.57 & 67491.31 & 106441.21 & -28519.38 \\ -38971.93 & -22501.36 & 0. & -180000. & 67941.31 & -116969.08 & -28519.38 & 319470.44 \end{bmatrix}$$

$$[M] = \begin{bmatrix} 0.000294 & 0. & 0. & 0. & 0. & 0. & 0.000147 & 0. \\ 0. & 0.000294 & 0. & 0. & 0. & 0. & 0. & 0.000147 \\ 0. & 0. & 0.000147 & 0. & 0. & 0. & 0.0000735 & 0. \\ 0. & 0. & 0. & 0.000147 & 0. & 0. & 0. & 0.0000735 \\ 0. & 0. & 0. & 0. & 0.0001697 & 0. & 0.0000849 & 0. \\ 0. & 0. & 0. & 0. & 0. & 0.0001697 & 0. & 0.0000849 \\ 0.000147 & 0. & 0.0000735 & 0. & 0.0000849 & 0. & 0.0006107 & 0. \\ 0. & 0.000147 & 0. & 0.0000735 & 0. & 0.0000849 & 0. & 0.0006107 \end{bmatrix}$$

By using the above-mentioned matrices, we establish the global matrix equation as

$$\left([K]-\omega^2[M]\right)\begin{Bmatrix} u_{1X} \\ u_{1Y} \\ u_{2X} \\ u_{2Y} \\ u_{3X} \\ u_{3Y} \\ u_{4X} \\ u_{4Y} \end{Bmatrix} = 0 \tag{12.46}$$

Step 5: Identify the essential boundary condition(s) and apply the boundary condition(s) to modify the global matrix equation.

With the configuration of the truss, we can identify the zero-displacement boundary conditions as

$$u_{1X} = u_{1Y} = u_{2X} = u_{2Y} = u_{3X} = u_{3Y} = 0 \tag{12.47}$$

The boundary conditions belong to rows 1 to 6. Ultimately, this means that only rows 7 and 8 (Equation 12.46) are needed to obtain the reduced mass and stiffness matrices required to find the natural frequencies. We employ this information in the next lines of code to obtain the natural frequencies.

```
vec_rows_of_knownloads=c(7,8);
ReducedK=DynamicPlaneTruss_ReducedK(GlobalK,vec_rows_of_
knownloads);
ReducedK

ReducedM=DynamicPlaneTruss_ReducedM(GlobalM,vec_rows_of_
knownloads);
ReducedM

naturalfrequencies=DynamicPlaneTruss_NaturalFrequencies
(ReducedM,ReducedK);
cbind(naturalfrequencies)
paste(round(naturalfrequencies/(2*pi)),"Hertz")
```

Running the code produces

```
> ReducedK
           [,1]      [,2]
[1,] 106441.21 -28519.38
[2,] -28519.38 319470.44
>
> ReducedM=DynamicPlaneTruss_ReducedM(GlobalM,vec_rows_
of_knownloads);
> ReducedM
             [,1]         [,2]
[1,] 0.0006107115 0.0000000000
[2,] 0.0000000000 0.0006107115
>
> naturalfrequencies=DynamicPlaneTruss_NaturalFrequencies
 (ReducedM,ReducedK);
> cbind(naturalfrequencies)
     naturalfrequencies
[1,]           12967.15
[2,]           23005.55
> paste(round(naturalfrequencies/(2*pi)),"Hertz')
[1] "2064 Hertz" "3661 Hertz"
```

From the preceding output, the two lowest natural frequencies of the truss are obtained as

$$\begin{Bmatrix} \omega_1 \\ \omega_2 \end{Bmatrix}^{\text{FEM}} = \begin{Bmatrix} 12967.15 \\ 23005.55 \end{Bmatrix} \text{rad/s} \tag{12.48}$$

Converting the previous unit into Hertz, we have

$$\begin{Bmatrix} \omega_1 \\ \omega_2 \end{Bmatrix}^{FEM} = \begin{Bmatrix} 2064 \\ 3661 \end{Bmatrix} \text{Hz} \tag{12.49}$$

The values in Hz match those obtained in Reference 7, in which the FORTRAN programing language was used for the FEM coding.

12.5.3 Example 12.3

By using the Euler–Bernoulli beam element, estimate the two lowest frequencies of the beams shown in Figure 12.3. Compare the results with the exact solution. Assume $L = 10\,\text{m}$; $A = 2.4 \times 10^{-3}\,\text{m}$; $I = 10 \times 10^{-6}\,\text{m}^4$; $E = 200\,\text{GPa}$, and $\rho = 7800\,\text{kg/m}^3$.

12.5.3.1 Solution

The two beams in Figure 12.3 are already discretized into four beam elements, and the connectivity of the elements is contained in Table 12.3.

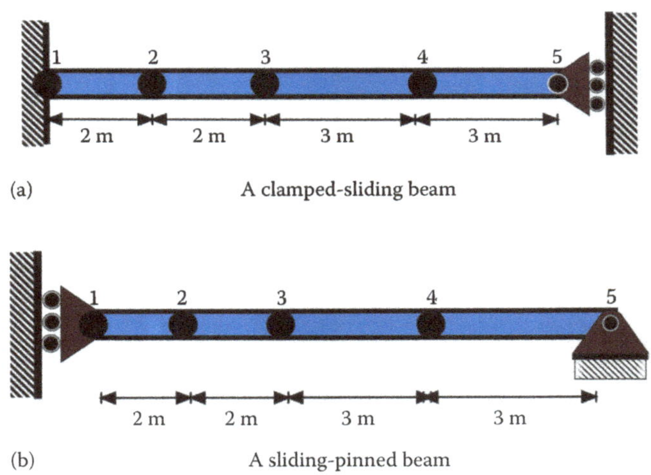

FIGURE 12.3
Discretized single-span beams with different boundary conditions.

TABLE 12.3
Element Connectivity and Properties

Element	Nodes (*i*, *j*)
1	1,2
2	2,3
3	3,4
4	4,5

With the data provided, the rest of the analysis steps is carried out using the R functions for the Euler–Bernoulli beams provided in Section 12.4.4.

Step 1: Specify the elements' properties.

```
YoungMod=2e11;A=2.4e-3;
Density=7800;MomentI=10e-6;
L1=L2=2;
L3=L4=3;
```

Step 2: Write the stiffness and mass matrices for each element.

```
k1=DynamicEulerBeam_StiffnessMatrix(4,YoungMod,MomentI,
L1);k1
k2=DynamicEulerBeam_StiffnessMatrix(4,YoungMod,MomentI,
L2);k2
k3=DynamicEulerBeam_StiffnessMatrix(4,YoungMod,MomentI,
L3);k3
k4=DynamicEulerBeam_StiffnessMatrix(4,YoungMod,MomentI,
L4);k4

m1=DynamicEulerBeam_MassMatrix(4,Density,A,L1);m1
m2=DynamicEulerBeam_MassMatrix(4,Density,A,L2);m2
m3=DynamicEulerBeam_MassMatrix(4,Density,A,L3);m3
m4=DynamicEulerBeam_MassMatrix(4,Density,A,L4);m4
```

The stiffness and mass matrices are obtained as

```
> k1=DynamicEulerBeam_StiffnessMatrix(4,YoungMod,MomentI,
L1);k1
         [,1]     [,2]     [,3]     [,4]
[1,]  3000000  3000000 -3000000  3000000
[2,]  3000000  4000000 -3000000  2000000
[3,] -3000000 -3000000  3000000 -3000000
[4,]  3000000  2000000 -3000000  4000000
> k2=DynamicEulerBeam_StiffnessMatrix(4,YoungMod,MomentI,
L2);k2
         [,1]     [,2]     [,3]     [,4]
[1,]  3000000  3000000 -3000000  3000000
[2,]  3000000  4000000 -3000000  2000000
[3,] -3000000 -3000000  3000000 -3000000
[4,]  3000000  2000000 -3000000  4000000
> k3=DynamicEulerBeam_StiffnessMatrix(4,YoungMod,MomentI,
L3);k3
```

```
          [,1]      [,2]       [,3]     [,4]
[1,]  888888.9   1333333  -888888.9  1333333
[2,] 1333333.3   2666667 -1333333.3  1333333
[3,] -888888.9  -1333333   888888.9 -1333333
[4,] 1333333.3   1333333 -1333333.3  2666667
> k4=DynamicEulerBeam_StiffnessMatrix(4,YoungMod,MomentI,
L4);k4
          [,1]      [,2]       [,3]     [,4]
[1,]  888888.9   1333333  -888888.9  1333333
[2,] 1333333.3   2666667 -1333333.3  1333333
[3,] -888888.9  -1333333   888888.9 -1333333
[4,] 1333333.3   1333333 -1333333.3  2666667
>
> m1=DynamicEulerBeam_MassMatrix(4,Density,A,L1);m1
          [,1]      [,2]      [,3]      [,4]
[1,] 14.013257  3.952457  4.850743 -2.335543
[2,]  3.952457  1.437257  2.335543 -1.077943
[3,]  4.850743  2.335543 14.013257 -3.952457
[4,] -2.335543 -1.077943 -3.952457  1.437257
> m2=DynamicEulerBeam_MassMatrix(4,Density,A,L2);m2
          [,1]      [,2]      [,3]      [,4]
[1,] 14.013257  3.952457  4.850743 -2.335543
[2,]  3.952457  1.437257  2.335543 -1.077943
[3,]  4.850743  2.335543 14.013257 -3.952457
[4,] -2.335543 -1.077943 -3.952457  1.437257
> m3=DynamicEulerBeam_MassMatrix(4,Density,A,L3);m3
          [,1]      [,2]      [,3]      [,4]
[1,] 21.019886  8.893029  7.276114 -5.254971
[2,]  8.893029  4.850743  5.254971 -3.638057
[3,]  7.276114  5.254971 21.019886 -8.893029
[4,] -5.254971 -3.638057 -8.893029  4.850743
> m4=DynamicEulerBeam_MassMatrix(4,Density,A,L4);m4
          [,1]      [,2]      [,3]      [,4]
[1,] 21.019886  8.893029  7.276114 -5.254971
[2,]  8.893029  4.850743  5.254971 -3.638057
[3,]  7.276114  5.254971 21.019886 -8.893029
[4,] -5.254971 -3.638057 -8.893029  4.850743
```

Step 3: Determine the global degree of freedom of the beam and expand the stiffness and mass matrix for each element to conform to the global degree of freedom.

The expansions of the matrices are done with the help of the function **DynamicEulerBeam_ExpandedMatrix()**.

```
n_nodes=5;Total_dof=n_nodes*2;
K1=DynamicEulerBeam_ExpandedMatrix(Total_dof,k1,1,2);
K2=DynamicEulerBeam_ExpandedMatrix(Total_dof,k2,2,3);
K3=DynamicEulerBeam_ExpandedMatrix(Total_dof,k3,3,4);
K4=DynamicEulerBeam_ExpandedMatrix(Total_dof,k4,4,5);

M1=DynamicEulerBeam_ExpandedMatrix(Total_dof,m1,1,2);
M2=DynamicEulerBeam_ExpandedMatrix(Total_dof,m2,2,3);
M3=DynamicEulerBeam_ExpandedMatrix(Total_dof,m3,3,4);
M4=DynamicEulerBeam_ExpandedMatrix(Total_dof,m4,4,5);
```

Step 4: Establish the global stiffness and mass matrices.

```
GlobalK=K1+K2+K3+K4;GlobalK
GlobalM=M1+M2+M3+M4;GlobalM
```

As the beam element is endowed with 2 degrees of freedom per node, and we have 5 nodes for the discretized structure, the assembled systems will have 10 total degrees of freedom. Hence, the output of the preceding lines of code produces the 10×10 global stiffness matrix $[K]$ and the 10×10 mass matrix M as

```
> GlobalK=K1+K2+K3+K4;GlobalK
         [,1]     [,2]     [,3]     [,4]       [,5]     [,6]       [,7]     [,8]       [,9]    [,10]
 [1,]  3000000  3000000 -3000000  3000000        0.0        0        0.0        0        0.0        0
 [2,]  3000000  4000000 -3000000  2000000        0.0        0        0.0        0        0.0        0
 [3,] -3000000 -3000000  6000000        0 -3000000.0  3000000        0.0        0        0.0        0
 [4,]  3000000  2000000        0  8000000 -3000000.0  2000000        0.0        0        0.0        0
 [5,]        0        0 -3000000 -3000000  3888888.9 -1666667  -888888.9  1333333        0.0        0
 [6,]        0        0  3000000  2000000 -1666666.7  6666667 -1333333.3  1333333        0.0        0
 [7,]        0        0        0        0  -888888.9 -1333333  1777777.8        0  -888888.9  1333333
 [8,]        0        0        0        0  1333333.3  1333333        0.0  5333333 -1333333.3  1333333
 [9,]        0        0        0        0        0.0        0  -888888.9 -1333333   888888.9 -1333333
[10,]        0        0        0        0        0.0        0  1333333.3  1333333 -1333333.3  2666667
> GlobalM=M1+M2+M3+M4;GlobalM
           [,1]      [,2]      [,3]      [,4]      [,5]      [,6]      [,7]      [,8]      [,9]     [,10]
 [1,] 14.013257  3.952457  4.850743 -2.335543  0.000000  0.000000  0.000000  0.000000  0.000000  0.000000
 [2,]  3.952457  1.437257  2.335543 -1.077943  0.000000  0.000000  0.000000  0.000000  0.000000  0.000000
 [3,]  4.850743  2.335543 28.026514  0.000000  4.850743 -2.335543  0.000000  0.000000  0.000000  0.000000
 [4,] -2.335543 -1.077943  0.000000  2.874514  2.335543 -1.077943  0.000000  0.000000  0.000000  0.000000
 [5,]  0.000000  0.000000  4.850743  2.335543 35.033143  4.940571  7.276114 -5.254971  0.000000  0.000000
 [6,]  0.000000  0.000000 -2.335543 -1.077943  4.940571  6.288000  5.254971 -3.638057  0.000000  0.000000
 [7,]  0.000000  0.000000  0.000000  0.000000  7.276114  5.254971 42.039771  0.000000  7.276114 -5.254971
 [8,]  0.000000  0.000000  0.000000  0.000000 -5.254971 -3.638057  0.000000  9.701486  5.254971 -3.638057
 [9,]  0.000000  0.000000  0.000000  0.000000  0.000000  0.000000  7.276114  5.254971 21.019886 -8.893029
[10,]  0.000000  0.000000  0.000000  0.000000  0.000000  0.000000 -5.254971 -3.638057 -8.893029  4.850743
```

For readability, the 10×10 global stiffness and global mass stiffness are reproduced in the following (rounded to the fourth decimal point):

$$
[K] = 10^6 \begin{bmatrix}
3. & 3. & -3. & 0. & 0. & 0. & 0. & 0. & 0. & 0. \\
3. & 4. & -3. & 0. & 0. & 0. & 0. & 0. & 0. & 0. \\
-3. & -3. & 6. & 0. & -3. & 3. & 0. & 0. & 0. & 0. \\
3. & 2. & 0. & 8. & -3. & 2. & 0. & 0. & 0. & 0. \\
0. & 0. & -3. & -3 & 3.88889 & -1.6667 & -0.8889 & 1.3333 & 0. & 0. \\
0. & 0. & 3. & 2. & -1.6667 & 6.6667 & -1.3333 & 1.3333 & 0. & 0. \\
0. & 0. & 0. & 0. & -0.8889 & -1.3333 & 1.7778 & 0. & -0.8889 & 1.3333 \\
0. & 0. & 0. & 0. & 1.3333 & 1.3333 & 0. & 5.3333 & -1.3333 & 1.3333 \\
0. & 0. & 0. & 0. & 0. & 0. & -0.8889 & -1.3333 & 0.88889 & -1.3333 \\
0. & 0. & 0. & 0. & 0. & 0. & 1.3333 & 1.3333 & -1\ 3333 & 2.6667
\end{bmatrix}
$$

$$
[M] = \begin{bmatrix}
14.0133 & 3.9525 & 4.8507 & -2.3355 & 0. & 0. & 0. & 0. & 0. & 0. \\
3.9525 & 1.4373 & 2.3355 & -1.0779 & 0. & 0. & 0. & 0. & 0. & 0. \\
4.8507 & 2.3355 & 28.0265 & 0. & 4.8507 & -2.3355 & 0. & 0. & 0. & 0. \\
-2.3355 & -1.0779 & 0. & 2.8745 & 2.3355 & -1.0779 & 0. & 0. & 0. & 0. \\
0. & 0. & 4.8507 & 2.3355 & 35.0331 & 4.9406 & 7.2761 & -5.255 & 0. & 0. \\
0. & 0. & -2.3355 & -1.0779 & 4.9406 & 6.288 & 5.255 & -3.6381 & 0. & 0. \\
0. & 0. & 0. & 0. & 7.2761 & 5.255 & 42.0398 & 0. & 7.2761 & -5.255 \\
0. & 0. & 0. & 0. & -5.255 & -3.6381 & 0. & 9.7015 & 5.255 & -3.3681 \\
0. & 0. & 0. & 0. & 0. & 0. & 7.2761 & 5.255 & 21.0199 & -8.893 \\
0. & 0. & 0. & 0. & 0. & 0. & -5.255 & -3.681 & -8.893 & 4.8507
\end{bmatrix}
$$

Consequently, the global matrix equation will be of the form

$$
\left([K] - \omega^2 [M]\right) \begin{Bmatrix} U_{1y} \\ \theta_{1z} \\ U_{2y} \\ \theta_{2z} \\ U_{3y} \\ \theta_{3z} \\ U_{4y} \\ \theta_{4z} \\ U_{5y} \\ \theta_{5z} \end{Bmatrix} = 0 \qquad (12.50)
$$

Step 5: Identify the essential boundary condition(s) and apply these to the global stiffness and mass matrices to obtain a reduced set of matrices.

For the beam in Figure 12.3a, the clamping on the left-hand side prevents both vertical translation and rotational movements, whereas the sliding prevents only rotational movements. Leading to the following boundary conditions:

$$U_{1y} = \theta_{1z} = \theta_{5z} = 0 \tag{12.51}$$

However, for the beam in Figure 12.3b, we have the following set of boundary conditions:

$$\theta_{1z} = 0; U_{5y} = 0 \tag{12.52}$$

On the basis of the boundary conditions mentioned earlier, we can obtain the reduced stiffness and mass matrices for the two beam cases in Figure 12.3. These reduced matrices are used to find the natural frequencies as shown in the following.

For the clamped-sliding beam

Given the boundary conditions in Equation 12.51, the rows 3–9 in Equation 12.50 will be the rows of known nodal forces that must be extracted to compute the natural frequencies. This is reflected in the variable **vec_rows_of_knownloads** in the next block of code.

```
vec_rows_of_knownloads=c(3:9);
ReducedK=DynamicEulerBeam_ReducedK(GlobalK,vec_rows_of_
knownloads);
ReducedK
ReducedM=DynamicEulerBeam_ReducedM(GlobalM,vec_rows_of_
knownloads);
ReducedM
naturalfrequencies=DynamicEulerBeam_NaturalFrequencies
(ReducedM,ReducedK);
cbind(naturalfrequencies[1:4])
```

By running the code, the following computed results are obtained:

```
> ReducedK
        [,1]     [,2]       [,3]     [,4]       [,5]     [,6]       [,7]
[1,]  6000000        0 -3000000.0  3000000        0.0        0        0.0
[2,]        0  8000000 -3000000.0  2000000        0.0        0        0.0
[3,] -3000000 -3000000  3888888.9 -1666667  -888888.9  1333333        0.0
[4,]  3000000  2000000 -1666666.7  6666667 -1333333.3  1333333        0.0
[5,]        0        0  -888888.9 -1333333  1777777.8        0  -888888.9
[6,]        0        0  1333333.3  1333333        0.0  5333333 -1333333.3
[7,]        0        0        0.0        0  -888888.9 -1333333   888888.9
>
> ReducedM=DynamicEulerBeam_ReducedM(GlobalM,vec_rows_of_knownloads);
> ReducedM
          [,1]      [,2]      [,3]      [,4]      [,5]      [,6]      [,7]
[1,] 28.026514  0.000000  4.850743 -2.335543  0.000000  0.000000  0.000000
[2,]  0.000000  2.874514  2.335543 -1.077943  0.000000  0.000000  0.000000
[3,]  4.850743  2.335543 35.033143  4.940571  7.276114 -5.254971  0.000000
[4,] -2.335543 -1.077943  4.940571  6.288000  5.254971 -3.638057  0.000000
[5,]  0.000000  0.000000  7.276114  5.254971 42.039771  0.000000  7.276114
[6,]  0.000000  0.000000 -5.254971 -3.638057  0.000000  9.701486  5.254971
[7,]  0.000000  0.000000  0.000000  0.000000  7.276114  5.254971 21.019886
> naturalfrequencies=DynamicEulerBeam_NaturalFrequencies(ReducedM,ReducedK);
> cbind(naturalfrequencies[1:4])
          [,1]
[1,]  18.21555
[2,]  98.76637
[3,] 246.09568
```

From the forgoing output, the three lowest natural frequencies of the clamped-sliding beam are obtained as

$$\begin{Bmatrix} \omega_1 \\ \omega_2 \\ \omega_3 \end{Bmatrix}^{\text{FEM}} = \begin{Bmatrix} 18.21555 \\ 98.76637 \\ 246.09568 \end{Bmatrix} \text{rad/s} \tag{12.53}$$

The exact closed form solution for the frequency $\omega_i (i = 1,2,3,\ldots,n)$ of a clamped-sliding beam is

$$\omega_i^{\text{exact}} = \frac{(\lambda_n)^2}{L^2}\sqrt{\frac{EI}{\rho A}}; \begin{Bmatrix} \lambda_1 \\ \lambda_2 \\ \lambda_3 \end{Bmatrix} = \begin{Bmatrix} 2.3650 \\ 5.4978 \\ 8.6393 \end{Bmatrix} \tag{12. 54}$$

Substituting the parameters (E, A, I, ρ, L) of the question into Equation 12.54 gives

$$\begin{Bmatrix} \omega_1 \\ \omega_2 \\ \omega_3 \end{Bmatrix}^{\text{Exact}} = \begin{Bmatrix} 18.21212 \\ 98.41833 \\ 243.03301 \end{Bmatrix} \text{rad/s} \tag{12.55}$$

From the earlier matrix, it is noticed that only very small deviation exists between the first, second, and third natural frequencies determined from the finite element calculations and the estimate of the exact solution.

For the sliding-pinned beam we have

Based on the boundary conditions in Equation 12.52, the applicable rows required to establish the reduced matrices are 1, 3, 4, 5, 6, 7, 8, and 10, as reflected in the variable **`vec_rows_of_knownloads`** in the next block of code.

```
vec_rows_of_knownloads=c(1,3:8,10);
ReducedK=DynamicEulerBeam_ReducedK(GlobalK,vec_rows_of_
knownloads);
ReducedK
ReducedM=DynamicEulerBeam_ReducedM(GlobalM,vec_rows_of_
knownloads);
ReducedM
naturalfrequencies=DynamicEulerBeam_NaturalFrequencies
(ReducedM,ReducedK);
cbind(naturalfrequencies)
```

Which produces the following output:

```
> ReducedK
        [,1]     [,2]     [,3]       [,4]     [,5]       [,6]    [,7]    [,8]
[1,]  3000000 -3000000  3000000        0.0        0        0.0       0       0
[2,] -3000000  6000000        0 -3000000.0  3000000        0.0       0       0
[3,]  3000000        0  8000000 -3000000.0  2000000        0.0       0       0
[4,]        0 -3000000 -3000000  3888888.9 -1666667  -888888.9 1333333       0
[5,]        0  3000000  2000000 -1666666.7  6666667 -1333333.3 1333333       0
[6,]        0        0        0  -888888.9 -1333333  1777777.8       0 1333333
[7,]        0        0        0  1333333.3  1333333        0.0 5333333 1333333
[8,]        0        0        0        0.0        0  1333333.3 1333333 2666667
>
> ReducedM=DynamicEulerBeam_ReducedM(GlobalM,vec_rows_of_knownloads);
> ReducedM
          [,1]      [,2]      [,3]      [,4]      [,5]      [,6]      [,7]      [,8]
[1,] 14.013257  4.850743 -2.335543  0.000000  0.000000  0.000000  0.000000  0.000000
[2,]  4.850743 28.026514  0.000000  4.850743 -2.335543  0.000000  0.000000  0.000000
[3,] -2.335543  0.000000  2.874514  2.335543 -1.077943  0.000000  0.000000  0.000000
[4,]  0.000000  4.850743  2.335543 35.033143  4.940571  7.276114 -5.254971  0.000000
[5,]  0.000000 -2.335543 -1.077943  4.940571  6.288000  5.254971 -3.638057  0.000000
[6,]  0.000000  0.000000  0.000000  7.276114  5.254971 42.039771  0.000000 -5.254971
[7,]  0.000000  0.000000  0.000000 -5.254971 -3.638057  0.000000  9.701486 -3.638057
[8,]  0.000000  0.000000  0.000000  0.000000  0.000000 -5.254971 -3.638057  4.850743
>
> naturalfrequencies=DynamicEulerBeam_NaturalFrequencies (ReducedM,ReducedK);
> cbind(naturalfrequencies)
     naturalfrequencies
[1,]           8.034228
[2,]          72.449828
[3,]         203.330389
```

From the output, the three lowest natural frequencies of the sliding-pinned beam are obtained as

$$\begin{Bmatrix} \omega_1 \\ \omega_2 \\ \omega_3 \end{Bmatrix}^{\text{FEM}} = \begin{Bmatrix} 8.034228 \\ 72.449828 \\ 203.330389 \end{Bmatrix} \text{rad/s} \tag{12.56}$$

The exact closed form solution for the frequency $\omega_n (n = 1,2,3,\ldots,N)$ of the sliding-pinned beam is [6]

$$\omega_n^{\text{exact}} = \frac{(\lambda_n)^2}{L^2}\sqrt{\frac{EI}{\rho A}};\ \lambda_n = \frac{(2n-1)\pi}{2} \quad \text{where } n = 1,2,3,\ldots,N \tag{12.57}$$

Substituting the parameters (E, A, I, ρ, L) of the question into Equation 12.57 gives

$$\begin{Bmatrix} \omega_1 \\ \omega_2 \\ \omega_3 \end{Bmatrix}^{\text{Exact}} = \begin{Bmatrix} 8.034112 \\ 72.307004 \\ 200.852789 \end{Bmatrix} \text{rad/s} \quad \text{where } n = 1,2,3,\ldots,N \tag{12.58}$$

where the values again show agreement with the FEM prediction.

In all, although the codes presented in this chapter and the other chapters of the book have been applied to only selected examples, they can handle many more variants of the common types of structural problems encountered in mechanics. The aim of the step-by-step solutions presented for all the examples is to make the codes and steps as transparent as possible to reduce mental work required for understanding. As a final point, it is possible to add several more layers of abstraction to the programing steps, for instance, automating the discretization and assembling steps. But that approach was jettisoned to give readers the freedom to extend the code in ways that suit his/her preference. Certainly, the author encourages the extension of the codes provided here.

12.6 Exercises

1. Determine the fundamental natural frequency of the cantilever beam shown in Figure 12.4.
2. Solve Example 12.2 by changing the support at point C to a roller support while keeping the support at points A and B fixed.
3. Find the natural frequencies of the truss structure shown in Figure 12.5. Let $A = 2.4 \times 10^{-3}$ m; $E = 200$ GPa, and $\rho = 7800\,\text{kg/m}^3$.

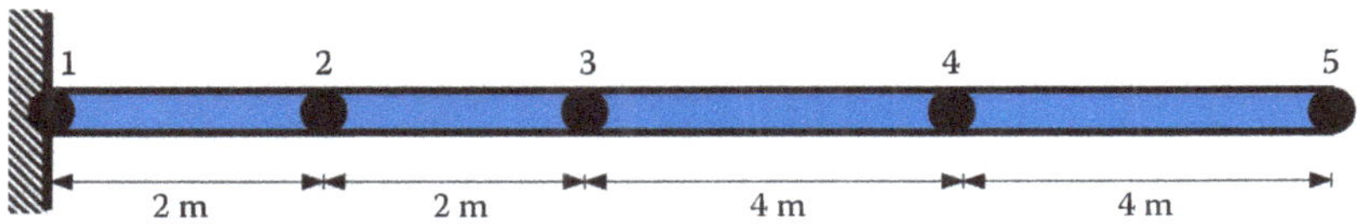

FIGURE 12.4
Figure 12Q1.

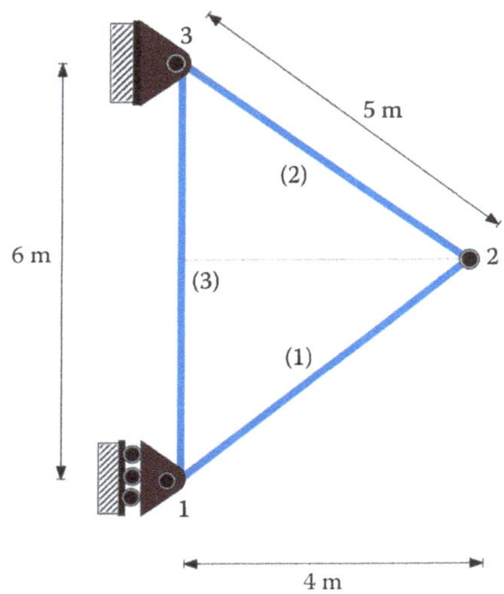

FIGURE 12.5
12Q 3.

References

1. M. Petyt, *Introduction to Finite Element Vibration Analysis*, Cambridge University Press, Cambridge, UK, 1998.
2. J. F. Doyle, *Static and Dynamic Analysis of Structures: with An Emphasis on Mechanics and Computer Matrix Methods*, Springer, Dordrecht, the Netherlands, 1991.
3. D. L. Logan, *First Course in the Finite Element Method*, Thomson, Canada, Ontario, 2007.
4. S. S. Rao, *The Finite Element Method in Engineering*: Elsevier Science, Burlington, NJ, 2010.
5. S. S. Rao, F. F. Yap, *Mechanical Vibrations*, Prentice Hall, Singapore, 2011.
6. R. D. Blevins, *Formulas for Dynamics, Acoustics and Vibration*, Wiley, Hoboken, NJ, 2015.
7. C. T. Ross, *Finite Element Programs for Structural Vibrations*, Springer Science & Business Media, London, UK, 2012.

Index

Note: Page numbers followed by f and t refer to figures and tables respectively.